FOOD INDUSTRY

FOOD INDUSTRY

DESIGN, TECHNOLOGY AND INNOVATION

Volume 1

Ethan Campbell
Lucy Wheeler
Eleanor Barnes

Kruger Brentt
Publishers
2026

Kruger Brentt Publishers UK. LTD.
Company Number 9728962

Regd. Office: 68 St Margarets Road, Edgware, Middlesex HA8 9UU

ISBN: 978-1-78715-643-2 (Vol. 1)

For information on all our publications visit our website at http://krugerbrentt.com/

PREFACE

The food industry is one of the most dynamic and rapidly evolving sectors worldwide, driven by changing consumer preferences, technological advancements, and the growing demand for safe, nutritious, and innovative food products. The design, technology, and innovation in food production play a pivotal role in ensuring quality, efficiency, and sustainability across the supply chain. From processing and packaging to quality assurance and product development, a deep understanding of modern food industry practices is essential for professionals, researchers, and students in food science and technology.

This book, Food Industry: Design, Technology and Innovation, provides a comprehensive exploration of the principles, technologies, and innovative approaches shaping the modern food industry. It is intended as a reference for students, researchers, industry professionals, and entrepreneurs seeking insights into food processing technologies, industrial design, and innovation management within the sector.

The initial chapters focus on the fundamental concepts of food industry design, including facility layout, equipment selection, workflow optimization, and regulatory considerations. Emphasis is placed on designing food production systems that ensure operational efficiency, safety, and compliance with hygiene standards.

Subsequent chapters explore technological advancements in food processing, preservation, packaging, and automation. The book covers traditional and emerging methods, such as thermal processing, high-pressure processing, extrusion, nanotechnology, and smart packaging, highlighting their applications, benefits, and limitations. Special attention is given to integrating sustainability and energy efficiency into food industry operations.

Innovation is a central theme throughout the book, with discussions on product development, sensory evaluation, consumer trends, and market-driven strategies. The role of research and development (R&D) in creating novel food products, improving nutritional quality, and enhancing shelf life is thoroughly examined.

Case studies from leading food companies illustrate how technology and innovation are successfully applied to meet market demands and regulatory standards.

The book also addresses critical challenges in the food industry, such as food safety, quality control, traceability, and the impact of globalization on supply chains. Strategies for managing these challenges using modern technologies, data analytics, and process optimization are highlighted to provide a practical perspective.

Food Industry: Design, Technology and Innovation serves as a vital resource for anyone engaged in food science, food engineering, or food business management. By combining theoretical principles with practical applications and contemporary case studies, this book equips readers with the knowledge and tools needed to innovate, optimize, and lead in the modern food industry.

Ethan Campbell

Lucy Wheeler

Eleanor Barnes

CONTENTS

VOLUME 2

SECTION 1
TYPES OF FOOD

Chapter 1

A COMPREHENSIVE EXAMINATION OF CAGAITA (EUGENIA DYSENTERICA DC) MACRO AND MICRO COMPONENTS FROM A TECHNOLOGICAL PERSPECTIVE

1. INTRODUCTION: AN OVERVIEW

Numerous fruit species indigenous to the Brazilian Cerrado provide significant economic and ecological potential, along with social relevance to the indigenous community (Bezerra, Silva, Ferreira, Ferri, & Santos, 2002). These fruits frequently augment the diet and serve as sources of medicine, textile fibers, construction materials, and fuel. The advancement of new technology may render these fruits possible avenues for economic exploitation. (Martinotto, Soares, Santos, & Nogueira, 2008).

The Cerrado region has an abundance of species of fruit, still underused by local communities for scientific unknown and lack of incentive for marketing (Veira, Costa, Silva, Ferreira & Sano, 2006). The sustainable use of these species can be an excellent alternative to add value to raw materials available in the Cerrado region and improve the health of the population, thereby contributing to the income of rural communities and encouraging the conservation of native species.

The cagaita tree, belongs to the Myrtaceae family of plants, consisting of 14 genera and represented by 211 species that naturally occur in the Cerrado. Myrtaceae is one of 10 plant families found in this biome or ecosystem that together contribute to more than 51% of its richness. The cagaita is found in the Brazilian states of Goias, Minas Gerais, São Paulo, To cantins and Bahia (Silva, Chaves, &Naves, & 2001). It occurs at highest densities in latosoil and is observed in areas with mean annual temperatures between 21.1°C and 25.5°C and at altitudes of 380 m to 1100 m (Souza, Naves, Carneiro, Leandro & Borges, 2002).

The cagaiteira is a medium-sized tree, reaching a height of 30 meters, with a cylindrical and twisted trunk that measures between 20 cm and 40 cm in diameter. The suberous bark and its fissures are distinctly unusual. The crown is elongated and dense, featuring square, hairless branches; aside from the buttons, the pedicels, leaves, and juvenile branches are puberulent. It is a deciduous plant and is selectively heliophytic and xerophilous. (Donadio, Môro, & Servidone, 2002).

Flowering occurs in the middle of the dry season, from mid-July to early August, with the simultaneous emergence of new leaves of the cuprea (Fig.1) (Brito, Pereira, Pereira, & Ribeiro, 2003). The cagaiteira's flowers are always axillary and are either singular or clustered in arrays of three. They are hermaphrodites, and complete, are from 1.5 to 2 cm in diameter, actinomorphic, dialipetalous, dialisepalous, tetramerous, and are endowed with white pet als (Lorenzi, 2000).

Figure 1. Cagaiteira flower and branches with flowers Source: www.plantasonya.com.br.

The cagaita tree can be used almost entirely, bringing its economic value, and the great po tential for sustained exploration (Table 1).

Table 1. Forms of exploitation and use of Eugenia dysenterica DC.

Feature	Utility	References
Tree	Ornamental landescape	Martinottoet al., 2008
Flowers	Apiculture	Lorenzi, 2002
Stalk	Construction, furniture, pallets, firewood and charcoal	Chaves & Telles, 2006; Martinotto et al., 2008
Shell	Tannery, antidiarrheal	Lorenzi, 2002; Chaves & Telles, 2006; Martinotto et al., 2008
Leaves	Lawn trees, antidiarrhoeal, antifungal, moluscocida and treatment of diabetes and jaundice	Chaves & Telles, 2006; Martinotto et al., 2008

The cagaiteira exhibits significant potential for agricultural production systems due to its high and relatively stable yield over the years, the fruit's suitability for processed products, compatibility with pasture, high drought tolerance, resilience to edaphic and biotic stress, fire resistance, and ease of propagation through seeds and seedling establishment in the field, among other factors. (Veira, Costa, Silva, Ferreira & Sano, 2006).

Zucchi, Brondani, and Pinheiro (2003) describe the cagaita fruit as a flattened, globular berry, pale yellow in color, measuring 2 to 3 cm in diameter, and containing 1 to 3 white seeds enveloped in a mildly acidic pulp (Fig. 2). The seeds are connected to the fruit by a dry, membrane mesocarp, whereas the endocarp is succulent. The seeds are round, pale yellow upon ripening, possess an acidic taste, and weigh from 14 and 20 grams. (Silva, Chaves, & Naves, 2001).

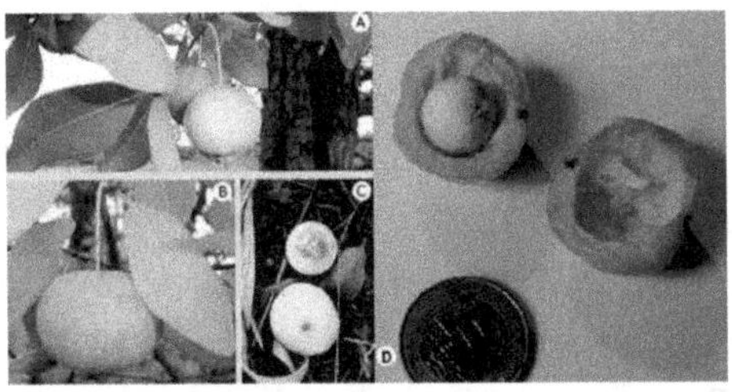

Figure 2. Cagaita fruit (Eugenia dysenterica DC): A – unripe; B – imature; C – ripefruit; D – Fruit with the seed (Source: Tatagiba, 2012; Stolfi, 2012).

2. CAGAITA PULP PROCESS

The mature fruits of cagaita (Eugenia dysenterica) are harvested by hand. After cleaning (im mersion in sodium hypochlorite 200 ppm) and selection, the fruits are depulped, packed in polyethylene bags, and freezing and stored at -18 oC (Fig. 3).

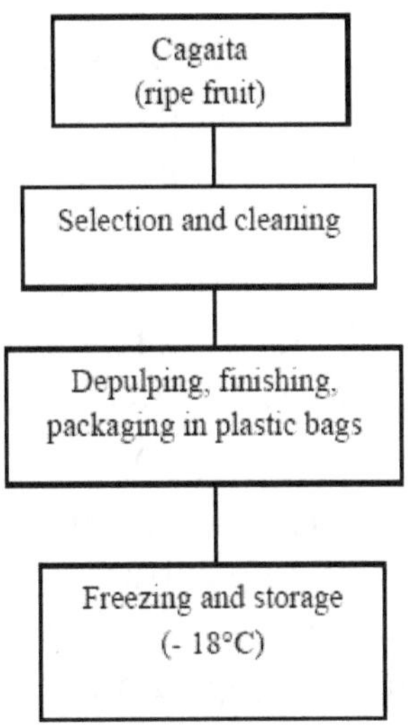

Figure 3. Whole cagaita pulp process (Cardoso et, 2011)

3. NUTRITIONAL AND PROXIMAL COMPOSITION

Studies have shown that the cagaita fruit's nutritional composition indicates a high water content (95.01%). It has the highest percentage of polyunsaturated fatty acids (such as linole ic (10.5%) and linolenic acids (11.86%)), surpassing corn, sunflower, peanut, soybean, olive and palm oils. Fatty acids play an important role in the human body, forming the basis of substances that are critical for developing cell membranes found in the brain, the retina and the reproductive system (Almeida, 1998).

Carvalho et al. (2010) found the moisture content of cagaita pulp to measure 94.12%, the ti tratable acidity at 13.78 g m-1 and a pH of 3.05. These values are higher than other fruits of the same genera, such as the pitanga and jambo. Conversely, the ash and protein contents are lower than the jambo and pitanga (Oliveira, Figueiredo, & Queiroz, 2006). Similarly, Ri beiro (2011) evaluated the proximal composition of the cagaita pulp that was extracted with and without peels. The test results for moisture, ash, protein, lipids and carbohydrates (by difference – NIFEXT) for this pulp (with and without peels) varied from 90.08 to 88.55 g·100g-1; 0.25 to 0.33 g·100g-1; 1.85 to 2.03 g·100g-1; 0.20 to 0.36 g 100g-1 and 7.62 to 8.73 g 100g-1, respectively. The titratable acidity, pH and soluble solids ranged from 13.78 to 14.63 g.100ml-1; 2.90 to 2.69 and 8.20 to 8.70°Brix, respectively, leading to the conclusion that re moving the peel results in a reduction of carbohydrate content such that some remains in it after extracting the juice. Silva, Lacerda, Santos, and Martins (2008) found 20.01 TEV (Ener gy Total Value), 94.34 (moisture); protein 0.82; lipids 0.44; carbohydrates 3.08 and ash 0.28 g m-1 in the cagaita pulp. The authors did not mention whether the pulp was obtained with or without peels.

In cagaita pulp extracted with the peels, Roesler et al., (2007) found 2.09 (proteins); 0.32 (lip ids); 0.23 (ash); 89.71 (moisture); 20.47 (total sugars) pH of 2.8 and 26.4 (total acidity). Cardo so et al., (2011) found 0.73 g of citric acid 100g−1, pH of 3.3 and soluble solids of 9.12°Brix in cagaita pulp from the Cerrado region of the state of Minas Gerais. Moisture content was 91.56 g 100g−1, with similar results found by Roesler et al. (2007) in cagaita pulp from the Cerrado region within the Goias state.

Silva, Santos-Junior and Ferreira (2008) investigated the cagaita fruit at different stages of maturation; however, the results for the moisture did not differ significantly, ranging from 92.77 to 93.21 g 100g-1.

From the results obtained by Ribeiro (2011), one can conclude that the cagaita fruit, with or without peels, is basically made up of carbohydrates and water. As expected, the moisture content in pulp extracted without peels was higher than the moisture content in pulp with peels. This was because the latter contained peels and the former was essentially pulp with a high water content. The value for cagaita pulp without peels was 90.08 g 100g-1 and the pulp with peels was 88.55

g 100g-1. However, no significant difference (P < 0.05) was found. Removing the peels yields a reduction in carbohydrate content, although some remains in it after extracting the juice.

Other researchers studying the cagaita fruit obtained similar results. Roesler et al., (2007) evaluated only the pulp, obtaining 89.71%. Silva, Santos-Junior & Ferreira (2008) investigat ed the fruit at different stages of maturation, however, the results for the moisture content did not differ significantly, ranging from 92.77 to 93.21 g 100 g-1.

Martins (2006) found a carbohydrate content of 5.4 g 100 g-1, which was lower than that re corded by Ribeiro (2011), at 7.62 and 8.73 g 100 g-1. These results may be related to the geo graphic location of the analyzed fruits. For example, the temperature, sun exposure and maturity, among other factors, may have had an effect on the results.

Due to its low lipid content, the cagaita fruit is recommended as part of a low calorie diet. The values found by Ribeiro (2011) varied from 0.20 to 0.36 g 100 g-1 for pulp extracted with and without peels. Those values were similar to those reported by Martins (2006) and Roes ler et al., (2007), being 0.20 and 0.32 g 100 g-1, respectively. It is worth noting that this was the only parameter that did not yield a significant difference in the 5% level of significance, showing a higher content of lipids in the peels of the fruit.

Vallilo, Garbelotti, Oliveira, and Lamardo (2005) evaluated other Myrtaceae fruits and found similar low values of lipids: 0.23 g 100 g-1 in Surinam cherry (Eugenia uniflora L), 1.53 g 100 g-1 in cambuci (Campomanesia phaea Berg), 0.80 g 100 g-1 pears in the field (Eugenia klotzchiana Berg) and 0.54 g 100 g-1 in guava (Psidium guajava).

The protein levels were low; although, as expected, they were higher in the fruit with their peels (2.03 g 100 g-1) than in peeled fruits (1.85 g 100 g-1). Some studies with the same result found similar values: 2.09 g 100 g-1 for the whole pulp (Roesler et al., 2007) and 0.99 g 100 g-1 in cagaita pulp (Martins, 2006).

In another study of guava, cited later in this paper, Gutiérrez, Mitchell, and Solis (2008), re viewed the fruit in relation to its protein content of 0.88%.

It can be concluded that the cagaita fruit is not high-caloric due to its low levels of protein, carbohydrates, and especially lipids.

4. GLUCOSE, FRUCTOSE AND SUCROSE

Carvalho et al. (2009) and Ribeiro (2011) found a high concentration of fructose (2.54 g 100 mL-1), followed by glucose (1.75 g 100 mL-1) and the lowest concentration of sucrose (0.59 g 100 mL-1) (P >0.05) (Fig. 4). The high fructose content can be explained by the fact that the cagaita fruit used in their study was fully ripened.

Many different factors could have contributed to the low soluble sugar content in the cagaita pulp. One factor is mineral fertilization, where potassium is the primary mineral element causing starch accumulation in Citrus leaves (Lavon, Goldschmidt, Salomon, & Frank, 1995). On the other hand, the shortage of free sugars may trigger ethylene synthesis because defoli ation, which drastically reduces sucrose transport to the fruit, increases ethylene synthesis (Ortolá, Monerri, & Guardiola, 2007) and 1-aminocyclopropane-1-carboxylic acid (ACC) ac cumulation (Gómez-Cadenas, Mehouachi, Tadeo, Primo-Millo, & Talón, 2000).

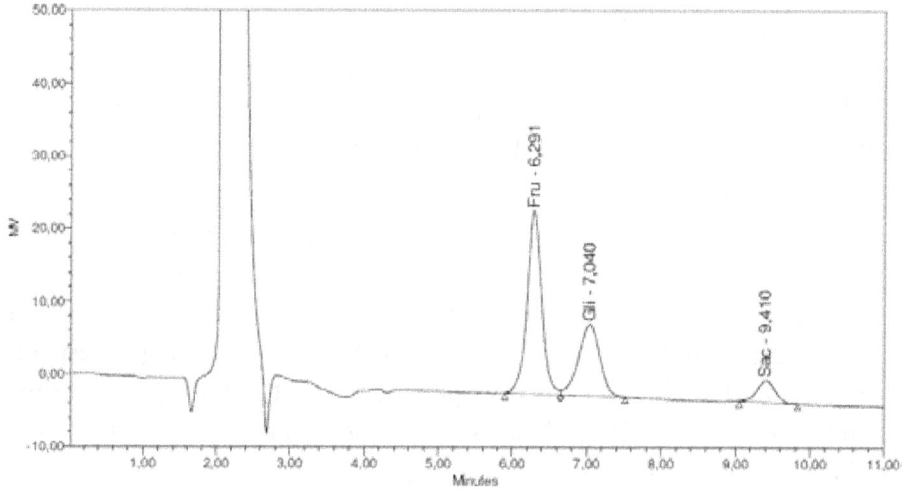

Figure 4. HPLC chromatogram of glucose, frutose and sucrose in whole cagaita pulp.

The low glucose, fructose and sucrose values in the cagaita indicate that this fruit is less sweet and contains less sugar than the guava, for example. This comparison has been veri fied by Lee & Kader (2000). Analyzing the fruits by High Performance Liquid Chromatogra phy (HPLC), their study found higher values in the ripe guava pulp (11.52 g.100 ml-1 of sucrose, fructose 11.37 g.100 ml-1 and, glucose 5.12 g.100 ml-1.

5. ASCORBIC ACID (VITAMIN C)

According to Andrade, Diniz, Neves & Nóbrega (2002), the sources of ascorbic acid are clas sified by different levels: high sources, such as strawberry, guava and pineapple, contain 100 to 300 mg·100 g-1; medium sources, such as orange, lemon and papaya contain an average of 50 to 100 mg·100 g-1; and low sources, such as lime, pear and mango, contain 25 to 50 mg·100 g-1. The vitamin C content in cagaita pulp as reported by Ribeiro (2011) was 56.66 mg 100 g-1 (Fig. 5) and by Cardoso et al. (2011), it was 34.11 mg 100 g-1, with 30.03 mg 100 g-1 of ascorbic acid and 4.08 mg 100 g-1 of de-hydro ascorbic acid. Therefore, the cagaita can be classified as a medium source of ascorbic acid. The pulp of the cagaita fruit has shown con siderable promise for its vitamin C content and is considered a source of that nutrient when compared to other fruit. Silva, Santos-Junior, and Ferreira (2008) found the level of vitamin C to be 27.46 mg 100 g-1 in cagaita pulp from the Cerrado region in the state of Goias.

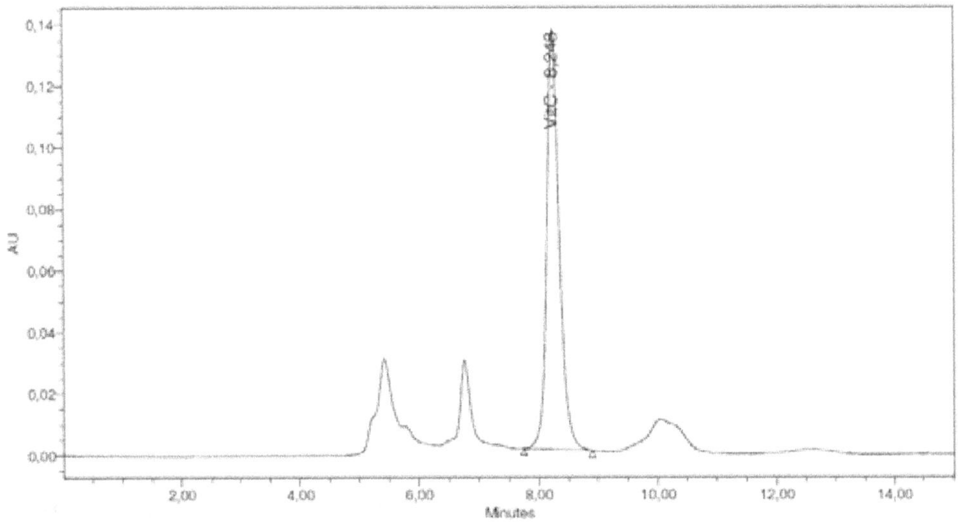

Figure 5. HPLC chromatogram of ascorbic acid in whole cagaita pulp.

On the other hand, the National Sanitary Surveillance Agency (ANVISA) legislation (Brazil, 1998) recommends that for a food to be considered a "source" of a certain vitamin, it should contain, at least, 15% of the Recommended Daily Intake (RDI) per 100 g of reference. To be considered "rich" in a vitamin, it should contain at least 30% of the RDI. Therefore, the cagai ta can be categorized as rich in vitamin C because it exceeds 30% of the RDI (US National Academy of Sciences, 2000).

The ascorbic acid content of 26 kinds of exotic fruits from a variety of species and families were evaluated by Valente, Albuquerque, Sanches-Silva and Costa (2011). The results ranged from 1.42 to 117 mg 100 g-1, and those fruits that had values similar to cagaita were guava (Psidium guajava) with 65.8 mg 100 g-1, kiwi (Actinidia chinensis Planch), cv. Hayward with 55.2 mg 100 g-1, papaya (Carica papaya), cv. Taiwan with 64.2 mg 100 g-1 and mango (Mangifera indica L), cv. Palmer, with 40.9 mg 100 g-1, among others.

6. POLYPHENOLS COMPOUNDS

In general, phenolic compounds behaving as antioxidants are multifunctional, achieving bi oactivity in several ways: fighting free radicals by donating a hydrogen atom from a hydrox yl group (OH) of their aromatic structure; chelating transition metals, such as the Fe^{2+} and Cu^{+}; interrupting the propagation reaction of free radicals in lipid oxidation; modifying the redox potential of the medium and repairing the damage in molecules attacked by free radi cals (Podsedek, 2007; Kyungmi & Ebel, 2008). These same phenolic compounds also block the action of specific enzymes that cause inflammation, modify the metabolic pathways of prostaglandins, permit platelet clumping and inhibit activation of carcinogens (Liu, 2005; Valko et al., 2007).

Historically, like tannins, phenolic compounds were classified as anti-nutrients, which have demonstrated adverse effects on human metabolism. However, identifying the specific properties of these phenolic compounds has stimulated the development of research aimed at identifying their potential health benefits (Kaur & Kapoor, 2001).

It is worth noting that a substance can be defined as polyphenolic antioxidant if it meets two conditions: (1) presence at a low concentration on the substrate to be oxidized (and this may delay or prevent oxidation), and (2) high stability of radicals formed after the reaction (Kaur & Kapoor, 2001).

Several spectrophotometric methods have been developed for the quantification of phenolic compounds in foods. The most commonly used by the scientific community is the Folin-Cio caltcau method, which involves the oxidation of phenol with a reagent and yellow phospho molybdate heteropolyacid phosphotungsten (Folin-Ciocalteau) and colorimetric measurement of W-Mo blue complex formed in reaction in an alkaline medium (Singleton, Orthof, & Lamuel-Raventos, 1999). The results are expressed in gallic acid equivalents.

Some results of the polyphenols content in ethanolic (18.38 g GAE kg-1) and aqueous (16.23 g GAE kg-1) extracts of cagaita pulp were reported by Roesler et al. (2007). The content of the total phenolics in cagaita pulp was evaluated by Ribeiro et al. (2011) who found 10.51 mg GA g-1 in pulp with peels and, in the pulp without peels, found 9.01 mg gallic acid g-1.

Therefore, no significant difference was found at a 5% level between them. Thus, the cagaita fruit was found to have high total phenolic compounds.

7. ANTIOXIDANT CAPACITY

Determining the antioxidant activity of foods, in addition to recognizing its antioxidant po tential before being consumed, is important to assess the defense against oxidation and deg radation reactions that can lead to the degradation of its quality and nutritional value (Lima, 2008). Currently, there are no approved or standard methods for the determination of anti oxidant activity. However, several in vitro methods have been and are being tested to evalu ate the total antioxidant activity of substances and foods, especially in complex matrices such as wine, fruits and other vegetables. These methods are necessary because of the diffi culty in comparing and measuring each compound separately and also because of the po tential interactions between different antioxidants in the system. (Cao & Prior, 1999; Kulkarni, Aradhya, & Divakar, 2004; Scherer & Godoy, 2009).

The methods most often cited in the literature include the antioxidant power in the reduc tion of iron (FRAP), DPPH (radical 2,2-diphenyl-1-picrihidrazil) Activity of OxygenRadicalAbsorption(ORAC),ABTS[acid2,2-Azin-bis(3-ethylbenzothiazoline) – 6 - sulfonic acid Spectrometry and Electron Spin Resonance (ESR) (Kulkarni, Aradhya and Divakar, 2004; Li ma, 2008). While evaluating the efficiency of using methanol and ethanol as solvents to determine the antioxidant activity in cagaita pulp (Ribeiro et al., 2011) found that the amount of ethanol ranged between 6.6% and 96.82% and that of methanol ranged between 11.20% and 92.60%, in different concentrations. It was also shown that the cagaita pulp reached its maximum value at a concentration of 500 µg ml-1, in both cases.

Roesler et al. (2007) found the antioxidant activity (IC50) in cagaita pulp extracted with peels to measure 387.47 mg ml-1 in the ethanolic extract and 879.33 mg ml-1 in the aqueous extract.

8. CAROTENOIDS

Gomes et al. (2011) measured the total carotenoid content in the whole cagaita pulp and also in the freeze-dried pulp and found 0.87 and 9.29 mg 100 g-1, respectively (Table 2 and Fig. 6). Lutein was the most abundant carotenoid in the whole and freeze-dried pulps (0.21 and 2.22 mg 100 g-1, respectively), followed by zeaxanthin (0.19 and 2.05 mg 100 g-1, respectively) and β-carotene (0.11 and 1.33 mg 100 g-1, respectively).

According to these results, cagaita may be a source of lutein and zeaxanthin (which are nat ural antioxidants), particularly in freeze-dried pulp. By microencapsulating the freeze-dried pulp, it can become a beneficial food additive because cagaita pulp is widely consumed in the Brazilian Cerrado.

Table 2. Carotenoids (µg/g) and isomers of saponified and not saponified cagaita pulp

Samples	Total Carote-noids	β-caro-tene	9-cis-β-carotene	13-cis-β-carotene	β- crip-toxantin	α-caro-tene	Lutein	Zeaxan-thin
Whole Pulp	8.22 ± 0.06	0.97 ± 0.08	Nd	Nd	0.35 ± 0.01	Nd	1.81 ± 0.12	1.99 ± 0.05
Saponified Whole Pulp	5.83 ± 0.52	1.70 ± 0.18	0.20 ± 0.01	0.09 ± 0.01	1.49 ± 0.11	0.18 ± 0.16	0.85± 0.01	0.79 ± 0.02

Source: Gomes, 2012

Lutein can be found in a variety of vegetables and is especially plentiful in cabbage (15 mg 100 g-1), parsley (10.82 mg 100 g-1), spinach (9.20 mg 100 g-1) and pumpkin (2.40 mg 100 g-1). However, it is found in lower concentrations in fruits such as peach and orange (0.02 and 0.35 mg 100 g-1, respectively).

Gomes (2012) identified α-carotene, β-carotene isomers and 9:13-cis β-carotene, β-cryptox anthin, lutein and zeaxanthin in the pulp produced in cagaita Damianópolis, Goias, Brazil (Fig. 6).

The β-carotene and β-cryptoxanthin the most abundant carotenoids, lutein and zeaxanthin and the carotenoids intermediate, and the α-carotene carotenoid the minority (Table 2).

There were significant differences in levels of total carotenoids according to the saponifica tion step. The hydrolysis step was necessary to facilitate the identification of different caro tenoids. The average concentration of total carotenoids found in the extracted pulp without the saponification step was 8.22 mg / g. There was a 29% decrease in total carotenoid content of the pulp subjected to saponification step (5.83 µ / g ± 0.18). This drop was expected and may occur as a function of temperature application of tests, and also by the exposure time of the pigment to the alkali (Mercadante, 1999; Penteado, 2003).

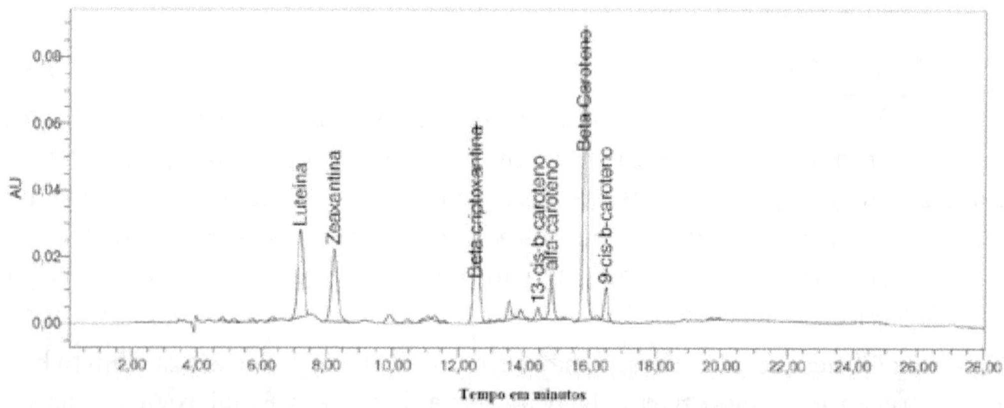

Figure 6. HPLC chromatogram of saponified cagaita pulp. Source: Gomes, 2012

Cardoso et al. (2011) found a lower total carotenoid content (0.77 mg100 g-1) in the cagaita pulp from the Cerrado in the state of Minas Gerais. The major carotenoids were the α--caro tene (0.31 mg 100 g-1) and β-carotene (0.39 mg 100 g-1) provitamin A carotenoids. They still found a small quantity of lycopene (0.06 mg100 g-1), however lutein and zeaxanthin were not found.

9. MINERALS

According to Carvalho et al. (2009), the most abundant mineral found in the cagaita pulp was potassium (75.83 mg 100 g-1), followed by sodium (6.80 mg 100 g-1), phosphorus (6.68 mg 100 g-1) and magnesium (5.92 mg 100 g-1). The levels of zinc were lower (0.23 mg 100 g-1), as were the levels of iron (0.06 mg 100 g-1) and calcium (0.65 mg 100 g-1) (Table 1). Higher values of calcium (0.8 mg 100 g-1) and, similarly, iron (0.04 mg 100 g-1) were found by Silva, Santos-Junior Junior, and Ferreira (2008) in the cagaita pulp, but zinc was not found at high er levels. Leterme, Buldgen, Estrada, and Londoño (2006), in analyzing the fruits of araçá- boi (belonging to the same family and genus as the cagaita), found similar values: 78 mg 100 g-1 (potassium), 7 mg 100 g-1 (phosphorus), 2 mg 100 g-1 (sodium) and 9 mg 100 g-1 (magnesi um), respectively.

Table 3. Minerals in the unpeeled cagaita pulp (Eugenia dysenterica DC).

Mineral	mg/100g	Mineral	mg/100g
Potassium	75.83 (± 0.43)	Aluminum	0.23 (± 0.06)
Phosphorus	6.68 (± 0.14)	Zinc	0.23 (± 001)
Sodium	6.80 (± 0.13)	Manganese	0.13 (± 0.01)
Magnesium	5.92 (± 0.08)	Iron	0.06 (± 0.01)
Calcium	0.65 (± 0.08)	Copper	0.01 (± 0.01)

Mean Value (± Standard deviatio/n (n = 3)). Source: Carvalho et al., 2009

Comparing the cagaita (Eugenia dysenterica DC) to the results of the study by Dembitsky et al. (2011), in which different fruits were analyzed, confirms that the acerola (Malpighia puni cifolia Linn) contains lower amounts of potassium (41 mg/100 g), zinc (0.09 mg/100 g) and manganese (0.7 mg 100 g-1) and much higher amounts of calcium (4 mg 100 g-1), iron (37 mg 100 g-1) and magnesium (22 mg 100 g-1).

While analyzing the fruits of guava-boi (Eugenia stipitata Mark Vaughn) that belong to the same family and genus as the cagaita, Leterme, Buldgen, Estrada, and Londoño (2006) found similar amounts: 78 mg 100 g-1 of potassium, phosphorus 7mg 100 g-1, mg 100 g-1, 2 mg 100 g-1 and 9 mg 100 g-1 of sodium and magnesium. These variations could be due to climatic conditions, soil type and the addition of fertilizers, for example.

10. VOLATILE COMPOUNDS

Volatile compounds are responsible for the aroma and flavor of foods. The same fruit, even if native to Brazil, can vary greatly from region to region, with different varieties having a dissimilar volatile composition (Alves & Franco, 2003). The methods used for the extraction of volatile substances are time-consuming, requiring large amounts of sample (Sánchez-Pal omo, Díaz-Maroto, & Pérez-Coello, 2005). Solid-phase Microextraction (SPME) is a fast, low- cost technique that allows the extraction of volatile substances that can then be analyzed by gas chromatography coupled to mass spectrophotometry (GC/MS). This technique replaces traditional extraction methods, avoiding the formation of artifacts without the need for sol vents, thereby minimizing artifact formation (Pawliszyn, 1997; Riu-Aumatell, Castellari, & López-Tamames, 2004).

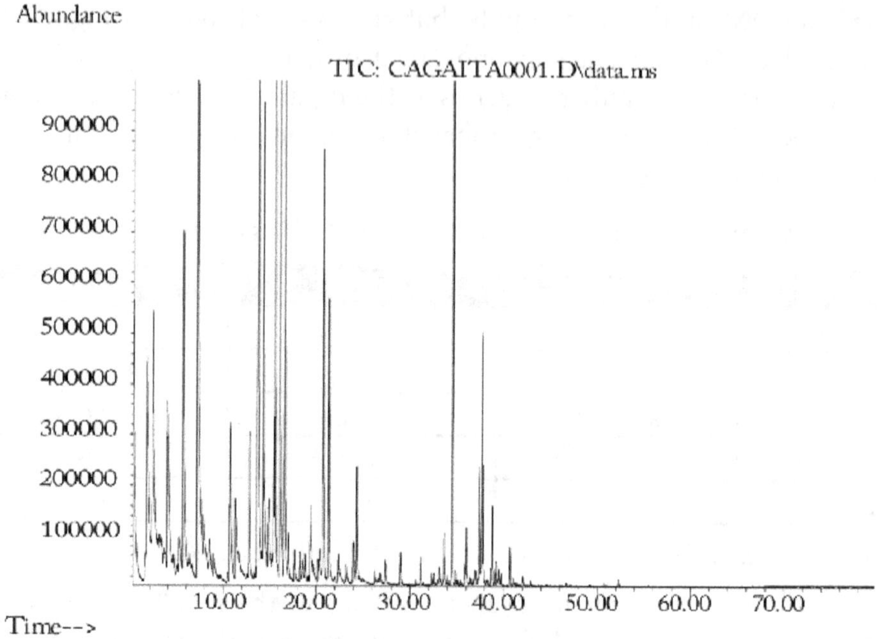

Figure 7. Chromatogram of the cagaita pulp volatile compounds. Source: Cardoso et al, (2011).

Fifty six volatile compounds were found in cagaita pulp extracted by solid phase micro-extrac tion and were analyzed by GC/MS. Among them, 19 could not be identified by Carvalho et al. (2009). Ethyl hexanoate was the most abundant compound in the cagaita pulp (51.4%), fol lowed by the ethyl butanoate (14.7%), which also imparts the fruity aroma of the fruit juices and pulp. The results revealed that a greater concentration of esters, mainly methyl, ethyl hex anoate (6.5%) and butanoate, are responsible for the fruity aroma. Alcohols and terpenes were present at low concentrations, with ethanol being the most abundant (3.0%). These volatile compounds were also found in pineapple, apple and papaya, among other fruits (Van Den Dool, & Kratz, 1963; Adams, 1972). Alves and Franco (2003) also identified some major compounds in murici, finding esters and alcohols. Ethanol (28.1%), ethyl hexanoate (25.1%) and methyl hexanoate (5.2%) were the major components. However, they reported that the high ethanol levels could be due to fermentation following maturation. Because no other authors re ported these compounds, it is not possible to compare the reported results. A typical total ion chromatogram obtained from the cagaita pulp analysis is presented in Figure 7.

It is noteworthy that this is the first time that volatile compounds have been found in cagaita fruit from the Cerrado region in Goias.

11 MEMBRANE PROCESSES APPLIED TO CAGAITA PULP

The consumption of fruit juice in Brazil and in the industrialized world has increased signif icantly in recent decades. Using fruit juice or pulp that has been clarified by the membrane processes of microfiltration is already a reality in the international market. The cagaita pulp can be introduced as a new product used in the formulation of carbonated beverages, ener gy and isotonic drinks. The demand for products with less nutritional and sensory changes led to the development of non-thermal preservation techniques such as the process of mem brane separation. The membrane separation process is based on the selective permeability of one or more components through a membrane. The determination of the hydraulic permea bility is an important tool in evaluating the permeate flux and the integrity of the mem brane. Cardoso et al., (2011) evaluated the cagaita pulp clarified by microfiltration with a tubular polyethersulfone membrane (0.3 μm) at 2 Bar (Fig. 8). A mean flux after 2 hours process was 20 L./m2 h. and the clarified juice yield was 43%. The results for the flux of the juice permeate were acceptable and the permeate was clear and translucent.

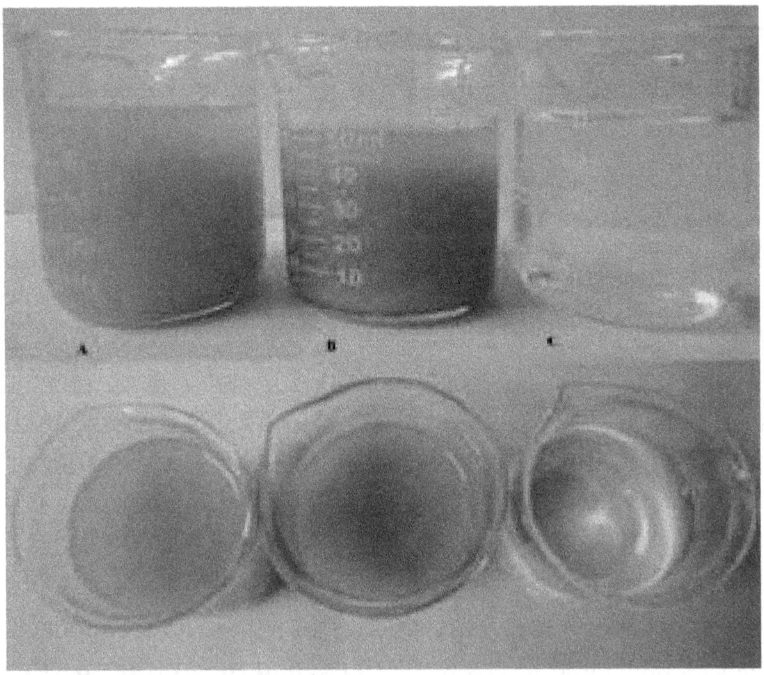

Figure 8. Cagaita pulps (Eugenia dysenterica DC): A – Whole, B – Concentrated e C – Clarified. (Cardoso et al., 2011).

12. MICROBIOLOGICAL QUALITY

Microbiological studies of cagaita pulp revealed no growth of microorganisms. Coliforms at 45°C, were indicative of its tolerance to sample 10^2 CFU (colony forming units) as was the absence of salmonella in 25 g of the sample (Carvalho et al., 2009). Therefore, the analyzed pulps were found fit for human consumption because they were in accordance with stand ards established by ANVISA (Brasil, 1998).

Table 4. Microbiological analysis of whole, retentate and clarified cagaita pulp.

Samples	Total Coliforms (UFC/mL)	Thermotolerant Coliforms (UFC/ mL)	Yeast and Mold (UFC/ mL)	Salmonella sp. (Absence 25 g or mL)
WCP	< 10	< 10	< 10	Absence
RCP	< 10	< 10	< 10	Absence
CCP	< 10	< 10	< 10	Absence

WCP: Whole Cagaita Pulp; RCP: Retentate Cagaita Pulp; CCP: Clarified Cagaita Pulp

13. PARTICLE SIZE OF THE CAGAITA PULP

Particle size analysis is an important tool to observe the enzymatic hydrolysis and the parti cle size reduction in order to optimize the membrane pore size before clarification processes.

Particle size analysis can be an useful tool to observe particle size reduction during enzy matic hydrolysis optimization to reduce juice viscosity. Few studies are found in the litera ture reporting the use of particle size analysis to observe viscosity decrease in fruit juices.

Laser diffraction analysis was used to evaluate the effects of cloud particle characteristics such as shape, volume fraction, and soluble pectin on the viscosity of cloudy apple juice. Cloudy apple juice results in a suspension of irregular-shaped particles ranging from 0.25 to 0.5 μm in size. Data indicate that the effect of nonspherical particles on cloudy apple juice viscosity can be neglected and soluble pectin can significantly increase the viscosity (Geno vese & Lozano, 2000).

The distribution of the average particle diameter, i.e., its frequency as measure by Carvalho et al. (2009 and 2011), was 12.11%, and the average particle diameter within cagaita pulp was 68.17 μm (Fig. 9). The presence of nanoparticles of less than 1 micrometers was still ob served, but in low frequency (0.1%).

After enzymatic hydrolysis of lemon juice at different incubation times, Carvalho et al., (2006) evaluated the particle size reduction in prior membrane microfiltration processes in order to obtain better permeate fluxes. The whole lemon juice showed a wide distribution of particle size ranging from 5 to 900 μm, and the greatest particle size reduction after hydroly sis ranged from 5 to 200 m. There were few particles above this size.

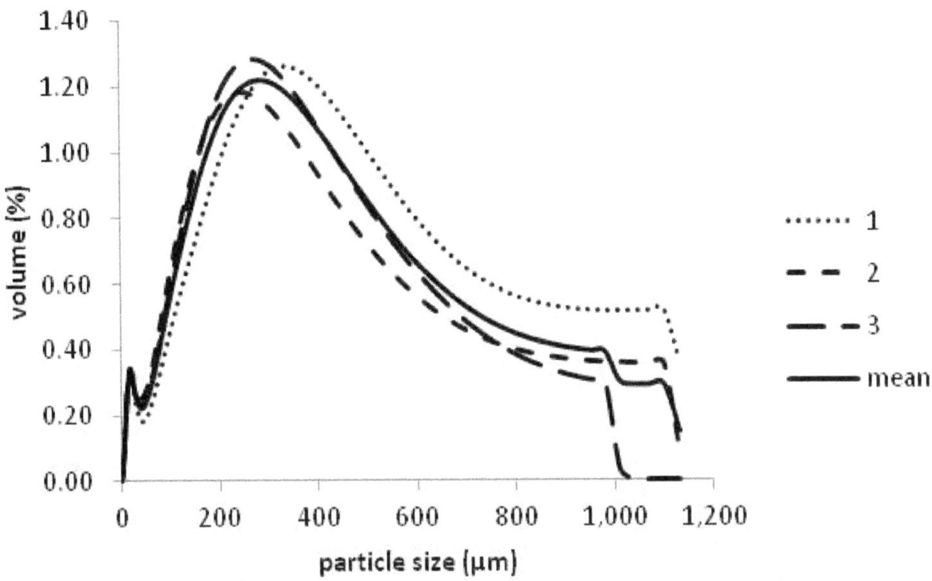

Figure 9. Particle size and frequency of cagaita pulp (Eugenia dysenterica DC).

14. CONCISE SUMMARY

The findings presented by several writers referenced in this study concerning the physical and chemical properties of the antioxidant effects of the cagaita fruit suggest promise for therapeutic and medical uses. Moreover, an array of novel products with advantageous attributes, including jams, juices, and energy drinks, can be derived from the fruit of the cagaita. Utilizing a proven technology like membrane processing to obtain clarified juice, followed by the addition of nutrients, presents the opportunity for a lucrative economic endeavor. The populace of the Brazilian Cerrado region consumes cagaita fruit both whole and processed by hand, making the industrial manufacturing of cagaita fruit products a potential economic option, particularly given that much of the current production remains underutilized.

REFERENCES

1. Adams, J. B. (1972). Changes in the polyphenols of red fruits during heat process ing:the kinetics and mechanism of anthocyanin degradation.Campden Food Preser vation Research Association of Technology. Bulletin, 22.

2. Almeida, S. P. (1998). Frutas nativas do Cerrado: caracterização físico-química e fonte potencial de nutrientes. In: Sano, S. M.; Almeida, S. P. Cerrado: ambiente e flora. Pla naltina: Embrapa-CPAC, , 247-285.

3. Alves, G. L., & Franco, M. R. B. (2003). Headspace gas chromatography/mass spec trometry of volatile compounds in murici (Byrsonima crassifolia L Rich). Journal of Chromatography A, , 985, 297-301.

4. Andrade, R. S. G., Diniz, M. C. T., Neves, E. A., & Nóbrega, J. A. (2002). Determina ção e distribuição de ácido ascórbico em três frutos tropicais. Eclética Química, 27, 393-401.

5. Bezerra, J. C. B., Silva, I. A., Ferreira, H. D., Ferri, P. H., & Santos, S. C. (2002). Mollus cicidal activity against Biomphalaria glabrata of Brazilian Cerrado medicinal plants. Fi toterapia, , 73, 428-430.

6. Brazil, Health., & Ministry, . National Agency of Sanitary Surveillance Ordinance SVS/MS nº 27, 01.13. (1998). Brasília, Brazil.

7. Brito, M. A., Pereira, E. B. C., Pereira, A. V., & Ribeiro, J. F. (2003). Cagaita: biologia e manejo. Planaltina. DF: Embrapa Cerrados. 80p.

8. Cao, G., & Prior, R. L. (1999). Measurement of oxygen radical absorbance capacity in biological samples. Methods of Enzymology, , 299, 50-62.

9. Cardoso, L. M., Martino, H. S. D., Moreira, A. V. B., Ribeiro, S. M. R., Pinheiro, Sant'., & Ana, H. M. (2011). Cagaita (Eugenia dysenterica DC) of the Cerrado of Minas Gerais, Brazil: Physical and chemical characterization, carotenoids and vitamins. Food Re search International, , 44(7), 2151-2154.

10. Cardoso, F. S. N., Carvalho, L. M. J., Gomes, P. B., Ramos, M. F. S., Tebaldi, N., , M., Deodoro, I. N., Ferreira, N. A., & Figueiredo, R. E. P. (2011). Clarification of cagaita(Eugenia dysenterica L) pulp by microfiltration. EFFoST 2011 Annual Meeting: Proc ess, Structure- Function Relationships. November, Berlin, Germany., 09-12.

11. Carvalho, L. M. J., Castro, I. M., Silva, C. A. B., Fonseca, R. B., & Silva, E. M. M. (2006). Effect of enzymatic hydrolysis on particle size reduction in lemon juice (Citrus limon L), cv. Tahiti. Brazilian Journal of Food Technology, , 9, 277-282.

12. Carvalho, L. M. J., Ribeiro, E. M. G., Moura, M. R. L., Viana, D. S., Motta, E. L., Barbi, N., & Figueiredo, V. (2009). Study of volatile compounds in cagaita. CIGR-VI Interna tional Symposium on Food Processing. Monitoring Technology in Bioprocesses and Food Quality Management. Potsdam. Germany. 31 August-02 September.

13. Carvalho, L. M. J., Ribeiro, E. M. G., Viana, D. S., Moura, M. R. L., & Vieira, A. C. R. A. (2010). Use of different solvents to determine cagaita antioxidant activity. IUFOSTth World Congress of Food Science and Technology. Shangay. China., 2010-15.

14. Chaves, L. J., & Telles, M. P. C. (2006). Capítulo 7: Cagaita in Frutas Nativas da Re gião Centro-Oeste. Brasília: Embrapa Recursos Genéticos e Biotecnologia, 2006. 320 p.

15. Dembitsky, V. M., Poovarodom, S., Leontowicz, H. M., Vearasilp, S., Trakhtenberg, S., & Gorinstein, S. (2011). The multiple nutrition properties of some exotic fruits: Biological activity and active metabolites. Food Research International, 44 (7), 1671-1701.

16. Donadio, L. C., Môro, F. V., & Servidone, A. A. (2002). Frutas brasileiras. Jaboticabal: Novos Talentos Ed.. 288p.

17. Genovese, D. D., & Lozano, J. E. (2000). Effect of cloud particle characteristics on the viscosity of cloud apple juice. Journal of Food Science, 65(4), 641- 645.

18. Gomes, P. B., Carvalho, L. M. J., Cardoso, F. N., Tebaldi, N., Ribeiro, E. M. G., & Car valho, J. L. V. (2011). Carotenoids in cagaita (Eugenia dysenterica DC): whole and lyophilized pulps. Book of Abstracts: EFFOST 2011 Annual Meeting-Process- Struc ture Function Relationships. November, Berlin, Germany., 9-11.

19. Gomes, P. B. (2012). Cagaita pulp clarification by microfiltration and evaluation of carotenoids losses after membrane process. MsC Thesis. Rio de Janeiro Federal Uni versity, Pharmaceutical Sciences MSc and Ph.D Programm, Rio de Janeiro, Brazil.

20. Gómez-Cadenas, A., Mehouachi, J., Tadeo, F. R., Primo-Millo, E., & Talón, M. (2000). Hormonal regulation on fruitlet abscission induced by carbohydrate shortage in Cit rus. Planta, , 210, 636-643.

21. Gutiérrez, R. M. P., Mitchell, S., & Solis, R. V. (2008). Psidium guajava: A review of its traditional uses, phytochemistry and pharmacology. Journal of Ethnopharmacology, , 117, 1-27.

22. Kaur, C., & Kapoor, H. C. (2001). Antioxidants in fruits and vegetables the Millenni um's Health. International Journal of Food Chemistry, , 36, 703-725.

23. Kyungmi, M., & Ebele, S. E. (2008). Flavonoid effects on DNA oxidation at low con centrations relevant to physiological levels. Food and Chemical Toxicology, , 46, 96-104.

24. Kulkarni, A. P., Aradhya, S. M. E., & Divakar, S. (2004). Isolation and identification of a radical scavenging antioxidant punicalagin from pith and capillary membrane of pomegranate fruit. Food Chemistry, , 87, 551-557.

25. Lavon, R., Goldschmidt, E. E., Salomon, E., & Frank, A. (1995). Effect of potassium, magnesium and calcium deficiencies on carbohydrate pools and metabolism in Cit rus leaves. Journal of American Society for Horticultural Science, , 120, 54-58.

26. Lee, S. K., & Kader, A. A. (2000). Preharvest and postharvest factors influencing vita min C content of horticultural crops. Postharvest Biology and Technology, 20, 207-220.

27. Leterme, P., Buldgen, A., Estrada, F., & Londoño, A. M. (2006). Mineral content of tropical fruits and unconventional foods of the Andes and the rain forest of Colombia. Food Chemistry, , 95, 644-652.

28. Lima, A. (2008). Caracterização química, avaliação da atividade antioxidante in vitro e in vivo, e identificação dos compostos fenólicos presentes no pequi (Caryocar brasi liense Cambi). Tese de Doutorado- Faculdade de Ciências Farmacêuticas, Universi dade de São Paulo, 219p.

29. Liu, F. (2005). Antioxidant activity of garlic acid from rose flowers in senescence ac celerated mice. Life Sciences, , 77, 230-240.

30. Lorenzi, H. (2000). Árvores brasileiras: manual de identificação e cultivo de plantas arbóreas do Brasil, 2, 3. Ed. Nova Odessa: Plantarum.

31. Martinotto, C., Paiva, R., Soares, F. P., Santos, B. R., & Nogueira, R. C. (2008). Cagai teira (Eugenia dysenterica DC). Lavras: MG. Boletim Técnico, Evaluation of physical, chemical and sensory properties of pawpaw fruit Assimina triloba as indicator of ripe ness. Journal of Agricultural and Food Chemistry, 42, 968-974., 78, 1-21.

32. Martins, B. A. (2006). Avaliação físico-quimica de frutos do cerrado in natura e proc essados para a elaboração de multimisturas. Dissertação de Mestrado-Programa de Mestrado em Ecologia e Produção Sustentável em Ecologia e Produção Sustentável, Universidade Católica de Goiás, 85p.

33. Mercadante, A. Z. (1999). Chromatographic separation of carotenoids. Archive Latin Am. Nutri. 49: 52S-57S.

34. Penteado, M. V. C. (2003). Vitaminas: aspectos nutricionais, bioquímicos, clínicos e analíticos. Barueri, São Paulo: Manole. 612 p.

35. Oliveira, F. M. N., Figueirêdo, R. M. F., & Queiroz, A. J. M. (2006). Análise comparati va de polpas de pitanga integral, formulada e em pó. Revista Brasileira de produtos Ag roindustriais, 8 (1), 25-33.

36. Ortolá, A. G., Monerri, C., & Guardiola, J. L. (2007). Fruitlet age and inflorescence characteristics affect the thimming and the increase in fruitlet growth rate induced by auxin applications in Citrus. Acta Horticulturae, , 463, 501-508.

37. Pawliszyn, J. (1997). Solid phase microextraction: theory and practice. New York: Wi ley VHC. Inc. 247 p.

38. Podsedek, A. (2007). Natural antioxidants capacity of Brassica vegetables: a review. LWT: Food Science and Technology, , 40, 1-11.

39. Ribeiro, E. M. G., Carvalho, L. M. J., Viana, D. S., Soares, A. G., Gomes, P. B., Barros,H. D., & Moura, M. R. L. (2011). Antioxidant capacity in cagaita fruit (Eugenia dysen terica DC) using the oxygen radical absorbance capacity

(ORAC) assay. Book of Ab stracts. ISEKI 2011, Milan, Italy. August, September, 02., 3.

40. Ribeiro, E. M. G. (2011). Atividade antioxidante e polifenóis totais da casca e da pol pa do fruto da cagaita (Eugenia dysenterica DC). Master Science Thesis. Rio de Janeiro Federal University. Pharmacy College. Pharmaceutical Sciences Pos Graduation Pro grame.
41. Riu-Aumatell, M., Castellari, M., & López-Tamames, E. (2004). Characterization of volatile compounds of fruit juices and nectars by HS/SPME and CG/MS. Food Chemis try, , 87, 627-637.
42. Roesler, R., Malta, L. G., Carrasco, L. C., Holanda, R. B., Sousa, , & , C. A. S. (2007). Ciência e Tecnologia de Alimentos,, , 27(1), 53-60.
43. Sánchez-Palomo, E., Díaz-Maroto, M. C., & Pérez-Coello, M. S. (2005). Rapid deter mination of volatile compounds in grapes by HS-SPME coupled with GC-MS. Talan ta, , 66(5), 1152-1157.
44. Scherer, R., & Godoy, H. T. (2009). Antioxidant activity index (AAI) by the 2,2 di phenyl-1-picrylhydrazyl method. Food Chemistry, , 112, 654-65.
45. Silva, R. S. M., Chaves, L. J., & Naves, R. V. (2001). Caracterização de frutos e árvores de cagaita (Eugenia dysenterica. DC.) no sudeste do estado de Goiás. Brasil. Ciência e Tecnologia de Alimentos, , 23(2), 330-334.
46. Silva, M. R., Santos-Junior, R. T. O., & Ferreira, C. C. C. (2008). Estabilidade da vita mina C em cagaita in natura e durante a estocagem da polpa e refresco. Pesquisa Agro pecuária Tropical, march., 38(1), 53-58.
47. Silva, M. R., Lacerda, D. C., Santos, G. G., & Martins, D. M. (2008). Caracterização química de frutos nativos do cerrado. Ciência Rural, , 38(6), 1790-1793.
48. Singleton, V. L., Orthofer, R., & Lamuela-Raventos, R. M. (1999). Analysis of total phenols and other oxidation substrates and antioxidants by means of Folin Ciocalteu Reagent. Methods in Enzymology, , 299, 152-178.
49. Souza, E. R. B., Naves, R. V., Carneiro, I. F., Leandro, W. M., & Borges, J. D. (2002). Crescimento e sobrevivência de mudas de cagaiteira (Eugenia dysenterica DC) nas condições do Cerrado. Revista Brasileira de Fruticultura, 24 (2), 491-495.
50. US National Academy of Sciences (USA).(2000). Dietary Reference Intakes for vita min C, vitamin E, selenium and carotenoids. Washington DC: National Academy Press, 506p.
51. Valente, A., Albuquerque, T. G., Sanches-Silva, A., & Costa, H. S. (2001). Ascorbic acid content in exotic fruits: A contribution to produce quality data for food composi tion databases. Food Research International, doi:10.1016/j. foodres.2011.02.012,

52. Valko, M., Leibfritz, D., Moncol, J., Cronin, M. T. D., Mazur, M., & Telser, J. (2007). Free radicals and antioxidants in normal physiological functions and human disease. International Journal of Biochemistry & Cell Biology, , 39, 44-84.

53. Vallilo, M. I., Garbelotti, M. L., Oliveira, E., & Lamardo, L. C. A. (2005). Característi cas físico-química dos frutos do cambucieiro (Camponesia phaea). Revista Brasileira de Fruticultura, , 27(2), 241-244.

54. Van Den, Dool. H., & Kratz, P. D. (1963). A Generalization of the retention index sys tem including linear temperature programmed gas-liquid partition chromatography. Journal of Chromatography, , 11, 463-471.

55. Veira, R. F., Costa, T. S. A., Silva, D. B., Ferreira, F. R., & Sano, S. M. (2006). Frutas Nativas da Região Centro-Oeste do Brasil. Embrapa Recursos Genéticos e Biotecnolo gia. Brasília, DF.

56. Zucchi, M. I., Brondani, R. P. V., & Pinheiro, J. B. (2003). Genetic structure and gene flow in Eugenia dysenterica DC in the brazilian cerrado utilizing SSR markers. Genetics and Molecular Biology, 26 (4), 449-457

2
Chapter

THE BEAN - INHERENTLY CONNECTING AGRICULTURE AND HUMAN WELFARE

1. INTRODUCTION: AN OVERVIEW

Human living necessitates a consistent provision of food rich in various vitamins, minerals, trace elements, amino acids, vital fatty acids, and starch. Improvements in crop production have predominantly taken place in cereals such as rice, wheat, and maize, while grain legumes like beans and lentils have only realized a fraction of these advancements.[1]. The transition has adversely affected human well-being [2], as cereals, following polishing or dehusking, retain minimal quantities of protein and minerals.

The Leguminosae family is particularly intriguing due to its high protein content and its capacity to fix atmospheric N2, rendering it independent of fuel-derived nitrogen fertilizers. The common bean (*Phaseolus vulgaris* L.) is consumed more than any other grain legume. Due to its significance, it is frequently referred to as the 'poor man's meat,' however this analogy may not adequately reflect the value of the bean [3]..Beans are abundant in the amino acids lysine and methionine, rendering them complementary to grains. Moreover, they are abundant in nutritional fiber and possess a low oil content. Beans exhibit significant genetic diversity, tailored to specific local conditions and culinary inclinations. An assessment of the diverse collections, specifically by CIAT and the USDA Plant Germplasm System for beneficial features, has commenced; nonetheless, advanced plant breeding of the bean remains limited. [e.g. 4, 5].

Beans are consumed as mature grain and immature seeds as well as green pods and leaves taken as vegetables [6]. As early as 1958, the UN organisation FAO organised a conference where the production and consumption of bean were discussed. In this context, [7] noted that data on production and consumption on grain legumes generally were incomplete. It seems plausible that this condition prevails till today given that a large proportion of the bean crops are produced for home consumption in backyards and small gardens and frequently it is also intercropped with maize by smallholders as a secondary crop. Consequently, reliable statistics may be difficult to obtain regarding production.

Bridging agriculture and human wellbeing is the answer to major challenges like world hunger, diminishing natural resources, and climate changes. The bridging can be done in two ways, either by enhancing the content of nutrients in the starch-rich stable food or by enhancing the accessibility of nutrient-dense food in the diet. Acknowledging beans importance in the diet of large segments of the world population, we will in this chapter explore possibilities to bridge the production side with the consumption side. This we will do by focussing on enhancing the amounts of important nutrients in our dominant diets.

Improving the nutritional content of available food can be achieved by traditional fortification by processing dietary components.Alternatively, it can be achieved by 'biofortification', which seeks to enhance the genetic foundation of plant foods to increase their nutritional value.

Enhancing our availability to nutrient-dense food necessitates a reevaluation, as these dietary items may already be integral to the traditional diet.This perspective necessitates that local production and productivity serve as our focal point, acknowledging that individuals' distinct preferences and cultures may affect their choices in producing various cultivars.This perspective necessitates the involvement of individuals in the process [8], and this chapter will explore this using the Phaseolus bean as a model for a nutrient-rich component of the diet.

2. THE BEAN

Improving the content in the starch-rich food elements like wheat, rice and maize is obviously possibly but the starting point is very low (Table 1). The grain legumes, on the contrary, have a high starting point from where to seek improvements. Beans are superior to cereals in their macro- and micronutrient content as demonstrated in Table 1, in agreement with [9] although trials with other pulses under farmers' conditions have demonstrated that genetic potential are not always expressed under more marginal conditions. Furthermore, the legumes holds a potential for entering the diet in a diversity of ways, ranging from the dry mature seeds, to green seeds and pods as well as leaves used as vegetables, see also [10]. An efficient bridge to human wellbeing can thus be established by enhancing the access to and intake of the beans with their high nutrient density.

The production and the uses of legumes decrease in some regions while it increases in others. Brazil and Argentina have become major producers and exporters of soya bean due to its value in the feed industry, while the production of grain legumes for home consumption decreases steadily in a country like Bangladesh [12]. A historical view since 1970 show however a consistent decline in the average annual consumption of grain legumes per capita from 9 to 7 kg per person [6].

Table 1. Nutrient content of two grain legumes (Cajanus cajan: pigeonpea and Phaseolus vulgaris L.: bean) and maize (with or without husk) cultivated under farmers' conditions in eastern and southern Africa. Included is also the content of rice, wheat and potato flour sampled from various shops. After [11] and Høgh-Jensen, unpublished data).

	Protein (%)	Mg (%)	P (%)	S (%)	K (%)	Ca (%)	B (ppm)	Na (ppm)	Cr (ppm)	Mn (ppm)	Fe (ppm)	Ni (ppm)	Cu (ppm)	Zn (ppm)	Mo (ppm)
Bean	25.0	0.171	0.396	0.178	1.450	0.177	11.0	23.0	0.00	18.0	65.0	1.00	3.00	38.0	28.0
Pigeonpea	23.6	0.157	0.370	0.126	1.710	0.110	11.4	10.1	0.14	14.0	29.9	3.69	11.8	23.2	1.22
Maize	8.4	0.122	0.380	0.206	0.430	0.005	-	-	-	7.9	33.2	0.43	2.84	29.0	0.34
Maize dehusked	1.1	0.002	0.024	0.113	0.015	0.011	0	96.3	0.40	0.15	3.58	0.29	0.18	0.9	0.02
Potato flour	0.7	0.005	0.009	0.097	0.121	0.017	0	41.0	0.37	0.5	7.5	0.10	0.09	0.6	0.01
Wheat flour	15.1	0.010	0.030	0.162	0.398	0.030	0	11.6	0.31	14.7	29.6	0.12	3.37	21.7	0.81
Basmati ris	4.2	0.030	0.012	0.162	0.132	0.043	0	9.0	0.36	12.9	5.2	0.25	2.05	23.5	0.55

In a trial with approx. 100 bean genotypes grown under relatively fertile one-site conditions in Malawi, an unexpected small variation was observed in terms of iron and zinc content of the grain. Mean contents of iron in the bean grains were 67.7 (± a SE of 0.95) and zinc were 33.6 (± a SE of 0.54) ppm (Høgh-Jensen and Chirwa, unpublished data). This demonstrated that genetic diversity may not be fully expressed when conditions are the same. However, seven of the best performing varieties were selected for subsequent trialling under varying local conditions in Malawi and Tanzania in the dry season of 2005 utilizing residual moisture. This trialling expressed on average over 230 plots selected for variation a content of 90 and 37 ppm iron and zinc, respectively (Table 2). The promising varieties consequently performed above expectations and certainly above average - even under fairly harsh conditions and less welcoming soils.

What varied the most was actually the yield between farmers. Consequently, the low hanging fruit is to focus on trialling and selecting the highest yielding varieties and to work with farmers to optimize the cultivation of beans (Table 3). Breeders have had some success by simply selecting for yields under conditions with semi-controlled drought periods [13,14] or across environments [15]. This approach does not disregard the more sophisticated breeding efforts like marker-assisted selection [e.g. 5]. However, the diversity seems yet only partly tapped, which means that local conditions to a large extent can be accommodated in a simpler trialling approach. The effect of this localness is expressed in the yield differences shown in a trialling of 6-8 bean varieties in Tanzania and Malawi, ranging from 100 kg grain per hectare to almost 3 tonnes (Table 3).

Table 2. Mean grain yield, individual grain weight, and iron and zinc content in dry matter for tested varieties in Malawi and Tanzania in the dry season of 2005.

Variety per country	Grain yield (kg DM ha-1)	Grain weight (g 1000 grains-1)	Iron (ppm)	Zinc (ppm)
Malawi				
101	1671	516	88	39
102	1410	444	88	39
103	1131	442	114	46
104	1470	478	88	38
108	1262	419	110	41
109	1749	503	91	39
Napilira	1280	408	104	42
Tanzania				
Jesca	782	324	61	33
Lyamungo85	860	358	81	32
Lyamungo90	1015	393	77	31
Selian94	1010	314	88	36
Selian97	1121	346	82	34
Uyole84	710	259	70	33
Uyole96	746	404	87	40
Wanja	924	385	78	34

Table 3. Average bean grain yield per farmer, who tested 6-8 varieties.

Farmer	Country	Grain yield (kg ha-1)	Country	Grain yield (kg ha-1)
1	Tanzania	296	Malawi	1129
2		740		1146
3		146		2006
4		634		1508
5		432		1696
6		155		1602
7		189		1600

8		99		1415
9		1924		844
10		1475		1068
11		2829		648
12		704		1518
13		1534		1241
14		1336		876
15		716		1573

3. INNOVATION IN A VALUE CHAIN THAT ALSO ACCOMMODATE HUMAN WELL-BEING

Documented trialling efforts have so far been dominated by the researchers and only including the farmers, processers, traders, etc. to a limited extend. This does not mean that actors of change like innovative farmers, NGO, etc., have not had such activities. Our experiences tell us however that many of these data are difficult to get access to as they appear in reports, notebooks, newsletters, and similar documents that are found on shelves and stores. It appears logical that such trialling efforts must be linked to a learning process. The localness must however not hinder that the conclusions from such learning processes to be made available to others. The increasing using of open online repositories of research documents, which often is termed "grey literature", is an important step to share knowledge. The increasing publication rate in open access literature is another that will bring actors of change into the knowledge stream and to our common building of joint research capacity [see e.g. 16].

Since the Second World War, the innovation model in science has been linear, although a new model - less linear - emerged in the 1990s, called the 'Triple-Helix model', based on interactions between policy, science and society [17]. Increasingly, this model is being seen as also having a fourth leg, namely that of business. The fairly sequential linear innovation approach where production >> processing >> retailing may be adequate when talking about industrialized agricultural commodities. However, when quality requirements are less standard, the development of the requested traits at the commodities at various steps along the chain may require quite different orchestrated processes [16].

Such a process have been depicted by [18], drawing on experiences from working with small and market-inexperienced farmers, small processers with limited financial and processing capacity, and more fragmented retailers where market requirements are only partly known. Due to the limited experiences and capacities along the chain, a number of learning loops are included where the various stakeholders interact. These interactions are centred on value chain forums and actions related to each transforming step in the chain. Such value chain forums are found very important to enable the adjustment and enabling of mutual learning. Included in the model are also the feedback loops and the transformation of the intelligence regarding market requirements (Figure 1).

The value chain forums can be regarded as the places that prototyping is taking place. Prototyping is an important step in the innovation process as this is where ideas are being presented, discussed and validated – or maybe even more importantly discharged. Prototyping is a very important mode of action to avoid mistakes that will be very expensive in the longer run, if the solutions are allowed to travel further up the value chain. Clearly such learning processes are a challenge to researchers as management is becoming management of the process and not management of the variables.

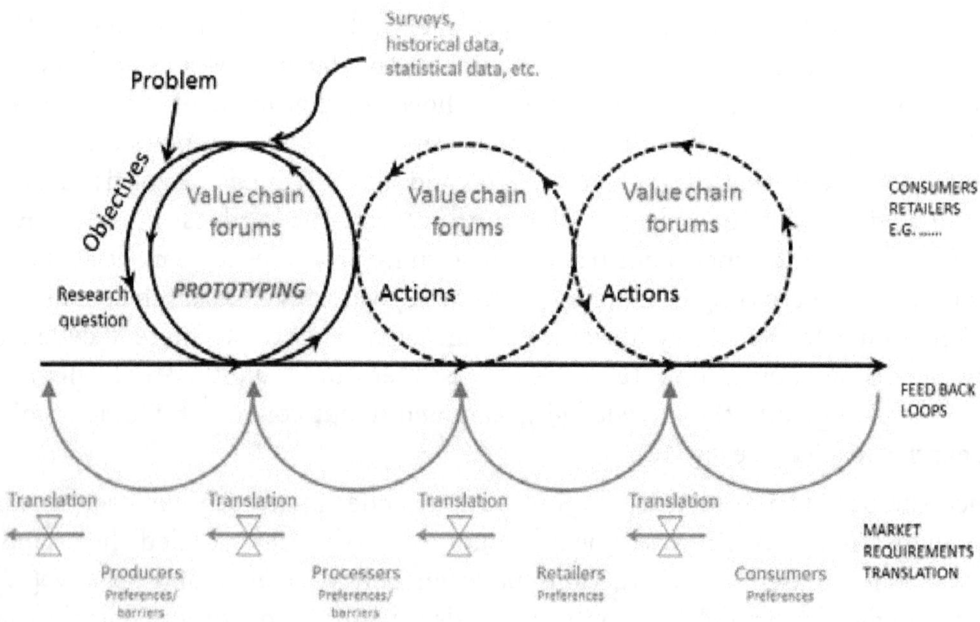

Figure 1. Prototyping in value chains innovation and development [modified after 19].

Prototypes are designed to answer questions. The prototypes need not to be sophisticated but should best be as simple as possible. Simplicity is important to keep costs down and to enable the question-solution discussion. At a moment where management wisdom insists that speed to market is the key to competitiveness, the maintenance of the learning loop is important – the cycle should be kept running to produce different ideas. Simplicity and differentiation is the two carrying principles here! But the circle MUST be stimulated by feeding in intelligence from the other actors along the chain, e.g. retailers and sellers, among others, to maintain chain agility. Consequently, innovation is not solely about technology.

Innovation in this context is more about means to obtain, consolidate, translate and manage knowledge, means to transform knowledge, and organisational learning. In that sense, innovation becomes a culture of prototyping [see e.g. 20,21]. The possibilities of including dietary requirements in the first learning loop (Figure 1) are good as long as these requirements can be quantified and described and as long as they are causal.

Nutrient dense diets can be sought in two ways. One is to find variation within one element of the diet that can form the basis for selecting the most promising in order to enhance the content. Or to seek a better production and/or access to the part of the diet that is particularly contributing with the nutrients. The later may be done by enhancing the production potential of bean varieties. It may however also be by promoting the use of the leaves for vegetable stews. [22] documented that the iron contents of the leaves compared to the mature grains could be 5-10 times higher on a dry matter basis. Leafy vegetables are indeed good sources of iron but they are mostly eaten for their vitamin-A and vitamin-C content. On a volume basis, the leafy vegetable and the boiled beans may provide similar amounts of iron. The boiled mature grain may however be a much better source of zinc [22].

4. NATURALLY BRIDGING AGRICULTURE AND HUMAN WELLBEING

To maintain productivity in agroecosystems, before the era of the fertilizer industry, humans traditionally have included animals in the systems and used their manures as fertilizers to drive the cereal production [23] in combination with grassland legumes to enhance the supply of nitrogen via symbiotic fixation [24]. Depending on locality, up to an average of 4-5 tonnes of manure could be applied per hectare in the UK [23] or as low as 1.5 tonnes per hectare in extensive mid-USA or north Spain [25].

The tropics have few examples where livestock is integrated in the agroecosystems in the same manner as frequently found under Northern temperate conditions [26). As fertilizer use in Africa is still a very modest proportion of worlds fertilizer use [27], the cereal yields per area unit has remained low (Figure 2).

The response of data like those of low yield levels (depicted in Figure 2) follows a paradigm development [26], which also is referred by [8]. During the 1960s and 1970s, an external input paradigm was driving the research and development agenda which later has been known as the 'Green Revolution'. In the early 1980s, the balance shifted from mineral inputs only, to low external input sustainable agriculture (LEISA) where organic resources were believed to enable sustainable agricultural production. During the 1990s, the Integrated Natural Resource Management research approach and ultimately the Integrated Soil Fertility Management paradigm emerged. Still it was however argued that Sub-Saharan

African farmers must use more fertilizer, improved germplasm, etc. to achieve a so-called “Second Green Revolution” [see e.g. 29].

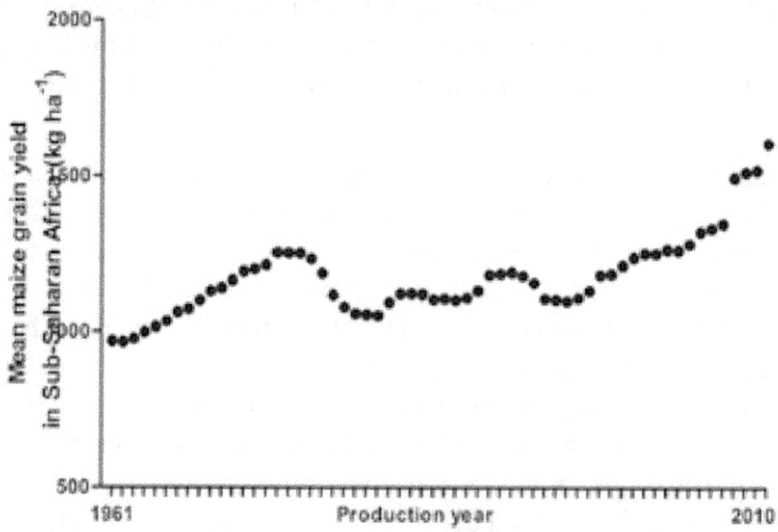

Figure 2. Official UN 5-year running average maize yield in the Sub Saharan African region between 1961 and 2010 [30].

A critical lesson from all this work is that a highly context-specific approach is required which takes into account the fertility status of the soil, the availability of organic inputs and the ability to access and pay for mineral fertilizers [28,31].

The response further assumes that the markets are perfect and that all agricultural commodities are entering a market. On one hand, large proportions of the diet of Africans are produced and consumed locally and may not enter the market. The part that enters the market may ignore the markets needs and preferences as it is sold as surplus on a local market. One commonly used model of innovation is the so-called value-chain model developed by Kline and Rosenberg, which emerged from studies of technological innovation. Modern innovation models must thus see many reverse processes and feedback loops in the incremental changes along the value chain, which further often has to include local conditions, cultural preferences, etc.

Elements in the bridge between agriculture and human wellbeing would thus be to trial for locally adapted bean varieties and to form a network among researchers that can promote a legume-based agriculture in these regions, in their particular social context. This approach would also recognize that a large proportion of the bean production occurs under conditions of significant drought stress [32], where agricultural inputs may not be an economically viable option. To overcome these particular stress conditions in combination with a vulnerable crop establishment phase, [10]) suggested investing in semi-perennial leguminous crops that has capacity to cope with short term weather variations. However, given the dominant role that beans have in nutrition in Africa and Latin America, robustness to environmental stress must be sought (Table 3) and combined with proper seed availability programmes [33].

The traditional plant-based diet is quite voluminous, i.e. it has high moisture content, with a limited protein and fat content. This is a particular challenge to children who require a diet of higher nutrient density than adults [34,35]. Some studies suggest that supplementary intake of animal protein, especially milk and fish, may stimulate childhood growth [e.g. 36]. However, some population segments may not have access to animal protein or cultural reasons limit their use of animal protein. Furthermore, dietary compositions vary over season in rural Africa and there may be temporal windows with surplus, adequate or lack of particular nutrients. Such windows may be influenced by reproduction cycles, health issues, harvest time and storage capacity, climate variability, household composition, among others [e.g. 37,38,39]. There is thus every reason to seek a higher density of nutrients in plant-based diets.

Cereals typical have a positive correlation between the nitrogen supply of the crop, thus the nitrogen content of the grain and the iron and zinc content [40]. Legumes are self-reliant on nitrogen through the biological fixation process. Consequently, correlations between nitrogen, iron and/or zinc content cannot be expected.

5. SEEKING THE NUTRIENT DENSE DIET – AN ADAPTABILITY ANALYSIS

It has frequently been assumed that farmers management and local growth conditions are fairly homogeneous and recommendations based on information generated on experimental stations dominate the extension services [e.g. 22]. However, homogeneity may be an illusion [e.g. 11]. Methods must thus be applied that allows for evaluation of performance under varying conditions.

Differentiating farmers are thus the approach in the so-called adaptability analysis [41]. This is an analysis that depicts the performance of the individual genotype across a wide range of environments versus the mean performance of the tested varieties can indicate if some varieties perform better or worse.

In terms of dry matter grain yield (Figure 3), there were no significant difference between the regression lines fitted to the observations in Malawi whereas the lines differed significantly ($p<0.05$) in Tanzania. In Tanzania, the slopes of the lines (Figure 3, right) had the following order in decreasing order: Selian97 > Selian94 > Lyamungo90 > Wanja > Lyamungo85> Uyole96 > Jesca > Uyole84.

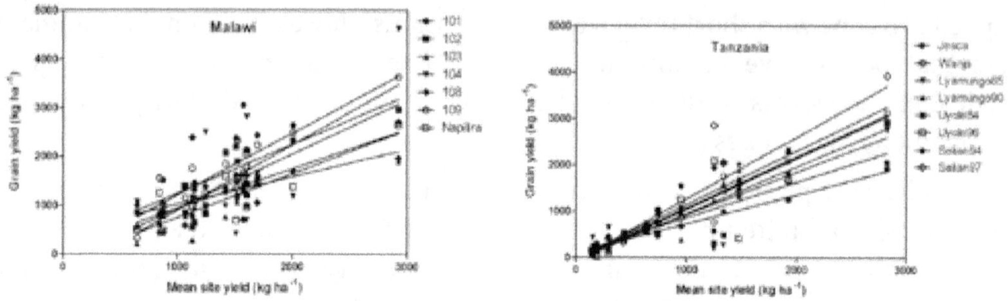

Figure 3. Individual observations and regression lines of grain dry matter yield of individual genotypes versus the mean site yield.

Phosphorus content in the grain follows pretty much a 1:1 ratio – so there is no effect of environment here as the slopes of the fitted regression lines did not differ ($p>0.05$). This is surprizing as the environment generally is considered P-scarce. The two environments clearly gave different proportions of phosphorus in the grain (Figure 4). And most observations from Malawi indicate that phosphorus in no way could be viewed as a limiting factor for beans at the current site with a mean site phosphorus proportion of 0.5% in the grain. Further, there seems no reason to believe that the individual genotypes could maintain a higher proportion of phosphorus in the gain across a phosphorus limiting environments as it appears to be the case in Tanzania.

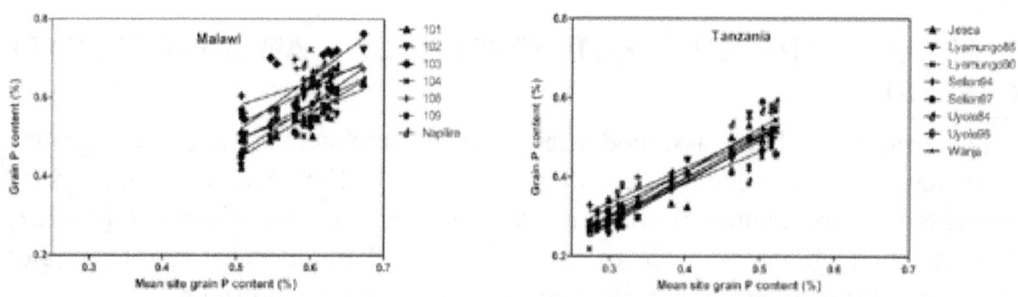

Figure 4. Individual observations and regression lines of the proportion (%) of phosphorus in the grain dry matter of individual genotypes versus the mean site yield.

A picture similar to phosphorus emerge (Figure 4) when plotting the proportions of iron in the grain (Figure 5). Obviously the two sites gave a different proportion of iron in the grain and there were tendencies to believe that some genotypes could be richer or poorer in iron than others. The 3 varieties with the highest proportion of grain iron content in Malawi were 103, 108 and Napilira while they in Tanzania were Selian94, Uyole96 and Selian97. In the Tanzanian case, the righest in iron thus seems to be the highest yielding across environments. An almost identical picture emerged regarding the proportion of zinc in the grain (Figure 6). The 3 varieties with the highest proportion of grain zinc content in Malawi were 103, 108 and Napilira while the 2 varieties with the richest zinc content in Tanzania were Selian94 and Uyole96.

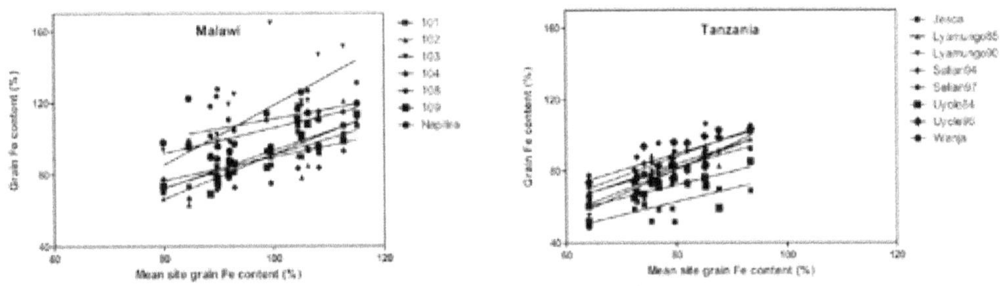

Figure 5. Individual observations and regression lines of the proportion (%) of iron in the grain dry matter of individual genotypes versus the mean site yield.

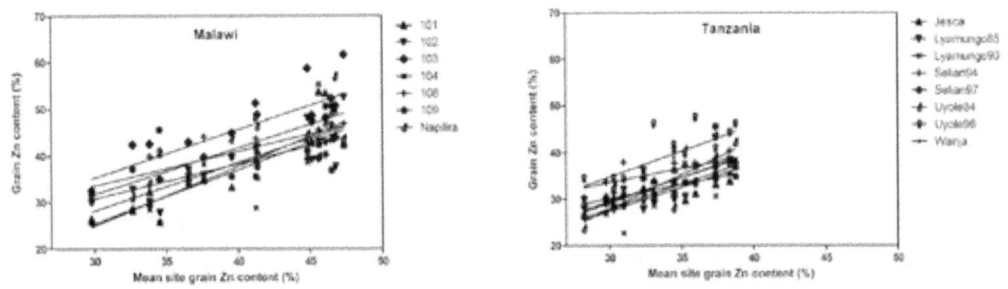

Figure 6. Individual observations and regression lines of the proportion (%) of zinc in the grain dry matter of individual genotypes versus the mean site yield.

Interestingly, however, is the fact that grain size did not appear to explain the differences between the element concentration as Malawi tended to have varieties that had individually larger grains (Table 2) and the grains with the highest proportion of phosphorus, iron and zinc in the grain dry matter. That eliminates a theory of element dilution at the end of the grain filling period which is often observed in bread wheat [42] but not always in other crops [40]. In other words, bean appears to continue to fill in elements in to the grain together with carbon while maturing.

The current data (Figures 4-5) demonstrate that efforts to find the genetic material that tend to accumulate elements, which are important for human wellbeing, in higher concentrations in the grains are justified. Naturally we may - from an evolutionary point of view - wonder what benefit the plant gets from this. But it should not stop us from utilizing this variation in modern plant breeding efforts.

However, we are in a situation where we rely on small scale farmers to increase their production substantially. This production is both for home or local consumption but even more also for industrial purposes because of the rapid urbanisation of Africa and Asia. Building on farmers' capability and knowledge of their own environments may be the best way to enhance output from agriculture. That requires innovative approaches at farm level to test and select the best suited genetic material (Figure 2) to that particular environment. This will further require new approaches to seed supply systems as "one type fits all" approach will not do the job. On the contrary, seed supply systems must build on an approach of "multiple types to fit any environment", which obviously is a major challenge to extension and research.

6. A BOWL OF BEANS

The complementarity in the amino acid composition among beans and maize has been recognized for long [7, and references herein]. Grain legumes are characterised by being markedly deficient in the essential amino acids of methionine and tryptophan but rich on lysine. Cereals normally hold more methionine than the grain legumes so a high complementarity and higher combined nutritional value could be expected. Indian scientists were front runners in documenting such efforts [e.g. 43,44]. In recent years there has been a change in the consumption of grain legumes in developed countries were they increasingly are viewed as "health foods".

The traditional plant-based diet in part of Africa and Asia can be quite voluminous, i.e. it have a high moisture content, and the protein and fat content may also be limited [35,44]. This pose a particular challenges to population segments that cannot ingest sufficient food to cover their needs, in shorter or longer periods of their lives [e.g. 35,36,37].

Dietary diversity is important for the wellbeing of humans [45,46]. An inexpensive bowl of beans or other grain legumes would benefit many people. Agriculture has the potential to supply this bowl. Here we argue that by accepting that conditions vary much locally, we will have to adapt a learning approach to selecting bean varieties based on local productivity of the various genotypes given the local pest and disease pressures, soil fertilities and soil fertility management practices, on local preferences for processing and eating the beans, on the beans role in the local cropping systems, on differentiated population and resource groups.

From the industry's perspective, enhanced yields will be advantageous as intensification will facilitate profitable production. This is exemplified by the case of soybean production in South America [47].Such instances underscore the anticipated future scenario[48] in which the industrial emphasis on certain functional qualities would intensify the concentration on the amalgamation of yields and distinct quality criteria.In a future where manufacturing must be augmented to accommodate an additional 2 billion global residents, quality attributes significant for human health and well-being may become paramount. These features must encompass iron and zinc.

Beans are predominantly propagated and replanted from prior harvests. Consequently, the locality is already reflected in the planting choices of communities. Distributing new enhanced seed varieties is, based on experience, exceedingly challenging when these crops are involved. The optimal approach for the food business to ensure ample bean supplies when collaborating with numerous smallholders is to establish contracts based on specific quality attributes. Outlet and market preferences have been shown to significantly influence farmers' behaviors.

At the time of composing this text, food prices appear to have permanently surpassed the relatively low levels established after the 2007-2008 price peak. Beans are predominantly cultivated by smallholders, hence possessing the potential to enhance family food security. This research contends that beans possess the capacity to connect agriculture and human well-being due to their nutritional value, genetic diversity, and management by local communities. The statistics indicate that farmers and change agents can enhance dietary quality by selecting the highest-performing varieties.

REFERENCES

1. Welch RM & Graham RD (2005) Agriculture: the real nexus for enhancing bioavailable micronutrients in food crops. Journal of Trace Elements in Medicine and Biology 18, 299-307.
2. Tontisirin K, Nantel G & Bhattacharjee L (2002) Food-based strategies to meet the challenges of micronutrient malnutrition in the developing world. Proceedings of Nutrition Society 61, 243-250.
3. Broughton WJ, Hermández G, Blair M, Beebe S, Gepts P & Vanderleyden J (2003) Beans (Phaseolus spp.) – model food legumes. Plant and Soil 252, 55-128.
4. Hillocks RJ, Madata CS, Chirwa R, Minja EM & Msolla S (2006) Phaseolus bean improvement in Tanzania, 1959-2005. Euphytica 150, 215-231.
5. Schneider KA, Brothers ME & Kelly JD (1997) Marker-assisted selection to improve drought resistance in common bean. Crop Science 37, 51-60.
6. Akibode S & Maredia M (2011) Global and Regional Trends in Production, Trade and Consumption of Food Legume Crops. Report March 27, 2011. Michigan State University.
7. Patwardhan VN (1962) Pulses and beans in human nutrition. American Journal of Clinical Nutrition 11, 12-30.
8. Høgh-Jensen H, Oelofse M & Egelyng H (2010) New challenges in underprivileged regions calls for people-centred research for development. Society and Natural Resources 23, 908-915.

9. Welch RM, House WA, Beebe S & Cheng Z (2000) Genetic selection for enhanced bioavailability levels of iron in bean (Phaseolus vulgaris L.) seeds. Journal of Agricultural and Food Chemistry 48, 3576-3580.

10. Høgh-Jensen H (2011) To meet future food demands we need to change from annual grain legumes to multipurpose semi-perennial legumes. In: Food Production – Approaches, Challenges and Tasks (ed.: A Aladjadjiyan). InTech – Open Access Publishers. ISBN 978-953-307-887-8. pp. 3-24.

11. Høgh-Jensen H, Myaka FA, Kamalongo D, Rasmussen J & Ngwira A (2006) Effect of environment on multi-element grain composition of pigeonpea cultivars under farmers' conditions. Plant and Soil 285, 81-96.

12. AO STAT (2012) http://faostat.fao.org/site/339/default.aspx, accessed August 2, 2012.

13. Beebe SE, Rao IM, Cajiao C & Grajales M (2008) Selection for drought resistance in common bean also improves yield in phosphorus limited and favorable environments. Crop Science 48, 582-592.

14. White JW & Singh SP (1991) Breeding for adaptation to drought. In: Common Beans- Research for Crop Improvement (eds.: A Schoonhoven & O Voyest). CIAT, Cali, Colombia. pp. 501-560.

15. Acosta-Gallegos JA & Adams MW (1991) Plant traits and yield stability of dry bean (Phaseolus vulgaris L.) cultivars under drought stress. Journal of Agricultural Science 117, 213-219.

16. The Royal Society (2011) Knowledge, Networks and Nations: Global Scientific Collaboration in the 21st Century. The Royal Society, RS Policy document 93/11.

17. Erzkowitz H & Leydesdorff L (eds.) (1997) Universities in the Global Economy: A Triple Helix of University-Industry-Government Relations. Cassell Academics, London.

18. Kline SJ & Rosenberg N (1986) An overview of Innovation. In: The Positive Sum Strategy: Harnessing Technology for Economic Growth (eds: R Landau & N Rosenberg). National Academy Press, Washington D.C. pp. 275-305.

19. Høgh-Jensen H et al. (2011) Innovation research in value chains. ICROFS Newsletter 2, 10-13.

20. Schrage M (1995) No More Teams. Doubleday Currency, New York.

21. Hughes PM & Cosier G (2001) Prototyping, people and culture of innovation. BT Technology Journal 19, 29-34.

22. Tryphone GM & Nchimbi-Mzolla S (2010) Diversity of common bean (Phaseolus vulgaris L.) genotypes in iron and zinc contents under screenhouse conditions. African Journal of Agricultural Research 5(8), 738-747.

23. Brassley P (2000) Plant Nutrition. In: The Agrarian History of England and Wales (ed.: J Thirsk). Part I, 1850-1914, vol. VII. Cambridge University Press, Cambridge. pp. 533-548.

24. Kjærgaard T (2003) A plant that changed the world: the rise and fall of clover 1000-2000. Landscape Research 28, 41-49.

25. Tello E, Garrabou R, Cussó X, Olariata JR & Galán E (2012) Fertilizing methods and nutrient balance at the end of traditional organic agriculture in the Mediterranean bioregion: Catalonia (Spain) in the 1860s. Human Ecology: e-print online.

26. Sumberg J (2003) Towards a disaggregated view of crop-livestock integration in Western Africa. Land Use Policy 20, 253-264.

27. FAO (2011) Current world fertilizer trends and outlook to 2015. FAO, Rome.

28. Bationo A (2009) Soil fertility – paradigm shift through collective action. Observatory on Science, Technology, and Innovation for ACP Agricultural and Rural Development. 25/9/2009.

29. Conway G & Toenniessen G (1999) Feeding the world in the twenty-first century. Nature 402, Supp: C55-C58.

30. FAO STAT (2012) http://faostat.fao.org/site/291/default.aspx. Data accessed October 4, 2012.

31. Bationo A, Waswa B, Kihara J and Kimetu J (eds.) (2007) Advances in Integrated Nutrient Management in Sub-Saharan Africa: Challenges and Opportunities. Springer, Dordrecth.

32. Graham PH & Ranalli P (1997) Common bean (Phaseolus vulgaris L.). Field Crops Research 53, 131-146.

33. David S, Mukandala L & Mafuru J (2002) Seed availability, an ignored factor in crop varietal studies: A case study of bseans in Tanzania. Journal of Sustainable Agriculture 21, 5-20.

34. Dewey KG & Brown KH (2003) Update on technical issues concerning intervention complementary feeding of young children in developing countries and implication for intervention programme. Food Nutrition Bulletin 24, 5-28.

35. Ogbonnaya JA, Ketiku AO, Mojekwu CN, Mojekwu JN & Ogbonnaya JA (2012) Energy, iron and zinc densities of commonly consumed traditional

complementary foods in Nigeria. British Journal of Applied Science and Technology 2(1), 48-57.

36. Hoppe C, Mølgaard C & Michaelsen KF (2006) Cow's milk and linear growth in industrialized and developing countries. Annual Review of Nutrition 26, 131-173.
37. Darmon N, Ferguson E & Briend A (2002) Linear and nonlinear programming to optimize the nutrient density of a population's diet: an example based on diets of preschool children in rural Malawi. Americal Journal of Clinical Nutrition 75, 245-253.
38. Ndekha M, Tulmala T, Vaahtera M, Cullinan T, Salin M-L & Ashorn P (2000) Seasonal variation in the dietary sources of energy for pregnant women in Lungwena, rural Malawi. Ecology of Food and Nutrition 38(6), 605-622.
39. Torheim LE, Ferguson EL, Penrose K & Arimonds A (2010) Women in resource-poor settings are at risk of inadequate intakes of multiple micronutrients. The Journal of Nutrition 140, 20515-20585.
40. Cakmak I, Pfeiffer WF & McClafferty (2010) Biofortification of durum wheat with zinc and iron. Cereal Chemistry 87, 10-20.
41. Hildebrand PE & Russell JT (1996) Adaptability Analysis: A Method for the Design, Analysis and Interpretation of On-Farm Research-Extension. Iowa State University Press, Ames IA.
42. Fan MS, Zhao FJ, Feirweather-Tait SJ, Poulton PR, Dumham S & McGrath SP (2008) Evidence of decreasing mineral density in wheat grain over the last 160 years. Journal of Trace Elements in Medicine and Biology 22, 315-324.
43. Phansalkar SV, Ramachandran M & Patwardhan VN (1957) Nutritive value of vegetable proteins. I. Protein efficiency ratios of cereals and pulses and the supplementary effects of the addition of a leafy vegetable. Indian Journal of Medical Research 45(4), 611-621.
44. Someswara Rao K, Swaminathan MC, Swarup S & Patwardhan VN (1959) Protein malnutrition in South India. Bulletin World Health Organisation 20(4), 603–639.
45. Ferguson E, Gibson RS, Opare-Obisaw C, Osei-Opare F, Lamba C & Ounpuu S (1993) Seasonal food consumption patterns and dietary diversity of rural preschool Ghanaian and Malawian children. Ecology of Food Nutrition 29, 219-234.
46. Moursi MM, Arimond M, Dewey KG, Tréche S, Ruel MT & Delpeuch F (2008) Dietary diversity is a good predictor of the micronutrient density of the diet of 6- to 23- month-old children in Madagascar. The Journal of Nutrition 138, 2448-2452.

47. Pengue WA (2005) Transgenic crops in Argentina: The ecological and social debt. Bulletin of Science Technology Society 25(4), 314-322.
48. Tomei J & Upham P (2009) Argentinean soy-based biodiesel: An introduction to production and impacts. Energy Policy 37(10), 3890-3898.
49. FEWS NET (2012) Price watch: May food prices, June 29, 2012

3 Chapter PRODUCTION OF SAKE IN THE JAPANESE ALCOHOLIC BEVERAGE SECTOR

1. INTRODUCTION: AN OVERVIEW

SAKE production is a significant segment of the Japanese culinary industry. It has preserved a robust connection with the culture in its production regions, like to other alcoholic beverages like wine, beer, and tequila in other countries. SAKE boasts a history spanning 1000 years, with brewers' abilities and procedures developed scientifically even prior to the establishment of the science of chemistry. Specifically, low-temperature sterilization of SAKE was performed in the 16th century, prior to Louis Pasteur's invention of pasteurization. The technique is meticulously delineated in ancient Japanese texts.

The importance of SAKE culture and its traditional brewing processes has been examined through contemporary scientific analysis and brewing research methodologies. Moreover, in SAKE brewing, distinctive approaches have been investigated, including low-temperature fermentation, attaining alcohol concentrations over 18% without distillation, open fermentation systems devoid of sterilization, and the development of a fruity scent in SAKE. Moreover, yeast, mold, and the raw material—rice—have been cultivated to be appropriate for SAKE production. Recent findings have clarified the preferences for SAKE among young customers aged 20 to 30, highlighting the possibility for fresh SAKE development. This article outlines the history of SAKE, its propagation methods, production materials, and latest research developments.

2. HISTORY OF SAKE

Cultivation of rice, the raw material for SAKE brewing, originated in China. Seed rice harvested more than 10 millennia ago have been found in Kiangsi province and Hunan province in China. Probably, Japanese rice was introduced from China, where rice was cultivated in dry fields using dry rice cultivation methods. Introduced from China early, rice was cultivated in dry fields in Japan also. However, the method of rice cultivation in paddy fields boosted yields to higher levels than those achieved in dry fields. The wet method brought social changes: a reliable labor force is necessary for cultivation by planting rice in paddy fields, harvesting it, and maintaining paddy fields, equipment, and irrigation. The labor force resources from families were limited. More labor was required from settlements. The settlements formed communities. Later communities formed ancient Japan. At that time, rice was of particular value: it was a divine food. SAKE, made from that divine food of rice, was also revered as blessed. It was used as a sacrifice to the gods. Moreover, people believed in a divine spirit indwelt in rice. Extending that belief, people believed that intoxication by alcohol beverages as SAKE made from rice brought gods into the human body. Furthermore, they solidified the community by sharing divine foods as rice and beverage as SAKE among members as they cooperated in rice cultivation [1].

Alcoholic beverages can be made from cereals as beer, SAKE, or whisky. Saccharification processing is extremely important. Generally, it is an important feature of Asian alcoholic beverage production that mold cultivate in cereal, so-called 'KOJI', is used for production. However, according to ancient literature, Osumi-no-Kuni-Fudoki, which recorded the culture and geography of Kagoshima in ancient times before production of SAKE using KOJI, alcoholic beverages were made with saliva as a saccharifying agent with a method of chewing rice in the mouth. It was produced by that method until the eighth century [1].

In China, ancient Chinese KOJI had been used from ancient times, as described in Chinese ancient texts such as the Chi-Min-Yao-Shu. Chinese KOJI is made from barley or wheat. It is kneaded cereal flour with water and hardened as a brick or cake. Modern KOJI is made from non-heated cereal flour or wheat as material. However, it is described that ancient KOJI made from a mixture of heat-treating material with mixed non-heated wheat flour, roasted wheat flour, and steamed wheat flour, and the mixture cultivated KOJI mold after kneading with water or extraction without inoculation of seed mold. Kanauchi and co-authors [2, 3] reported their features. Results show that Aspergillus spp. was grown on and within steamed cereal cake as the dominant KOJI mold, Rhizopus spp. was grown on and within a non-heating cereal cake as the dominant KOJI mold. Furthermore, both Aspergillus spp. and Rhizopus spp. as dominant KOJI molds were grown on and within a roasted cereal cake and a cereal cake mixed with heat-treating of cereal materials of three kinds [2,3]. Mold strains were dominant selectively in cereal cake because denatured protein was impossible to decompose by Rhizopus spp., but Aspergillus oryzae was impossible to assimilate non-heated starch in wheat flour [2,3]. In modern China, non-heated cereals such as barley or peas are used, whereby Rhizopus spp. or similar physiological features have Mucor spp., which can grow on non-cereals, predominant in it. It is difficult to decompose denatured cereal protein to enhance their uptake for nutrition of micro-organisms, Rhizopus spp. has a weak protease or peptidase to grow on steamed cereals [4].

It remains unclear whether Chinese type KOJI was used for SAKE production or not by ancient Japanese. However, steamed rice is used for SAKE production where Aspergillus spp. has been used since ancient times.

Between the Asuka-Nara Era and Heian Era (5th – 8th century), imperial families and courtiers established huge craft factories of which wide areas were dedicated to crafts. Many technicians and workers were employed in them, monopolizing practical technologies of all areas. Brewery also continued in such factories, and SAKE was brewed using advanced technologies during those eras.

During the Heian period (8th century and thereafter), SAKE rose to importance for use in regional ceremonies or banquets. SAKE brewing by SAKE-NO-TSUKASA was both a SAKE brewery and a supervisory office of the imperial court. Some kinds of SAKE were brewed for emperor, imperial family, and the aristocracy for use at ceremonies or banquets. Brewing methods were described in ENGISHIKI, an ancient book of codes and procedures related to national rites and prayers. For example, GOSYU was specially brewed for the Emperor using steamed rice, KOJI, and mother water. The mash was fermented using wild fermentative yeast for ten days; then the mash was filtrated. The resultant SAKE was used for subsequent brewing as mother water. The SAKE brewed steamed rice, KOJI, and strained SAKE brewing were repeated four times to produce SAKE with a very sweet taste [1]. The literature in this period described SAKE of more than two kinds. The minor aristocracy and many people were not able to drink SAKE because it was extremely expensive.

Between the later Heian Era and Muromachi Era via the Kamakura Era (12th–16th century) SAKE was produced and sold at Buddhist temples and private breweries. During that period, it was a popular alcoholic beverage. In the 12th century, the feudal government issued alcohol prohibition laws many times to maintain security. Officers destroyed SAKE containers throughout cities.

At the beginning of the Muromachi Era, according to the 'GOSHU-NO-NIKKI', SAKE was brewed already using a modern process in which rice-KOJI and steamed rice and water were mashed successively step-by-step. Moreover, the techniques applied lactic-acid fermentation, which demonstrates protection of the mash from bacterial contamination and dominant growth of yeast during SAKE production [1, 5].

During the 16th century, the TAMON-IN Diary was written for 100 years. TAMON-IN were small temples belonging to the KOFUKUJI temple in Nara. The diary described heating methods used to kill contaminated germs already in this century. In Europe, Louis Pasteur announced low-temperature pasteurization of wine and milk in 1865. However, Japanese brewers had acquired experimentally pasteurized SAKE during the 16th century [5].

During the Edo Era, the brewing season extended from the autumnal equinox to the vernal. However, results show good tasting SAKE brewing conducted in midwinter using a method called 'KANZUKURI'. The brewing techniques of those brewers in the Ikeda, Itami, and Nada districts (Osaka City and Hyogo Prefecture) held the leadership in SAKE brewing at that time. During the Genroku period (end of the 17th Century), the total number of breweries was reported as greater than 27,000 [5].

After the Meiji Era, SAKE brewing methods changed drastically based on European science. Many improvements of SAKE brewing were accomplished by applying beer-brewing methods directly. However, many special techniques are used in SAKE production. For example, mold culture is not required in beer brewing. Japanese brewers built the technology of SAKE production which mixed European beer brewing and old Japanese traditional techniques during the Meiji Era [5].

3. SAKE MATERIALS

3.1 Water

Water is an important material used in SAKE brewing, accounting for about 80% (v/v) of SAKE. It is used not only as the material but also in many other procedures such as washing and steeping of rice, washing of bottles or SAKE tanks, and for boiling. Generally, approx. 20–30 kl of water is necessary to process one ton of rice for SAKE brewing [5, 6]. The water for SAKE brewing must be colorless, tasteless and odorless; it must also be neutral or weakly alkaline, containing only traces of iron, ammonia, nitrate, organic substances, and micro-organisms. In particularly, iron ions are injurious to SAKE, giving it a color and engendering deterioration [5, 6]. Therefore, iron in brewing water is removed using appropriate treatments such as aeration, successive filtration, adsorption (with activated carbon or ion-exchange resins) and flocculation (with a reagent of alum) [7, 8].

3.2 Rice

The quality of rice, the principal raw material of SAKE, strongly affects the SAKE taste, but details of its effects are not clearly elucidated. Contrary to the other Asian alcohol production, Japonica short-grain varieties are used for SAKE production. In Korea and Taiwan, other short-grain varieties might also be used for alcoholic beverages.

3.2.1 Grain size

Large grains are suitable for SAKE production. Figure 1 shows rice grain size. The grain size is generally reported as the weight of 1,000 kernels. A weight of more than 25.0 g has been quoted as a mean value of 101 selected varieties by scholars [5, 9]. The selected varieties have a white spot in the center known as SHINPAKU, which contains high levels of starch.

3.2.2 Chemical constituents

Rice contains 70–75% carbohydrates, 7–9% crude protein, 1.3–2.0% crude fat, and 1.0 ash, with 12–15% water. Other components such as proteins or lipids in rice, excepting starch, are unnecessary for SAKE production. In fact, SAKE produced with rice having excessive proteins or lipids does not have good flavor or taste. Their compounds exist on the endosperm surface, mainly around the aleurone layer. Therefore they are removed by rice polishing. Moreover, the following have close correlations among the weight of 1,000 kernels: crude protein contents, speed of adsorption of water during steeping, and formation of sugars by saccharification of rice with amylases [5, 10].

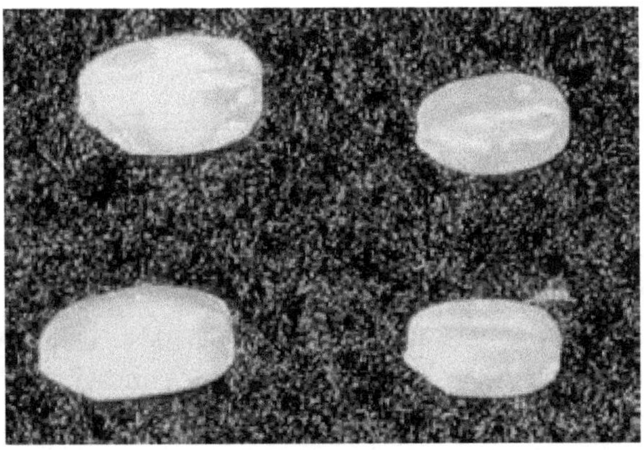

Figure 1. Rice grain size. Left side shows YAMADANISHI for variety of SAKE brewing rice. Right side shows HITOMEBORE for variety of diverting rice.

4. MICROORGANISMS

4.1 KOJI mold (Aspergillus oryzae)

The scientific name of Japanese KOJI mold is Aspergillus oryzae. It grows on and within steamed rice grains. The mold accumulates various enzymes for SAKE production. Enzymes of about 50 kinds have been found in KOJI, the most important of which are amylases. α- Amylase (Endo- α- amylase, EC.3.2.1.1) and saccharifying amylase (Exo-α-glucosidase; E.C. 3.2.1.20) play important roles in amylolytic action [5, 11]. Furthermore, proteases of some kinds are also important enzymes: acid-proteases and alkaline-proteases are found in KOJI. In SAKE mash, the enzymes decompose protein to form amino acids and peptides (oligo- amino acid) at low pH values such as pH 3–4 [5]. Furthermore, amino acids or peptide-supported yeast grow with food or nutrition. The enzyme acts indirectly, decomposing rice protein while combining to an active site of the α-amylase [12].

The taxonomy of mold was studied for Aspergillus oryzae by Ahlburg and Matsubara (1878) and Cohn (1883). A report by Wehmer (1895) was published, describing KOJI mold class A. oryzae in detailed mycological studies as an A. flavus-oryzae group.

They are slight graded on variations in morphological and physiological properties [5]. Murakami et al. identified and reported that KOJI mold strains used for SAKE brewing belonged to A. oryzae and not A. flavus. Two species were distinguished based on mycological characteristics of each authentic type culture of the two species [13, 14]. It is noteworthy that no Japanese industrial strain of KOJI mold is capable of aflatoxin production.

In SAKE brewing, conidiospores produced over bran rice, so-called TANE-KOJI, are sprayed and inoculated on steamed rice. KOJI is prepared in an incubation room, a so-called KOJI- MURO.

4.2 Yeast

4.2.1 Physiology of SAKE yeast

Fermentative multi-budding yeast, Saccharomyces cerevisiae, which has been used not only in SAKE brewery, but also in beer brewry, winery and bakery, was discovered in ca. 1830 by J. Meyen; it was named by E.C. Hansen in 1882 [5]. SAKE yeast is classified taxonomically in the Saccharomyces cerevisiae group [15]. However, the yeast was distinguished from other strains of S. cerevisiae by additional properties such as vitamin requirements [16, 17], acid tolerance, sugar osmophilic character, and adaptability to anaerobic conditions. Additionally, SAKE yeast has advantageous features that enable its growth under high sugar contents and low pH conditions, to produce SAKE under open system fermentation.

SAKE yeast formed a large amount of foam during main mash fermentation. Because one- third of the capacity of the fermentation vessel is occupied by foam during usual main fermentation, preventing foam formation would be greatly advantageous to breweries to save space occupied by the foam and scaling up the amount of MOROMI produced. Some large- molecular-weight compounds that arise from steamed rice grains are also regarded as taking part in foam formation. Recently, foam formation has involved existing proteins, with foam formation on the yeast surface.

Figure 2. TAKA-AWA foam. (Photograph by Shiraki Tunesuke Co., Ltd.)

Ouchi and Akiyama obtained foam-less mutants that have the same characteristics as the parent yeast except for foam-formation [18, 19]. A foam-less mutant of SAKE yeast, a favorite strain of Saccharomyces cerevisiae (The Brewing Society of JAPAN is distributing it as SAKE yeast), has become available for SAKE brewing. Recently, foam protein in SAKE yeast, AWA 1, was cloned. TAKA-AWA foam has been obvious molecular biologically [20].

4.2.2 Aroma production by SAKE yeast

The SAKE aroma is produced by yeast mainly because rice, as a SAKE material, has weaker aroma than materials used for wine or beer. Furthermore, SAKE contains ethanol, higher concentrations of alcohol, and many aroma-producing compounds. Aromatic compounds are an important factor used to characterize SAKE. Recently, a flavor wheel for SAKE was produced similar to existing ones used for wine and beer [21,22].

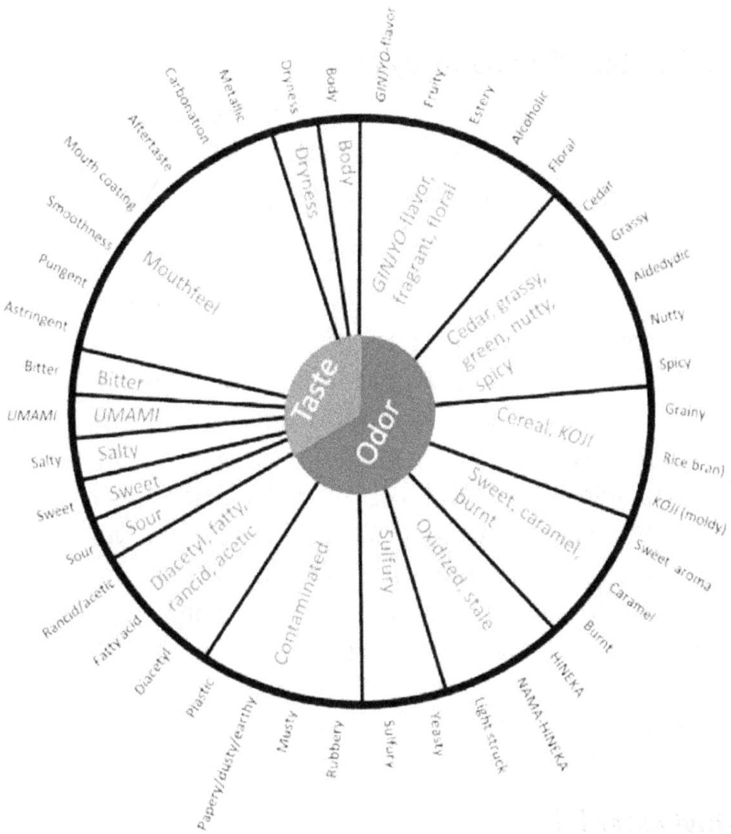

Figure 3. Flavor wheel of SAKE.

Typical yeast metabolic processes producing aromatic compounds are shown in Fig. 3.

According to this wheel, the aromas can be categorized as floral aroma, fruit-like nutty, caramel-like, and lipid-like. A fruit-like flavor is imparted to SAKE from yeast production because many Japanese consumers favor SAKE that has a fruit-like aroma. A yeast mutant producing fruity aromas was isolated for SAKE brewing. Their typical chemical components are ethyl caproate, which gives an apple-like aroma, and iso-amyl acetate or iso-amyl alcohol, which give a banana-like aroma. Before development of methods of breeding yeast to produce aromas, it was not easy for aromatic SAKE to be brewed and supplied stably for customers. Some competent SAKE brewers had controlled temperature severely to adjust enzymes that produced KOJI mold as amylase. Controlling the amounts of sugars as nutrient elements produced by amylase in mash adjusts the metabolisms of yeast growth and production of SAKE aromas as ethyl caproate and iso-amyl acetate. However, it is readily apparent that SAKE aroma synthesis by metabolic pathways or control mechanisms. The yeast producing fruity aroma was bred for use in commercial brewing [1, 5].

Higher alcohol metabolism pathway [1]

The higher alcohols as aromatic compounds are iso-amyl alcohol and iso-butyl alcohol. The alcohols are produced by two pathways by yeast as shown below.

1. RCHNH2COOH⇔RCOCOOH⇔RCHO⇔RCH2OH
2. C6H12O6⇔ RCOCOOH⇔RCHO⇔RCH2OH

In these two pathways, 2-oxo acid is produced as a precursor. In the 1 pathway, 2-oxo acid is produced by deamination reaction between Ehrlich pathways. In the 2 pathway presented above, 2-oxo acid is produced between production of amino acid pathway. Oxo acid is produced by decarbonylation reaction and reduction reaction between both pathways, similarly as ethanol is produced from acetoaldehyde via pyruvic acid as oxo acid of one kind. For example, lacking amino acids as leucine and valine in SAKE mash, the yeast produces leucine and valine in SAKE mash. Furthermore, 2-oxo acid was transaminated from other amino acids. It is controlled by the amount of amino-acid-based amino bonds. Therefore, lacking extremely amino acid in mash, 2-oxo acid is converted to higher alcohol as iso-butyl alcohol and iso-amyl alcohol. Sufficing amino acid as leucine and valine in SAKE mash, the reaction of the 2 pathway inhibited by native feedback control and uptaken amino acid are converted by the 1 pathway.

Fatty acid ethyl ester [1]

Ethyl caproate is a favorite flavor providing an apple-like aroma for Japanese consumers. This compound is produced by esterification from caproic acid as a precursor. Caproic acid is synthesized by fatty acid synthase between fatty acid synthesis pathway from acetyl-CoA and malonyl-CoA in SAKE yeast. Their synthase composes FAS 1 (Fas1p; β-subunit) and FAS2 (Fas2p:α-subunit), which are hexamer proteins (α6β6 subunit) [23]. Ichikawa reported a yeast breeding method that produces high levels of ethyl caproate that high levels of precursor of ethyl caproate were producing in yeast cells [24]. Cerulenin, an antifungal antibiotic produced by Cephalosporium caerulens, inhibits beta-ketoacyl-ACP synthase as fatty acid synthetase. A mutant of cerulenin-resistant yeast strain decreases synthesis of long-chain fatty acids by mutating Gly1250 Ser in the gene. The strain can produce high levels of caproic acid [25].

Ethyl acetate group

Higher alcohol and esterified fatty acid produce a fruity aroma in SAKE. Usually, SAKE has 0.1 ppm or higher concentrations of ester compounds. That slight amount of ester produces a fruity aroma and intensifies the SAKE flavor. Excessive esters destroy the balance of the SAKE flavor. Many ester compounds produced mainly by yeast are acetate ester groups that react and which are produced by an alcohol–acetyl transferase reaction that transfers an acetyl bond from acetyl CoA to alcohol. Alcohol acetyl transferase (AATFase; E.C. 23.1.84) catalyzes the following reaction.

Acetyl CoA + Alcohol ⇔ Acetyl ester + CoA-SH

This enzyme, a microsomal enzyme, is an endogenous membrane protein dissolving by surfactant. Furthermore, more than 70% of the activity exists in it. AATFase has two isozymes of molecular weight 56 k Da. Isozyme P1 is reacted mainly in the yeast cell. Its activity has ca. 70–80% overall activity. Its optimum temperature is 25°C (Isozyme P2 is 40°C), and the optimum pH is 8.0. The pH range of its reaction is pH 7.5–8.5 (Isozyme P2 is pH 7.0–8.5). Their enzyme inhibited phosphatidylserine and phosphatidylinositol, having interfacial activity, and oleic acid and linoleic acid. Accordingly, this phenomenon showed that this enzyme has a hydrophobic active site in it [26].

4.3 Lactic acid bacteria (Lactobacillus sakei) [5, 27]

Lactic acid bacteria are the most important bacteria in SAKE brewing. Lactic acid bacteria are defined as listed below.

1. Bacteria ferment glucose and producing more than 50% lactic acid per 1 molar of glucose.
2. Bacteria is Gram positive. Their shapes are cocci or bacci.
3. They are facultative anaerobic bacteria.
4. They have no mobility.
5. They produce no spores.

Their fermentation types are two. One is homo type, 2 molar of lactic acid fermenting from 1 molar of glucose. The other is hetero type, 1 molar of lactic acid, 1 molar of ethanol and 1 molar carbon dioxide from 1 molar of glucose. Typical lactic acid bacteria for food processing are shown as the following: Leuconostoc spp. is a hetero-type cocci lactic acid bacteria, and Pediococcus spp. is a homo type cocci lactic acid bacteria. Lactobacillus spp. belongs to both types of bacci lactic acid bacteria.

In SAKE brewing, lactic acid bacteria are used in traditional seed mash, KIMOTO production for without sterilization safety open fermentation system without sterilization. In traditional seed mash, MOTO, production, it is known that Leuconostoc mesenteroides as hetero- lactic acid fermentation grows the MOTO preparation earlier under extremely low temperatures of less than 5°C. L. sakei as a hetero-lactic acid fermentation grows in it.

It is rarely that lactic acid bacteria spoil commercial SAKE. The bacteria are called HIOCHI bacteria, and have resistance to ethanol concentrations higher than 18% in SAKE. SAKE- grown HIOCHI bacteria have turbidity and an uncomfortable cheese-like smell from diacetyl [28]. Two types of lactic acid bacteria might be involved: one is L. homohiochi (homo lactic acid fermentation type); the other is L. heterohiochi (hetero-lactic acid fermentation type). Both bacteria have resistance to ethanol. The coefficient for growth of two bacteria in SAKE is mevalonic acid, which is produced by KOJI mold. Recently, mevalonic acid nonproductive mutants have been bred for SAKE-KOJI production [1, 17, 29].

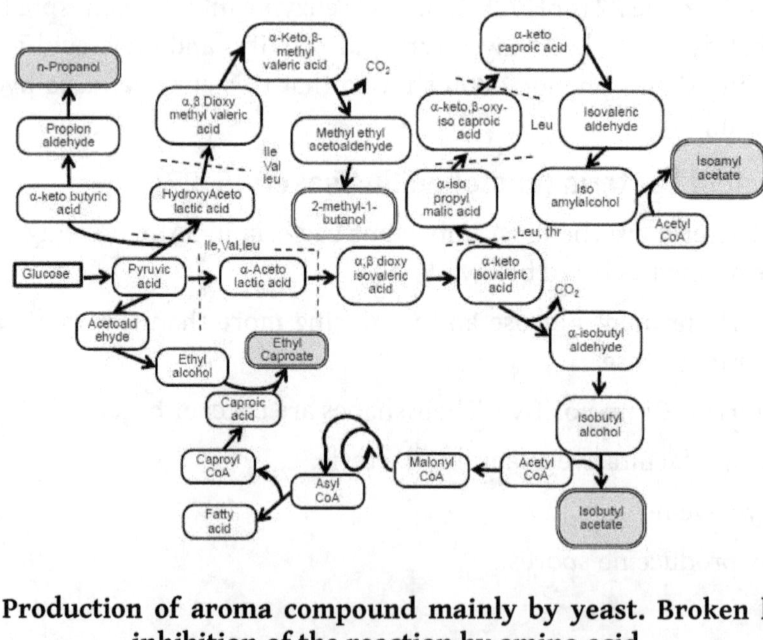

Figure 4. Production of aroma compound mainly by yeast. Broken line shows inhibition of the reaction by amino acid

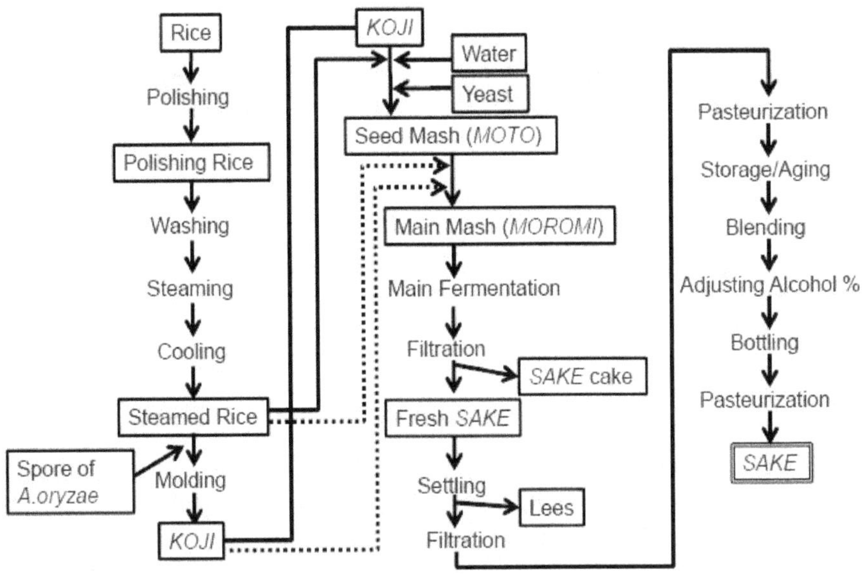

Figure 5. SAKE production.

5. SAKE PRODUCTION

5.1 Rice treatment (polishing, washing, and steeping)

In contrast to the use of malt in brewing beer or producing spirits, in SAKE brewing, polished rice is used. The main purpose of polishing is to remove unnecessary substances in rice aside from the starch, which are regarded as undesirable in SAKE brewing. Polishing re moves surface layers of the rice grains, which contain proteins, lipids, and minerals. The ratio of percentages by weight of polished rice to the original brown rice is defined as the polishing ratio. Changes in the amounts of some constituents of the processed grain with various polishing ratios are presented in Table 1 (Research Institute of Brewing, Japan, 1964). Crude fat and ash contents decrease most rapidly, whereas the protein content decreases gradually until the polishing ratio reaches 50%, after which it remains practically constant. In contrast to changes in the crude fat content, the lipid content (by hydrolysis) does not change with increase of the polishing ratio [30].

The lowest polishing ratio is strictly regulated under the Liquor Tax Law. In general, polished rice of 75–70% ratio is used for reasonably priced SAKE brewing. In contrast, polished rice of a 60% ratio is used for special brewing brands such as GINJYO-SHU, and rice of less than 50% polishing ratio is used for Grand grade SAKE, DAIGINJYO-SHU. The latter is a prestige class of SAKE. Sometimes, the SAKE is brewed using rice of a 30% polishing ratio.

Table 1. Changes in the contents of some rice grain components after polishing [5]

	Polishing ratio (%)			
	100	80	60	50
Moisture	13.5	13.3	11.0	10.5
Crude protein	6.55	5.12	4.06	3.8
Crude fat	2.28	0.11	0.07	0.05
Ash	1.00	0.25	0.20	0.15
Starch	70.9	74.3	76.3	77.6

The rice polisher depicted in Fig. 6 is used for SAKE brewing. The roller made of carborundum and feldspar rotates around a vertical axis, and scrapes the surface of grains. Rice grains supplied from the hopper are polished and fall to the bottom of the basket conveyer. The grains go through the sieve to remove the rice bran. The rice is carried by the basket conveyer to the hopper. The operation continues until the grains are polished to the required ratio [5].

Generally, with a mill having a roller that is 40 cm in diameter, average times for polishing are 6–8 h for 89%, 7–10 h for 75%, 10–13 h for 70%, and 16–20 h for 60% polishing ratio [5].

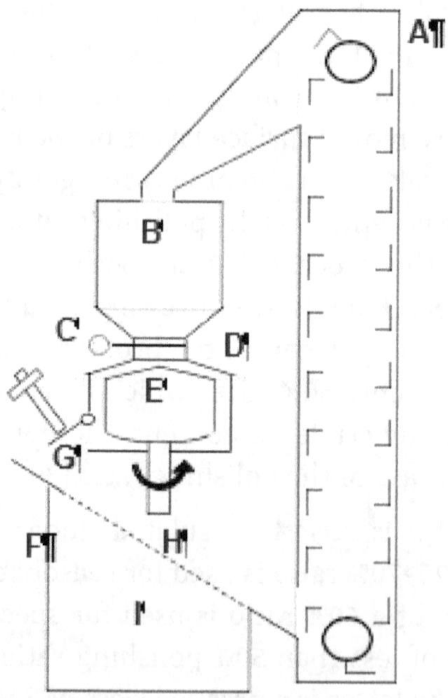

Figure 6. Diagram of a vertical type rice mill used in SAKE brewing [5]: A, the basket conveyor; B, rice hopper; C, the rice flow adjusting bulb; D, the polishing chamber; E, the roller; F, a resistance; G, exit; H, sieve and I, bran reservoir.

5.2 Washing and steeping

Rice is washed and steeped in water before steaming. During washing, the grains are polished further by collision of rice grains in water. During processing, the surface parts of the grains are removed, eliminating approx. 1–3% of the total grain weight [5]. Washed rice grains are passed into a vat and are steeped immediately in water. In washing and steeping procedures, the grains absorb water to about 25-30% of their original weight. The moisture promotes penetration of heat into the grain center during steaming and accelerates gelatinization of starch in the grains. Absorption of water is extremely important for preparing properly steamed rice, and controlling KOJI making and fermentation. The water absorption into grains differs according to the variety of rice and the polishing ratio [5, 10, 30]. Generally, rice grains are steeped in water for 1–20 h, and soft rice absorbs water within 1-3 h. Highly polished rice grains absorb water more rapidly. During washing and steeping, potassium ions and sugars are eluted from the grains [1, 31], whereas calcium and iron ions are absorbed onto the grains [5]. After steeping, excess water is drained off from the grains for about 4–8 h before steaming.

5.3 Steaming

Starch is changed to the α-form, and protein is denatured by the steaming process. Moreover, the grains are sterilized by steaming. The grains are usually steamed for 30–60 min, although previous reports show that steaming for as little as 15–20 min is sufficient to modify the starch and protein of rice produced in Japan [1]. During steaming, the grain moisture is absorbed to the extent of 7–12% of the weight of the starting rice grains, namely total water gain of about 35–40%. Historically, at many breweries, steaming processes usually generated steam from water in a large pot. Today, boilers are often used in many breweries for steaming. A steamer is a shallow and wooden tub in which is bored a hole (1/20 diameter of bottom) at bottom. The steamer is put above the 1.5–2.0 kl caldron, and rice is permeated by large amounts of steam from the caldron. Recently, a modern apparatus for steaming rice as belt conveyor type apparatus is used in automated modern breweries. The steamed rice is cooled to nearly 40°C for KOJI production, and the rice used for preparing MOTO and MOROMI-mash is cooled to less than 10°C. Breweries usually use machines to cool the steamed rice with a draft of air as it moves on the screened belt. A pneumatic conveyer system is often used to transfer steamed rice [1].

5.4 KOJI preparation [5]

A KOJI cultivates the KOJI mold, Aspergillus oryzae on and in steamed rice grains, and which accumulates various enzymes for SAKE production. For the preparation of KOJI, seed-molds are used at all breweries. The Aspergillus oryzae strains are cultivated in steamed bran rice dredging wood ash at 34–36°C for 5–6 days. This process results in abundant spore formation. Cultivation conditions influence the enzyme production. In general, higher cultivation temperatures (approx. 42°C) develop the activities of amylases. Lower temperatures (approx. 30°C) activate protease activities.

As cultivation times lengthen, more enzymic activities appear in the KOJI [32]. Nitrogenous substances and acids are accumulated more in KOJI that has been prepared from steamed rice of higher moisture contents [33]. They are regarded as related to the flavors and tastes of SAKE. After the steamed rice has been cooled to about 35°C by going through a cooling apparatus, it is transferred into the KOJI-MURO, a large incubating room, where temperature (26–28°C) and humidity are controlled at suitable levels to grow KOJI mold.

After inoculating or spraying TANE-KOJI as seed mold in the proportion of 60–100 g/1,000 kg of rice, then the mixture is heaped in the center of a table for KOJI preparation. At this stage, the temperature of the material is 31–32°C. As the spores germinate and mycelia develop, the rice begins to smell moldy like sweet chestnut. After incubation for 10–12 h, the heap of rice grains is mixed to maintain uniformity of growth, temperature, and moisture contents. After another 10–12 h, with growth of the mold, mold mycelia can be observed distinctly as small white spots on the grains. Furthermore, the material temperature has risen to 32–34°C. It is dispensed into wooden boxes, each with 15–45 kg of the grain. To control the rise in temperature and the moisture in the grain mass, the bottom of the box is made of wooden lattice or wire mesh. Temperature and moisture contents are also controlled by the thickness of the heaped grain layer in the box: 8 cm at the beginning, 6 cm at the first mixing, and 4 cm at the second mixing. Thereafter, at intervals of 6–8 h, the material is mixed and heaped again in the box. After incubation for about 40 h, the temperature of the material rises to 40–42°C. The mycelium develops to cover and penetrate the grains which have sufficient enzymes, vitamins and various nutritive substances for mashing and growth of SAKE yeast. Then the KOJI is taken out of the room and spread on a clean cloth to be cooled until it is used for mashing. α-amylase and acid-protease activities increase during KOJI making. Carbohydrates are decomposed finally to water and carbon dioxide, which engenders the production of energy for growth of the mold.

5.5 SAKE mash fermentation

5.5.1 'MOTO' as yeast starter

In SAKE brewing, MOTO is important as a yeast starter for the fermentation of MOROMI. MOTO is necessary to provide a pure and abundant yeast crop, and to supply sufficient lactic acid to prevent contamination of harmful wild yeast or bacteria during MOTO production or in the early stages of main fermentation.

In traditional MOTO preparation, lactic acid is produced by lactic-acid bacteria in the mash. In the modern method, pure lactic acid is added to the mash at the beginning of MOTO preparation. Lately, compression yeast cultivated using a method similar to that for baker's yeast used to ferment main mash safely with this yeast instead of MOTO. The amount of rice used for MOTO preparation is usually 7% of the total rice used for the entire SAKE mash.

Traditional Seed Mash

KIMOTO is a traditional MOTO. Actually, MOTO has been handed down from early times, and the MOTO was modified to be simple and convenience by Kagi et al. [34]. The modified MOTO is called KIMOTO. The YAMAHAIMOTO is based on the same microbiological principle as that of KIMOTO, and has practically replaced KIMOTO because the related procedure is simpler [5].

Steamed rice (120 kg) is mixed with 60 kg of KOJI and 200 L of water in a vessel at an initial temperature of 13–14°C. It is then kept for 3-4 days with intermittent stirring and agitation. During this period, the rice grains are partially degraded and saccharified, and the temperature falls gradually to 7–8°C. The mash is then warmed at a rate of 0.5–1.0°C/day by placing a wooden or metal cask filled with hot water in the mash after warming for an additional 10–15 days, after which the temperature reaches 14–15°C. In KIMOTO mash, some micro- organisms grow successively to each other as Fig. 7, and mash brings acid condition to grow SAKE yeast easily without contamination [5].

In early stages, contaminating wild yeast or germs disappears within the first two weeks as a result of the toxic effect of nitrite, produced by nitrate-reducing bacteria from nitrate contained in or added to the water. Slight nitrate contained in the mother water is converted to nitrite, which has toxicity for micro-organisms by nitrate-reducing bacteria such as Achromobacter spp., Flavobacterium spp., Pseudomonas spp., and Micrococcus spp. (derived from KOJI and water). A toxic substance, nitrite, yeast of one kind from KOJI as Pichia angusta [35] was assimilated after oxidating nitrite during MOTO mash. Nitrite is toxic for lactic acid bacteria and fermentative yeast in traditional MOTO. Their utility micro-organisms are able to grow under the MOTO mash containing nitrite. After removing nitrite, lactic-acid bacteria including Leuconostoc mesenteroides and Lactobacillus sakei (derived from KOJI) can grow in the MOTO mash. These bacteria multiply to reach a maximum count of about 107–108/g. However, other gram disappear before fermentation by SAKE yeast begins because of the accumulation of high concentrations of sugar and because of acidification resulting from the growth of lactic-acid bacteria [5].

The mixing process helps to dissolve the nutrients contained in the KOJI and steamed rice, and mash promotes the growth of lactic-acid bacteria in the early stages [36].

Convenient MOTO preparation as SOKUJYO-MOTO

Recently, SOKUJYO-MOTO is popular for use in SAKE brewing. It was devised by Eda [37]. It is based on the principle that addition of pure lactic acid to MOTO can prevent contamination by wild microorganisms. It takes a short time (7–15 days) to produce MOTO because of the time-saving lactic-acid formation by naturally occurring lactic acid bacteria, and saccharification of the mash proceeds quickly with the high initial mashing temperature (18–22°C). In this production, commercial lactic acid (75%, 650–700 m1/100 L of water) is added to the mash to adjust the pH value to 3.6–3.8. Although pure culture yeast is used as the inoculums, yeast grows more advantageously than do wild yeasts from KOJI. Furthermore, the latter eventually predominate during the MOTO process [5].

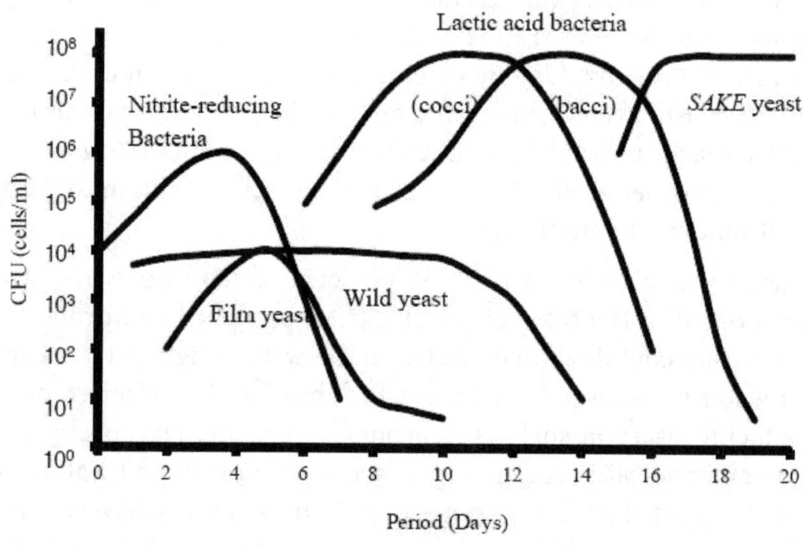

Figure 7. Changing numbers of micro-organisms in KIMOTO mash.

This predominance might be ascribed to the fact that the high mashing temperature and acidic conditions are close to the optimum for multiplication of both culture and wild yeasts. In addition, as opposed to the behavior in the classical process, no natural selection of wild yeasts by the toxic effect of nitrite occurs because the presence of lactic acid inhibits nitrate- reducing bacteria.

An example of the preparation of SOKUJYO-MOTO is the following: KOJI (60 kg) is added with 200 L of water and 140 ml of lactic acid (75%). A pure culture of SAKE yeast is inoculated to the mash (105 – 106/g). Its temperature is about 12°C. Steamed rice (140 kg) is added to the mixture, cooling it sufficiently to give a temperature of about 18–20°C. After keeping the mash for 1–2 days with intermittent stirring and agitation, it is warmed gradually in the same way as YAMAHAI-MOTO by increasing the temperature at a rate of approx. 1.0–1.5°C/day. As the temperature rises to about 15°C, SAKE yeast reaches its peak and fermentation begins.

The cultivation period can be shortened further by starting the mashing at 25°C and by keeping the temperature of MOTO over 18°C. Moreover, the variety of SOKUJYO-MOTO as KOONTOKA-MOTO (hot-mashed MOTO) is used by Japanese brewers. This mashing method is conducted at 56–60°C during several hours with subsequent inoculation of pure cultured SAKE yeast. To prevent excessive accumulation of sugars and the development of a high viscosity, the ratio of water to rice used is raised to 150–160 L/100 kg [5].

5.5.2 Main fermentation [5]

MOROMI, as main mash, is fermented in a large open vessel with a capacity ranging from 6–20 kl without special sterilization, in an open fermentation system. The weight of polished rice (1.5 t) was used for mashing one lot as standard. However, recently, larger vessels as 3– 7 tons or sometimes over 10 tons have been used for mashing one lot. The MOROMI mash is brewed steamed rice, KOJI and water. Table 2 shows proportions of various raw materials used for a typical MOROMI mash. The preparation of stepwise mashing as three steps is one characteristic of MOROMI mash production. First, steamed rice, KOJI and water are added to the MOTO. Consequently, the total acid and yeast population in MOTO are diluted to about one-half. The temperature of the first mash is about 12°C, and the yeast propagates gradually. After two days, the yeast grows until 108/g, which reaches the same order as that in MOTO. As a second addition, the materials are added in an amount that is nearly twice as much as the first addition. The yeast population and total acids are diluted by about half too. The temperature of the second addition is lowered to 9–10°C. In a third addition, materials are added in a larger amount.

Table 2. Proportions of raw materials used in a typical SAKEMOROMI [1]

	1st addition	2nd addition	3rd addition	4th addition	Total
Total rice (kg)	140	280	890	160	2000
Steamed rice (kg)	95	200	720	160	1580
KOJI rice (kg)	45	80	170		420
Water (liter)	155	250	1260	160	2460

The amount of MOROMI bring 14 folds as same as MOTO mash. Whereby yeast cells are diluted. This stepwise addition of material plays an important role in suppressing the invasion of wild micro-organisms together with lowering the mashing temperature in each addition. In SAKE brewing, temperature control is also extremely important to balance saccharification and fermentation, both of which occur simultaneously in MOROMI. Therefore, we call it 'Parallel Fermentation'. Small quantities of sugars released from steamed rice and KOJI are fermented gradually by SAKE yeast until the alcohol content reaches nearly 20% (v/v). Accumulated alcohol of 20% v/v in the mash from 40% (w/v) of sugars. If such a high concentration of sugars is supplied at once, then SAKE yeast would not ferment alcohol in the mash. Instead, the mash fermentation at a low temperature (below 10–18°C) is also a characteristic of SAKE brewing which gives the mash a balanced flavor and taste as well as a high alcohol concentration. After the third addition of materials, the mash is agitated, usually twice a day. The mash density then reaches maximum levels 3–4 days later.

A foam resembles soap suds. Furthermore, it spreads gradually over the surface, and subsequently increases to form a thick layer. A fresh fruit-like aroma at this stage indicates healthy fermentation. The fermentation gradually becomes more vigorous with a rise in mash temperature, and a rather viscous foam rises to form TAKA-AWA (a deep layer of foam, shown in Fig. 2), which reaches to the brim of the vessel. In some breweries, it is broken down with a small electric agitator. At this stage, the yeast cell count reaches a maximum of about 2.5 × 108/g [38]. Because the alcohol concentrations increase, the foam becomes less dense, and is easily dispersed. The fermentation finishes usually during 20–25 days. In some breweries, pure alcohol (30–40%) is added to the mash to adjust the final concentration to about 20–22% (v/v).

Quite often, to sweeten the mash, 7–10% of the total amount of steamed rice is added during the final stage of the MOROMI process to produce glucose from starch by the saccharifying action of KOJI that accumulates in the mash.

5.5.3 Filtration [1,5]

After alcohol fermentation, the mash is divided into SAKE and solids by filtration. The mash is poured into bags of about 5 L capacity made of synthetic fiber, which are laid in a rectangular box. SAKE is squeezed out under hydraulic pressure. After complete filtration, the solids pressed in a sheet are stripped out of the bags. Recently several automatic filter presses for filtering MOROMI mash have been used. The SAKE lees or SAKE cake, residue squeezing SAKE as cake was called 'SAKE-KASU', contains starch, protein, yeast cells and various enzymes, SAKE lees is used traditionally for making foodstuffs such as pickles and soup. In general, regarding 3 kl of SAKE containing 20% ethanol and 200–250 kg of KASU are obtained from one ton of polished rice. The slightly turbid SAKE is clarified to separate lees by standing in a vessel for 5–10 days at a low temperature.

5.5.4 Storage (aging) and bottling [5]

After settling the clarified SAKE for a further 30–40 days, The SAKE is pasteurized, killing yeasts, harmful lactic acid bacteria, and enzymes. The SAKE is heated to 60–65°C, passing it through a helical tube type heat exchanger for a short time. Recently plate-type heat exchangers with high efficiency of heat transfer have become available.

Figure 8. SAKE squeezer (Photograph by Hamada Co. Ltd)

As described in this chapter, the history of SAKE pasteurization began in the 16th century, before Pasteur's discoveries. After pasteurization, SAKE is transferred to sealed vessels for storage with or without addition of activated carbon. Pasteurization and the high content of alcohol in SAKE (usually 20%) prevent microbial infection. The blended SAKE is diluted with water to the appropriate alcohol content, usually 15.0–16.5% (v/v), and is filtered through activated carbon to improve the flavor and taste and to adjust the color and clarity. In modern procedures, filtration through activated carbon is followed by filtration through membranes or sheets having numerous pores of micrometer size, thereby removing minute particles including micro-organisms if any are present. This procedure enables the SAKE producer to omit pasteurization in the bottling procedure and therefore to prevent deterioration of quality caused by heating SAKE. The spoilage of SAKE is sometimes encountered, off-flavors and tastes are attributed mainly to the formation of diacetyl and acetic acid by HIOCHI bacteria.

SAKE is usually sold in a pale blue bottle of 1.8 l capacity, which is pervious to short and medium wavelengths in sunlight, as are beer, wine, and other alcoholic beverages. Coloring is spoilage of SAKE by sunlight, deferriferrichrysin precipitates, and tyrosine or tryptophan, kynurenic acid or flavin precipitates as precursors of colorants. Usually SAKE is aged and stored for a short time. It does not age for a long time of several years or longer. Vintage wine is aged much longer than SAKE. During storage, SAKE matures gradually. The maturation process is probably the result of oxidation reactions and physicochemical changes. SAKE changes and adopts a smoother taste. The storage temperature should be maintained carefully at 13–18°C, with consideration being devoted to the rate of maturation and the time of bottling.

SAKE is browned not only by amino-carbonyl reactions but also by still unknown reactions during aging. Long-aged SAKE has a sherry wine-like aroma that is attributable to furfurals produced in SAKE during aging. Furthermore, the SAKE taste is smooth and less stimulated by ethanol because of molecules of ethanol and water flocculate in the SAKE during aging. However, research of SAKE aging has been conducted by many researchers [1].

Recently, aging of SAKE to add value has been attempted by some breweries with so-called KOSYU as old vintage SAKE. KOSYU-SAKE has rich and complex flavors and tastes like those of cherry wine and a brown color by amino-carbonyl reaction as shown in Fig. 9. Aged SAKE can even have a chocolate color.

Figure 9. Changes of color of sake during aging.

6. VARIETIES OF SAKE [39]

In Japan, SAKE production and labeling are regulated strictly by the Liquor Tax Law. According to this law, SAKE is made from defined raw materials and methods of production as follows: 1, SAKE is an alcoholic beverage produced by fermenting materials such as rice, rice-KOJI, and water, with subsequent filtering of the material mixture. 2, SAKE is an alcoholic beverage fermenting a material such as rice, water, SAKE lees, rice-KOJI, and other material as authorized by government ordinance and filtering the material mixture. 3, SAKE is an alcoholic beverage filtrate of a mixture of SAKE and SAKE lees. Moreover, SAKE has been categorized as grand, first and second class, by alcohol concentration, and sensory evaluation by officers until 1992. However, labels and names of SAKE have not been regulated by law. For various reasons, many commercial products, SAKE which labels producing method or excessive name, was sold in the market and low-quality SAKE also was sold. Furthermore, many consumers were confused and purchased it mistakenly. Whereby, they were regulated by law in 1992.

Table 3. Classification of SAKE types by law [39]

SAKE type	Used material	Used rice Polishing ratio	Requirement
GINJYO -SYU	Rice, rice KOJI and pure distilled alcohol	Less than 60%	Fermentation at low temperature; fruity- flavor; good and clear appearance
DAI-GINJYO-SYU	Rice, rice KOJI and pure distilled alcohol	Less than 50%	Fermentation at low temperature; fruity- flavor; good and clear appearance
JYUNMAI-SYU	Rice, rice KOJI	-	Good flavor; good and clear appearance
JYUNMAI-GINJYO-SYU	Rice, rice KOJI	Less than 60%	Fermentation at low temperature; fruity- flavor; good and clear appearance
JYUNMAI-DAI-GINJYO-SYU	Rice, rice KOJI	Less than 50%	Fermentation at low temperature; fruity- flavor; good and clear appearance
TOKUBETSU-JYUNMAI	Rice, rice KOJI	Less than 60%	Especially good flavor; good and clear appearance
HONJYOZO-SYU	Rice, rice KOJI and pure distilled alcohol	Less than 70%	Good flavor; good and clear appearance
TOKUBETSU-JYUNMAI	Rice, rice KOJI and pure distilled alcohol	Less than 60%	Especially good flavor; good and clear appearance

Instead of SAKE grades such as grand grade, fiesta grade and second grade that had been used until 1992, SAKE is categorized as DAIGINJYO-SHU, GINJYO-SHU, JYUNMAI-SHU, JYUNMAI-DAIGINJYOU-SHU, JYUNMAI-GINJYOU-SYU, TOKUBETSU-JYUNMAI-SYU, HONJYOZO-SYU, and TOKUBETSU-HONJYOZO-SHU, and the labeling SAKE is regulated by the law as shown in Table 3.

The polishing rice ratio and using KOJI ratio regulated by the law to sell their categorized SAKE. Then they must be shown on the label. JYUNMAI means that SAKE is brewed using only rice and rice-KOJI and mother water, and GINJYO means special brewing. DAI-GINJYO means special brewing and prestige class in the SAKE brewery. Consequently, DAI-GINJYO tends to be expensive, but the price of SAKE is decided by the policy of the brewery. Additionally, KOSYU as aged vintage SAKE or NAMAZAKE as non-pasteurized SAKE is displayed on the SAKE label. It is necessary that some method or public organization manage other SAKE label items.

7. SAKE TASTES

As explained in this chapter, SAKE is a favorite food and beverages and individual favored SAKE and components of SAKE are important factors for purchasing SAKE. Over 500 chemical compounds exist in SAKE, producing a complex flavor and taste in SAKE.

SAKE consumption has decreased since the 1970s, it is 1.7 million kL. Recently, in 2009, SAKE consumption is about one-third that of the 1970s. According to a survey of household spending conducted by the Public Management Ministry in Japan [40], consumers in their 20s spend 1100 yen per month for SAKE, those in their 30s spend 2500 yen per month, and those in their 60s spend 3800 yen per month. Elder consumers spend three times as much as young consumers. To examine favorite tastes of young consumers (20s–30s) play a role to development of new SAKE for them and to increase SAKE consumption in Japan.

Table 4. Correlations for Evaluating SAKE using Sensory Evaluation Methods

	Appearance	Intensity of aroma	Appeal of aroma	Intensity of sourness	Intensity of bitterness	Intensity of sweetness	Appeal (Balance) of tastes	Preference of consumer
Appearance	1.000							
Intensity of aroma	0.450	1.000						
Appeal of aroma	-0.280	0.000	1.000					
Intensity of sourness	-0.590**	-0.300	0.100	1.000				
Intensity of bitterness	-0.200	0.160	-0.430*	0.120	1.000			
Intensity of sweetness	0.030	-0.380 **	0.210	0.030	-0.570	1.000		
Appeal (Balance) of tastes	-0.050	-0.190	0.550 *	0.070	-0.860 **	0.660**	1.000	
Preference of consumer	-0.110	-0.450 *	0.650 **	0.120	-0.800 **	0.670**	0.880**	1.000

*(Correlations are significant; **, $P < 0.01$, *, $P < 0.05$)*

Suzuki and co-authors [41] investigated the opinions and preferences of panelists (22 persons, 20s–30s) to conduct a SAKE sensory evaluation for research into favorite tastes and consumer preferences. The correlation of sensory evaluations of SAKE are presented in Table 4. Correlation was found between 'Intensity of aroma' and 'Balance of taste', for which the correlation factor is 0.55, and the relation between 'Appetite of consumers', with a correlation factor of 0.65. However, it showed a negative relation with 'bitterness', and the correlation factor was -0.430. Young consumers hope to buy and drink SAKE having a favorite flavor. Furthermore, correlation was found between 'Preferences of consumers', and 'Balance of taste', with a correlation factor is 0.88. 'Preferences of consumers', showed a relation with 'Intensity of bitterness', with close correlation factor of -0.80. Furthermore, a negative and close statistical correlation was found between 'Balance of SAKE tastes' and 'Intensity of bitterness', for which the correlation factor was -0.86. These data show that consumers hope to buy or drink SAKE with no bitter taste. 'Bitter' is a decreased balance of SAKE test and consumer appetite. Although bitter taste has played an important role in giving richness-taste to SAKE for a long time, young consumers are sensitive to bitter tastes in SAKE. It is therefore considered that the control of bitter taste must be undertaken in brewing processes.

8. CONCISE SUMMARY

SAKE brewing requires the application of high-quality processes that have been established experimentally, independent of the scientific method. Moreover, distinctive methods have been investigated, including low-temperature fermentation, achieving over 18% high alcohol concentration without distillation, utilizing an open fermentation system without sterilizing, and producing a fruity scent in sake. SAKE brewing utilizing solely rice as an ingredient can still yield fruity fragrances reminiscent of apple, melon, or banana. Specially produced SAKE for Japanese SAKE contests has 6–7 ppm of ethyl caproate, a notably high concentration for alcoholic beverages. This is one reason why the production of ethyl caproate yeast has been promoted and supported by governmental institutions across several Japanese prefectures. Nevertheless, significant skepticism remains over the adequacy of their SAKE adaptation to benefit customers. In a questionnaire survey, young consumers (ages 20–30) perceive SAKE as a beverage intended for the elderly [42]. This contributes to the perception that sake is an inexpensive alcoholic beverage. SAKE consumption is anticipated to decline due to the declining Japanese population, attributed to the nation's low birthrate and high life expectancy. All brewers and researchers in the SAKE industry must strive to produce high-quality SAKE that meets customer preferences, both domestically in Japan and for export to overseas markets. Moreover, SAKE can be esteemed by aficionados, like to 'Châteaux' wines, Grand cru, in European nations.

REFERENCES

1. Yoshizawa K. SAKE NO KAGAKU (Science of Alcoholic Beverages). Asakura, Tokyo, Japan; 1995.
2. Kanauchi, M., Shindo, H., Suzuki, M., Kakuta, T., Yoshizawa, K., & Koizumi, T. Characteristics of traditional wheat-qu (koji) described in the classic literature, "Chi min yao shu", of the ancient Chinese. (in Japanese) J. Brew. Soc. Jpn. (1998)., 93, 721-729.
3. Kanauchi, M., Shindo, H., Suzuki, M., Kakuta, T., Yoshizawa, K., & Koizumi, T. Role of extract from cockleburr [Xanthium strumarium] leaves used for wheat-qu (koji) making described in Chinese old literature Chi min yao shu. (in Japanese) J. Brew. Soc. Jpn. (1998)., 93, 910-915.
4. Tanaka, T., Okazaki, N., & Kitani, M. Comparison of Growth and Enzyme Production between A. oryzae and Rhizopus spp. Growth of Mold on Uncooked Grain (II), (in Japanese) J. Brew. Soc. Jpn. (1982)., 77(11), 831-835.
5. Rose AH. Economic Microbiology. Vol. 1, Alcoholic Beverages. Academic Press, New York; 1977.
6. Nojiro K, Kamata K, Tadenuka M, Yoshizawa K, Mizunuma T. JOZO NO JITEN (Encyclopedia of Brewage, Fermentation and Enology), Asakura, Tokyo, Japan; 1988.
7. Nanba Y, Momose H, Ooba T. SEISYU SEIZOU GIJYUTU (Techniques of SAKE Brewing), (in Japanese) Society of Brewing, Japan Tokyo; 1979.
8. Totuka A, Namba Y, Kobuyama Y. Removal of Metal ion from water by Poly Aluminum Chloride. (in Japanese) J. Brew. Soc. Jpn. 1971;67: 162-166.
9. Sato J, Yamada M. Research of Brewer's Rice using Physical and Chemical Methods, (in Japanese) Reports of the Research Institute of Brewing, Japan 1925;93: 506-642.
10. Yoshizawa K, Ishikawa T, Hamada Y. Sutadied on brewer's Rice (III), (in Japanese) J. Soc. Brew. Jpn. 973;68: 767-771.
11. Shimada S, Sugita O. in Mizumoto K. Studies of Aspergillus oryzae Strains for Sake- Brewing (IV) : On he Amylase Actions of "Koji" for Sake-Brewing. (in Japanese) Journal of Fermentation Technology. Osaka 1953;31: 498-501.
12. Siinoki T. The absorbing and decomposing Steamed rice by a-maylase in SAKE mash, (in Japanese) J. Brew. Soc. Jpn. 1984;79: 840-845.
13. Murakami H. Classification of the koji mold (19-23), (in Japanese) Reports of the Research Institute of Brewing, Japan 1972;144: 1-25.

14. Murakami H, Makino M. Classification of the koji mold (9). (in Japanese) Reports of the Research nstitute of Brewing, Japan 1968;140: 4-11.

15. Kodama K. In 'The Yeasts', (A. H. Rose and J. S. Harrison, eds.), volume 3, Academic Press, London, UK; 1970, pp. 225-282.

16. Takeda M, Tsukahara T. The Characteristics of SAKE Yeast (I). (in Japanese) Journal of the Fermentation Association of Japan 1965;23(8): 352-360.

17. Nojiro K, Kosaki M, Yoshii H. JOZOGAKU (Brewing, Fermentation and Oenology), Kodansha Tokyo, 1993) (in Japanese)

18. Ouchi K, Akiyama H. Non-foaming mutants of sake yeasts. Selection by cell agglutination method nd by froth flotation method. Agr. Biol. Chem. 1971;35: 1024-1032.

19. Nunokawa Y, Ouchi K. Sake brewing using foamless mutantsof sake yeast Kyokai No 7. (in Japanese) J. Brew. Soc. Jpn. 1971;66: 512-517.

20. Shimoi H, Sakamoto K, Okuda M, Atthi R, Iwashita K, Ito K. The Awa1 gene is required for the foam-forming phenotype and cell surface hydrophobicity of sake yeast. Appl Environ Microbiol. 2002;68(4): 2018-2025.

21. National Research Institute of Brewing; [http://www.nrib.go.jp/data/pdf/seiyoutai01.pdf] (in Japanese)

22. Japan Sake and Shochu Makers Association; [http://www.japansake.or.jp/sake/ english/pdf/no_4.pdf]

23. Stoops JK, Wakil SJ. The yeast fatty acid synthetase. Structure–function relationship and the role of the active cysteine-SH and pantetheine-SH. J. Biol. Chem. 1981;256: 8364-8370.

24. Ichikawa E, Hcsokawa N, Hata Y, Abe Y, Suginami K, Imayasu S. Breeding of sake yeast with improved ethyl caproate productivity. Agric. Biol. Chem. 1991;55: 2153-2154.

25. Tsutsumi H. Studies on producing aroma compound by SAKE yeast. (in Japanese) J. Biosci. and Bioeng. Jpn. 89: 717-719.

26. Minetoki T, Bogaki T, Iwamatsu A, Fujii T, Hamachi M. The purification, properties and internal peptide sequences of alcohol acetyltransferase isolated from Saccharomyces cerevisiae Kyokai No. 7. Biosci. Biotechnol. Biochem. 1993;57(12) :2094-8.

27. Soomro AH, Masud T, Kiran A. Role of Lactic Acid Bacteria (LAB) in Food Preservation and Human Health – A Review, Pakistan Journal of Nutrition 2002;1(1): 20-24.

28. Tomiyasu S. The flavor of spoilage HIOCHI-SAKE. (in Japanese) Journal of Fermentation Technology 933;10: 515-518.

29. Sugama S, Iguchi T. A study of the prevention of sake spoilage - Development of S.I. medium and its applications to prediction of hiochi phenomena. (in Japanese) J. Brew. Soc. Jpn. 1970;65: 720-725.

30. Yoshizawa K, Ishikawa T, Noshiro K. Studies of brewer's Rice (I). (in Japanese) J. Brew. Soc. Jpn. 1973; 68: 614-617.

31. Yoshizawa K, Ishikawa T, Unemoto F, Noshiro K. Studies of Brewer's Rice (II), (in Japanese) J. Brew. Soc. Jpn. 1973b;68: 705-707.

32. Suzuki M, Nunokawa Y, Imajuku I, Teruuchi Y, Uruma M. Studies of brewage KOJI - Comparison of temperature, period and each enzyme activity Preparation of KOJI at Laboratoy Scale –. (in Japanese) J. Brew. Soc. Jpn. 1956;51: 318-322.

33. Nunokawa Y. Studies of protease in Koji (IIII): the specificity of substrate of acid protease and lkaline protease. (in Japanese) J. Agri. Chem. Soc. Jpn. 1962;36: 884-890.

34. Kagi K, Otake I, Moriyama Y, Ando F, Eda K, Yamamoto T. Research of Brewing YAMAHAIMOTO, (in Japanese) Reports of the Research Institute of Brewing. Japan 1909;29: 1-38.

35. Shimaoka Y, Kanauchi M, Kasahara S, Yoshizawa K. The Elimination of Nitrite by Pichia angusta Y-11393 Isolated from Sake Koji (in Japanese) J. Brew. Soc. Jpn. 2012;107(7): 517-528.

36. Ashizawa C. Studies on Micro-flour in YAMAHAI MOTO (10) - Cocci and Bacci Lactic acid Bacteria –. (in Japanese) J. Brew. Soc. Jpn. 1965;60(10): 900-903.

37. Eda K. SOKUJYOMOTO. J. Brew. Soc. Jpn. 1909;4: 5-12.

38. Nojiro K. Sprinkle of SAKE yeast in SAKE mash and growth of the yeast in it. (in Japanese) J. Brew. Soc. Jpn. 1959;54: 658-661.

39. National Tax Agency; [http://www.nta.go.jp/shiraberu/senmonjoho/sake/hyoji/ seishu/gaiyo/02.htm]

40. Public Management Ministry :(http://www.e-stat.go.jp/SG1/estat/List.do?lid=000001070349)

41. Suzuki Y, Kanauchi-Kamiya H, Kanauchi M, Ishido T, Morita A, Tsubota Y. The factors of taste determining consumer preference for Sake by consumers in their 20s or 30s J. Soc. Brew. Jpn. 2012;107(9): 699-705.(in Japanese)

42. Sudo S., (http://www.nrib.go.jp/kou/pdf/46kou06.pdf

4 Chapter

STRUCTURING FAT FOODS

1. INTRODUCTION

1.1 Fat roles

Food fat provides taste, consistency, and helps us feel full. Fat is a major source of energy for the body, and aids in the absorption of lipid soluble substances including vitamins A, D, E, and K. Dietary fat is essential for normal growth, development, and maintenance, and serves a number of important functions. Increasing evidence indicates that fatty acids and their derived substances may mediate critical cellular events, including activation and ex pression of genes, and regulation of cellular signaling [1].

When and how humans learned to use fats and oils is unknown, but it is known that primi tive people in all climates used them for food, medicine, cosmetics, lighting, preservatives, lubricants, and other purposes. The use of fats as food was probably instinctive, whereas the other applications most likely resulted from observations of their properties and behavior under various environmental conditions. More than likely, the first fats used by humans were of animal origin and were separated from the tissue by heating or boiling. Recovery of oil from small seeds or nuts required the development of more advanced methods of proc essing, such as cooking, grinding, and pressing processes [2].

The total global oil and fat market is a huge economic factor. The rise of affluence in devel oping countries, this market is increasing and can be expected to increase further. The main fats commonly consumed are vegetable oils and fats, dairy fat and fats derived from ani mals, e.g. lard, tallow and fish oil [3].

Refining edible oils such as neutralization, bleaching, and deodorization, has been practiced for just over a century, but it has had a great impact on eating habits. Whereas the refining processes have increased the availability of sufficiently palatable oils, the oil modification processes (hydrogenation, interesterification, and fractionation) have increased the useful ness of edible oils by increasing their interchangeability [4].

2. FATS AND OILS

Fats and oils are water insoluble substances that are a combination of glycerin and fatty acids called triacylglycerols. Fats appear solid at ambient temperatures and oils appear liquid. Seeds, fruits, animal, and marine sources provide oils and/or fats; however, only a few of these sour ces are of economic importance. Fats and oils are the most concentrated source of energy of the three basic foods (carbohydrates, proteins, and fats), and many contain fatty acids essential for health that are not manufactured by the human body. Fats and oils are commonly referred to as triacylglycerols because the glycerin molecule has three hydroxyl groups where a fatty acid can be attached. The triacylglycerol structure is affected by the present and the position of at tachment (alpha, sn-1; middle, sn-2; outer, sn-3) of each fatty acid to the glycerin. The chemical and physical properties of fats and oils are largely determined by the fatty acids that they con tain and their position within the triacylglycerol molecule [2].

2.1 Fatty acids

Fatty acids consist of elements, such as carbon, hydrogen, and oxygen, which are arranged as a linear carbon chain skeleton of variable length with a carboxyl group at one end. Fatty acids can be saturated (no double bond), monounsaturated (one double bond), or polyunsa turated (two or more double bonds), and are essential for energetic, metabolic, and structur al activities. An unsaturated fatty acid with a double bond can have two possible configurations, either cis or trans, depending on the relative positions of the alkyl groups.

2.1.1 Fatty acids occurrence

The fatty-acid carbon-chain lengths vary between 4 and 24 carbon atoms with up to three dou ble bonds, with C18 the most common. Over 1000 fatty acids are known with different chain lengths, positions, configurations and types of unsaturation, and a range of additional sub stituents along the aliphatic chain. However, only around 20 fatty acids occur widely in nature; of these, palmitic, oleic, and linoleic acids make up ~80% of commodity oils and fats[4].

The most prevalent saturated fatty acids are lauric (C-12:0), myristic (C-14:0), palmitic (C-16:0), stearic (C-18:0), arachidic (C-20:0), behenic (C-22:0), and lignoceric (C-24:0). The most important monounsaturated fatty acids are oleic (C-18:1) and erucic (C-22:1). The es sential polyunsaturated fatty acids are linoleic (C-18:2) and linolenic (C-18:3) [2].

2.1.2 Saturated fatty acids

Saturated fatty acids contain only single carbon-to-carbon bonds and are the least reactive chemically [2]. Saturated acids with 10 or more carbons are solids, and melting points increase with chain length. Melting points alternate between odd and even chain length, with odd chain lengths having a lower melting point than the preceding even chain acid [4]. Most of the satu rated fatty acids occurring in nature have unbranched structures with an even number of car bon atoms. These acids range from short-chain-length volatile liquids to waxy solids having chain lengths of ten or more carbon atoms. Fatty acids from 2 to 30 carbons (or longer) do occur, but the most common and important acids contain between 12 and 22 carbons and are found in many different plant and animal fats.Saturated fatty acids are also functionally divided into short- and long-chain acids and are most widely known by their trivial names. The short-chain saturated acids (4:0–10:0) are known to occur in milk fats and in a few seed fats [1]. Medium chain fatty acids (8:0, 10:0, 12:0, and 14:0) occur together in coconut and palm kernel oils, both tropical commodity oils. In both of these oils, lauric acid (12:0) predominates (45 to 55%). Pal mitic acid (16:0) is the most abundant and widespread natural saturated acid, present in plants, animals, and microorganisms. Palm oil is a rich commodity oil source and contains over 40% of palmitic acid. Stearic acid (18:0) is also ubiquitous, usually at low levels, but is abundant in co coa butter (~34%) and some animal fats, e.g., lard (5 to 24%) and beef tallow (6 to 40%). A few tropical plant species contain around 50 to 60% of 18:0 [4]. The long-chain saturated acids (19:0 and greater) are major components in only a few uncommon seed oils.

2.1.3 Unsaturated fatty acids

Unsaturated fatty acids contain one or more carbon-to-carbon double bonds and are liquid at room temperature with substantially lower melting points than their saturated fatty acid counterparts. Monounsaturated fatty acids have only one double bond in the carbon chain and polyunsaturated fatty acids have two or more double bonds in the carbon chain [2]. Pol yunsaturated fatty acids, sometimes referred to as PUFAs or polyalkenoic acids, can be div ided into conjugated (double-bonded carbon atoms alternate with single bonds) and unconjugated (double bonds are separated by one or more carbon atoms with only single bonds) [1].

The most common monounsaturated is oleic acid (18:1 9c). Oleic acid is found in most plant and animal lipids and is the major fatty acid in olive oil (70 to 75%) and several nut oils, e.g., macadamia, pistachio, pecan, almond, and hazelnut (filbert) contain 50 to over 70%. High oleic varieties of sunflower and safflower contain 75 to 80% oleic acid. Cis-monounsaturated with 18 or less carbons are liquids at room temperature or low-melting solids; higher homo logues are low-melting solids. Trans-monounsaturated are higher melting, closer to the cor responding saturated acids. Double bond position also influences the melting point; both cis- and trans-C18 monounsaturated are higher melting when the double bond is at even po sitions than at odd positions [4]. Saturated fatty acids are very stable, but unsaturated acids are susceptible to oxidation; the more double bonds the greater the susceptibility. Unsaturat ed fatty acids, therefore, have to be handled under an atmosphere of inert gas (e.g. nitrogen) and kept away from oxidants or substances giving rise to free radicals [5].

2.1.4 Trans fatty acids

Monosaturated and methylene-interrupted polyunsaturated fatty acids are predominantly cis. Trans isomers, mainly monosaturated, are produced during catalytic partial hydrogenation, and can be present in substantial amounts in hardened fats, generally as a mixture of positional isomers. Heat treatment during deodorization of commodity oils may result in low levels of trans isomers, particularly of polyunsaturated. The undesirable nutritional properties of trans fatty acids have led to alternative ways of producing hardened fats, such as interesterification or blending with fully saturated fats, and to the use of milder deodorization procedures [4]. It is important to note that trans double bonds do occur in natural fats, as well as in industrially processed fats, but generally much less abundantly than cis bonds. Thus some seed oils have a significant content of fatty acids with trans unsaturation [5]. Saturated and trans fatty acids have a higher melting point than unsaturated and cis fatty acids [1].

2.1.5 Health problems

Concerns about possible toxic effects of fatty acids with trans-unsaturation began with the publication of results of experiments with pigs given diets containing hydrogenated vegeta ble fat for 8 months. They had more extensive arterial disease than those given otherwise equivalent diets devoid of trans-unsaturated fatty acids. Subsequently, numerous animal feeding trials, epidemiological studies of human populations and controlled dietary experi ments with human subjects have been reported [5]. In January 2003 the US Food and Drug Administration (FDA) instituted a requirement to list trans fat content as a separate item on the Nutrition Facts label on packaged foods from 2006. This change in labeling requirements has served as a catalyst to accelerate food product reformulation. On a voluntary basis, many food manufacturers and restaurants have reformulated their products and modified their operations to reduce trans fats in their offerings [6, 7].

Consumption of trans fatty acids raise the level of low density lipoprotein (LDL) cholester ol and decrease the level of high density lipoprotein (HDL) cholesterol. Based on results of epidemiological and intervention studies it is clear that these changes in blood profiles in crease the risk of coronary heart diseases. The main food sources for trans fatty acids are cookies and confectionary, snacks, and frying fats [8]. Consumption of significant amounts of trans fatty acids has been a major health concern for consumers and regulatory agen cies over the past decade. The major dietary sources of trans fatty acids are products for mulated with partially hydrogenated fats. Examples include margarines, shortenings, bakery products, and fast foods. The regulatory mandate from FDA and consumer con cerns have led to the development of alternative processes to produce foods with zero or reduced trans fatty acids contents [7, 9].

Fats and oils can be formulated as trans-acid-free products, but saturates are required for the solids contents that provide the functionality for plastic and liquid products. Reduced satu rates may be an option in some cases, but a saturate-free product is probably impossible if functionality is to be maintained [2]. Obviously, if the fat is completely hydrogenated there will be no double bonds and hence no problem; however, partially hydrogenated fats have trans double bonds. Trans double bonds are rare in naturally occurring fats, the major natu ral source is milk fat because they are formed by bacterial action in the rumen. So, most nat urally occurring oils and fats have cis double bonds; however, some trans double bonds are found in milk fat and some marine oils.

2.1.6 Low-trans

The new rules about trans fatty acids promise to strongly affect what is acceptable to con sumers and food manufacturers. It will be difficult to meet all the demands for low-trans fats and other traits that are important to consumers with the current technology, especially for frying fats and oils. Seed suppliers are busy trying to furnish seeds with compositions that will meet these needs and find farmers to grow these crops. Contracts with oilseed process ors have been made to process the harvest. Some food companies have pledged to use only trans-free fats and oils in their products [1].

Trans-free fat blends can be constructed by blending oils with fully hardened oils, or indeed where the entire blend has been randomized through interesterification. Blending vegetable oil types from different sources is an efficient alternative to hydrogenated vegetable oils, and still provides the appropriate physicochemical properties and nutritional requirements demand ed. Such fat blends can also be rich in polyunsaturated fatty acids as well as being trans- free. Trans-free options are commercially available in the form of a blend of tailored emulsifiers and oil blends where they meet demands for shelf-life, processing and distribution requirements. These trans-free options are available for a wide range of products covering, snacks, cakes, breads, tortillas, nutrition bars, cookies and breakfast cereals. Trans-free oil blends are also rou tinely designed for margarines, where they impart structure and texture, and shortenings where they provide firmness and contribute to crumb structure [10]. The combination of trans- free modification techniques (full hydrogenation, interesterification and fractionation) and the availability of a variety of different feedstocks can be used to produce virtually trans-free hard stocks with a range of physical properties such as solid phase lines determining melting per formances. Liquid seed oils, low in solids, are first fully hydrogenated to generate solids combined with a very low trans level (< 1.25%). These fully hydrogenated oils may subsequent ly be interesterified with non-hydrogenated liquid oil to reduce the solid fat content at high temperature (> 40°C). This solid fat content can be further reduced by fractionation [11].

2.2 Structural characteristics

Fats are the main structural components in many food products such as chocolate, confec tionery coatings, dairy products, butter, cream shortenings, margarine, and spreads. The sensory characteristics of fat-structured materials such as spreadability, hardness, and mouth feel are highly dependent on the structure of the underlying fat crystal network. This fat crystal network is built by the interaction of polycrystalline fat particles. The amount, ge ometry, and spatial distribution of solid fat crystals as well as their interactions at different levels within the network all affect the rheological properties of fats and fat-structured food products. Fat crystallization largely determines consistency, physical stability, visual ap pearance, and eating properties [4, 12, 13].

2.2.1 Crystals

A crystal consists of a material in a solid state in which the building entities—molecules, atoms or ions—are closely packed so that the free energy of the material is at minimum. As a result the entities are arranged in a regularly repeating pattern or lattice and are affected by the following points [14]: the molecules, or atoms, or ions are subject to heat motion; only the average positions will be fixed; diffusion can occur in a crystalline material, but the time scales involved are centuries rather than seconds; incorporation of a foreign molecule lead ing to a dislocation in the crystal lattice (Figure 1b); some solid materials are "polycrystal line", i.e., they are composites of many small crystalline domains of various orientations (Figure 1c).

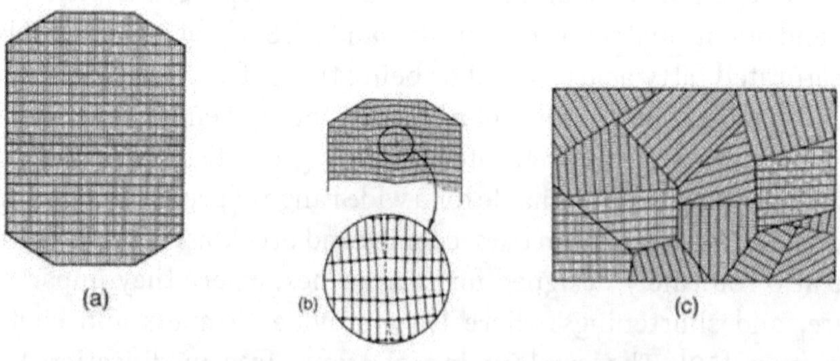

Figure 1. Two-dimensional illustration of crystalline order: (a) crystal lattice with perfect order, (b) a defect in the crystal leading to a dislocation in the lattice, (c) a polycrystalline material [14].

Different lattice arrangements and unit cells (Figure 2) can be constructed in terms of the lat tice parameters, also known as Bravais lattices: three spatial dimensions - a, b, and c; and three angles - α, β, and γ. For example, cubic systems all must have equal lengths (a=b=c) and angles equal to 90° [15].

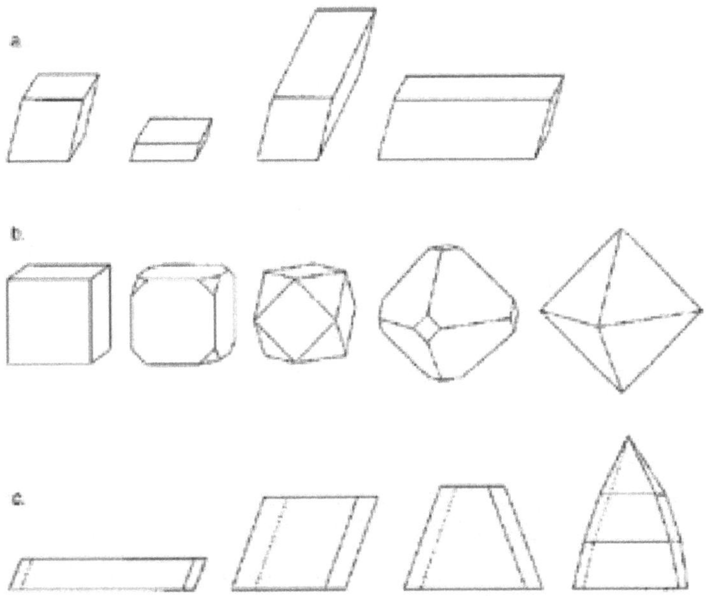

Figure 2. Variation in crystal morphology for identical unit cells: (a) rhombohedral, (b) cubic, and (c) monoclinic [15].

Crystals show enormous variation in external shape or habit caused by variation in the growth rate of the various faces of a crystal, which rates often depend on the composition of the solution. Corners and edges are rounded; curved faces can appear in large crystals; some needlelike crystals have a slight twist; some faces can grow faster than others (Figure 4). A noncrystalline solid is often referred to as an amorphous solid. Whether a material is crystal line or not can be established by x-ray diffraction. X-rays have a very small wavelength, of the order of 0.1 nm, which implies that individual atoms may cause scattering. If the atoms (or small molecules) occur at regular distances, sharp diffraction maxima occur, and the crystal structure can be derived from the diffraction pattern [14].

2.2.2 Crystallization

Many of the sensory attributes such as spreadability, mouthfeel, snap of chocolate, texture, etc., are dependent on the mechanical strength of the underlying fat crystal network. In ad dition to this obvious industrial importance, fat crystal networks form a particular class of soft materials, which demonstrate a yield stress and viscoelastic properties, rendering these plastic materials. The levels of structure in a typical fat network are defined as the fat crys tallizes from the melt.

The growth of a fat crystal network can be visualized thus: the triacyl glycerols present in the sample crystallize from the melt into particular polymorphic/ polytypic states. These crystals grow into larger microstructural elements (≈ 6 mm) which then aggregate via a mass- and heat-transfer limited process into larger microstructures (≈ 100 mm). The aggregation process continues until a continuous three-dimensional network is formed by the collection of microstructures. Trapped within this solid network structure is the liquid phase of the fat [12].

The crystallization process consists of two steps: nucleation and crystal growth. Nucleation can be described as a process in which molecules come into contact, orient and interact to form highly ordered structures, called nuclei. Crystal growth is the enlargement of these nu clei. Nucleation and crystal growth are not mutually exclusive: nucleation may take place while crystals grow on existing nuclei [16].

Nucleation can only be achieved via supersaturation or supercooling. A solution is supersa turated if it contains more of a component than can be theoretically dissolved within it at a particular temperature. Supercooling refers to the degree to which the solution is cooled with respect to the melting temperature of the crystallized solution. It is very difficult to de termine the parameters of supersaturation and supercooling for a crystallizing system, and therefore, as a good approximation, in practice only supercooling is usually considered for crystallization of triacylglycerol molecules from the melt [17].

When the temperature of a fat melt is decreased below its maximum melting temperature, it becomes supersaturated in the higher-melting triacylglycerol species present in the mixture. This so-called undercooling or supercooling represents the thermodynamic driving force for the change in state from liquid to solid. Fats usually have to be undercooled by at least 5-10ºC before they begin to crystallize. For a few degrees below the melting point, the melt exists in a so-called metastable region. In this region, molecules begin to aggregate into tiny clusters called embryos. At these low degrees of undercooling, embryos continuously form and breakdown, but do not persist to form stable nuclei. The energy of interaction between triacylglycerol molecules has to be greater than the kinetic energy of the molecules in the melt so as to overcome Brownian effects. For these flexible molecules, it is not sufficient to simply interact; molecules have to adopt a specific conformation in order to form a stable nucleus. The adoption of this more stable conformation is relatively slow, thus explaining the existence of a metastable region. As the undercooling is increased (i.e., at lower tempera tures) stable nuclei of a specific critical size are formed [18].

2.2.3 Polymorphism

An important way to characterize fats and oils is through the predominant crystalline phase, or polymorph, that tend to form upon crystallization. When the same ensemble of molecules can pack in different arrangements on crystallization, depending on the processing condi tions, the substance is said to demonstrate polymorphism. The different polymorphic states of a particular substance often demonstrate quite different physical properties (such as melt ing behavior and hardness), but on melting yield identical liquids [17].

Polymorphism is the ability of long-chain compounds such as fatty acids to exist in more than one crystal form, and this results from different patterns of molecular packing in the crystals. Triacylglycerols may occur in three main forms, namely, α, β', and β in order of in creasing stability and melting point. When fats are cooled, crystals of a lower melting form may be produced. These may change slowly or rapidly into a more stable form. The change is monotropic, that is, it always proceeds from lower to higher stability. Polymorphism re sults in the phenomenon of multiple melting points. When a fat is crystallized in an unstable form and heated to a temperature slightly above its melting point, it may resolidify into a more stable form [1]. The polymorphs differ in stability, melting point, melting enthalpy, and density. The α-polymorph is the least stable and has the lowest melting point, melting enthalpy, and density. The β-polymorph is the most stable and has the highest melting point, melting enthalpy, and density. The β'-polymorph has intermediate properties [4].

Under rapid cooling conditions, triacylglycerol molecules usually crystallize in metastable polymorphic forms, which subsequently transform into polymorphs of higher stability. On the other hand, at slow cooling rates, triacylglycerol molecules of similar chain lengths have time to associate with each other in more stable geometrical arrangements, resulting in the formation of a more stable polymorphic form. Due to the dependence of fat crystallization on the degree of undercooling and the cooling rate used, different results will be observed when using different cooling rates [18].

2.2.4 Tempering

Before its solid fat content can be determined, the fat must be exposed to a prescribed tem perature profile: first it has to be melted completely to destroy all traces of crystals, and then cooled to achieve virtually complete crystallization, and finally it has to be held at the meas uring temperature to come to equilibrium at that temperature. Sometimes, depending on the fat used, an extra step is introduced where the fat is held at a particular temperature, which is not the measuring temperature. This step is referred to as a tempering step. For confec tionery fats, a tempering step of 40 hours at 26°C is mentioned in the standard methods to ensure that cocoa butter and similar fats like cocoa butter equivalents (CBEs) are converted to their β-polymorph before the SFC is measured [4].

Tempering is a technique of controlled pre-crystallization employed to induce the most sta ble solid form of cocoa butter, a polymorphic fat in finished chocolates. The process consists of shearing chocolate mass at controlled temperatures to promote crystallization of triacyl glycerols in cocoa butter to effect good setting characteristics, foam stability, demoulding properties, product snap, contraction, gloss and shelf-life characteristics. Time–temperature protocols and shearing are employed to induce nucleation of stable polymorphs with the formation of three-dimensional crystal network structure influencing the microstructure, mechanical properties and appearance of products. The crystal network organization and the polymorphic state of the triacylglycerols crystals as affected by the crystallization condi tions are major factors determining rheological and textural properties of crystallized tria cylglycerols systems [19].

2.2.5 Solid fat content

The solid fat content (SFC) is a measure of the percentage of solid, crystalline fat in a sample at a selected temperature. Often, the SFC is measured at selected points within a temperature range. A measure of the SFC can be determined by a variety of methods: dilatometry, pulsed nuclear magnetic resonance (p-NMR), or differential scanning calorimetry (DSC). The method used and differences in the way it is executed can seriously affect the final result [4].

2.3 Fat design

Each application area requires its proper fat. The specifications of the fat depend on: recipe, equipment, procedure, temperature of fat and other ingredients, ambient temperature, stor age and distribution temperature of the final product. Some conditions to attend a satisfato ry fat design must be the compatibility among the components of the mixture: equivalent thermal properties (solid fat content, melting point and range); similar molecular size, shape and packing (to allow isomorphous replacement or formation of a single lattice unit in mix tures); similar polymorphism (transformation from stable to unstable forms should occur as readily for binary mixtures as with individual components) (Figure 3).

2.3.1 Processing

Edible fats and oils have been separated from animal tissues, oilseeds, and oil-bearing fruits for thousands of years. The combined largest source of vegetable oils is the seeds of annual plants grown in relatively temperate climates. The oilseeds are processed by expeller or screw press extraction, by prepress solvent extraction, or bay expander–solvent extraction. A second source of vegetable oil is the oil-bearing tree fruits and kernels. Oil-bearing fruits are pressed to obtain oil, sometimes after drying or sterilizing, or are cold pressed to preserve flavor and odor. Animal tissues may be wet- or dry-rendered (cooking processes) to separate the fats. Edible meat fats are supplied by lard from pigs, tallow from cattle and sheep, and milk fat or butter from cows. After recovering, fats and oils can be physically and/or chemically refined. Chemical refining removes most impurities with an alkaline solution, whereas physical refining removes them by distillation [2].

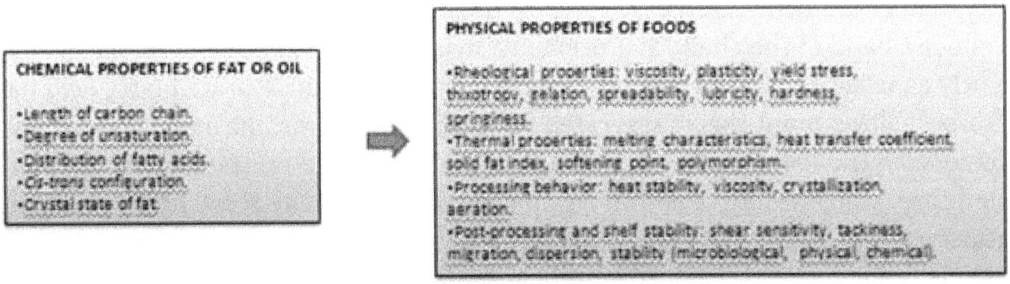

Figure 3. Physical and chemical functions of fats.

2.3.1.1 Industrialization

Searching for fat substitutes started in France during the Industrial Revolution. Large popu lation shifts from farms to factories and, in France, a depression and an imminent war with Prussia, created a demand for butter that the milk supply could not meet, escalating butter prices. The first acceptable butter substitute, named "margarine", was produced by the French chemist Mege Mouries in 1869, on commission from Emperor Napolean. Soon after the introduction of the first butter substitute on the market, several inventors patented vari ous modifications of Mouries' process [2, 4, 17]. Before 1900, animal fats were used as sour ces of fat with a high content of solids in margarine production. This led to a shortage of animal fats since they were also the main feedstock for soap making [3]. The best known modification processes applied today in the edible oil industry are hydrogenation, interes terification (chemical or enzymatic) and fractionation. The main purpose of these processes is to change the physicochemical properties of the oil or fat, by reducing the degree of unsa turation of the acyl groups (hydrogenation), by redistributing the fatty acids chains (interes terification) or by a physical separation of the triacylglycerols through selective crystallization and filtration (fractionation) [20].

2.3.1.2 Hydrogenation

Based on work done by the French chemist Paul Sabatier on the metal-catalyzed hydrogena tion of unsaturated organic compounds, German chemist Wilhelm Normann developed the method for hydrogenation of edible oils in 1903. Chemically, the hydrogenation of oils is the reduction of the double bonds in unsaturated fatty acids to single saturated bonds, by the reaction of hydrogen gas in the presence of a metal catalyst. The metal catalyst used at the time was nickel, and it has practically remained the same in the current hydrogenation pro cedures. Complete reduction of all double bonds in the oil would yield 100% saturated fatty acids, whereas reduction of only a fraction of the double bonds results in partially hydro genated fats. During the process of hydrogenation the cis double bond can open up and re form into a trans double bond, as well as shift positions along the fatty acid carbon chain. Structurally, cis double bonds in unsaturated fatty acids produce a bend in the chain that prevents unsaturated fatty acids from packing as tightly as saturated fatty acids. As a conse quence, a cis unsaturated fatty acid has a lower melting point than a saturated fatty acids with the same molecular weight. Conversely, the trans double bonds do not create a bend on the fatty acid chain. Therefore, trans unsaturated fatty acid chains are virtually straight, re sembling saturated fatty acids, and display higher melting points than the corresponding cis isomers [21].

The aim of the hydrogenation process is the total or partial saturation of the double bonds of unsaturated fats to obtain hard or plastic fats or to improve the stability to oxidation of an oil. The obtained product depends on the nature of the starting oil, the type and concentra tion of the catalyst used, the concentration of hydrogen, and the experimental conditions un der which the reaction takes place. Nickel catalyst was reported to catalyze undesirable side reactions such as cis, trans isomerization and positional isomerization of double bonds. The position of the double bonds affects the melting point of the fatty acid to a limited extent. The presence of different geometric isomers of fatty acids influences the physical character istics of the fat to a greater extent [22].

2.3.1.3 Interesterification

Interesterification has been developed as an alternative to hydrogenation, with the specific aim of eliminating the formation of trans fatty acids. The process rearranges the distribution of the fatty acids either chemically or enzymatically, within and between the triacylglycer ols, thus the fatty acid distribution is altered, but the fatty acid composition remains un changed – this rearrangement can be done either in a random or controlled manner. The technique is effective and can be used to produce fat products for spreads that are soft and spreadable and also trans-free. Interesterification is nothing new, having been around for some time, and the basic principles were first documented in 1969 [10].

2.3.1.4 Fractionation

Fractionation is a fully reversible modification process; it is basically a thermo-mechanical separation process in which a multi-component mixture is physically separated into two or more fractions with distinct physical and chemical properties. The separation can be based on differences in solidification, solubility, or volatility of the different compounds: fractional crystallization, fractional distillation, short-path distillation, supercritical extraction, liquid- liquid extraction, adsorption, complexation, membrane separation, etc. are the main techni ques practiced. Fractional crystallization refers to a separation process in which the fatty material is crystallized, after which the liquid phase is separated from the solid. It is based on differences in solubility of the solid triacylglycerol in the liquid phase, depending on their molecular weight and degree of unsaturation; this is a consequence of the ability of fats to produce crystals. On an industrial scale, crystals can be obtained according to three main technologies: detergent fractionation, solvent fractionation and dry fractionation [20].

2.3.2 Fat replacers

Fat replacers are called by many synonyms with various nuances in their usage: fat replacers can provide some or all of the functions of fat; fat substitutes resemble conventional fats and oils and provide all food functions of fat; fat analogs provide food with many of the charac teristics of fat; fat extenders optimize the functionality of fat; fat mimetics mimic one or more of the sensory and physical functions of fat in the food.

Fat replacers are most frequently used to replace fat in products with a high fat content and are used in a variety of food products, including frozen desserts, processed meats, cheese, sour cream, salad dressings, snack chips and baked goods. At the height of the interest in low-fat foods, more than 1000 fat-modified foods were introduced, with fat modified snacks being the fastest growing category of products in supermarkets at the time [11]. Normal fat contains nine calories per gram compared with five calories per gram for the sugar and pro tein components. If the proportion of fat is reduced the calorific value will fall. Corn starch, maltodextrin, pectin, gelatin, xanthan gum, guar gum, carrageenan, and soy protein were all commonly used ingredients in reduced fat products launched in the period 2008–10. Low in saturated fatty acids, sunflower oil was commonly used in new reduced fat foods. Fat re placers of the future will need to meet some important criteria, including reducing or replac ing the target fat effectively, being available at a cost appropriate to the benefits provided, and being safe and legal with no appreciable side effects.

2.3.3 Shortening

Shortening was the term used to describe the function performed by naturally occurring sol id fats such as lard and butter in baked products. These fats contributed a "short" (or ten derizing) quality to baked products by preventing the cohesion of the flour gluten during mixing and baking. Shortening later became the term used by all-vegetable oil processors when they abandoned the lard-substitute concept. Shortening has become virtually synony mous with fat and includes many other types of edible fats designed for purposes other than baking. Currently, a description for shortening would be processed fats and oils products that affect the stability, flavor, storage quality, eating characteristics, and eye appeal of pre pared foods by providing emulsification, lubricity, structure, aeration, a moisture barrier, a flavor medium, or heat transfer [2].

Fats and oils added to breads, cakes and similar baked goods are often referred to as shortenings that contribute to tenderness, improve volume gain of bread dough, enhance texture, crumb structure and shelf-life of the products. In order to produce a satisfactory shortening, one has to pay specific attention to the crystal structure, and similarly the con sistency of the shortening will depend on the ratio of solid to liquid fat present at differ ent temperatures [10].

Plastic shortening describes fats that are readily spread, mixed or worked. The property of plasticity is highly important in fats used as shortening agents in baked products. Com mercially, these are prepared by hydrogenation of oils, during which, some of the double bonds are isomerised into trans fatty acids from their cis configuration. Trans fatty acids have higher melting points and greater stability against oxidative rancidity than their cis- isomers and are important contributors to the functional properties of hydrogenated prod ucts. To meet the requirements of health-conscious consumers fats having a wide melting range which crystallize in the b' polymorphic form without the formation of trans fatty acids are needed [23].

Palm oil, because of its naturally β' tending nature, is favoured for shortening applications, such that it can impart stability to the emulsion, smooth consistency and provide good aera tion properties [10].

3. MECHANICAL PROPERTIES

When triacylglycerols are cooled from the melt to a temperature below their melting point, i.e., when they are supercooled, they undergo a liquid–solid transformation to form primary crystals with characteristic polymorphism. These primary crystals aggregate, or grow into each other, to form clusters, which further interact, resulting in the formation of a continu ous three-dimensional network. The mechanical properties of a fat, can be influenced by all these levels of structure; however, most directly by the level of structure closest to the mac roscopic world, namely the microstructure [24].

It is during crystallization that the template for the final physical properties of the result ing fat crystal network is created. Hence, the mechanical properties of a fat crystal net work are determined by the different levels of structure, such as chemical composition, solid fat content (SFC), and crystal habit (polymorphism and microstructure). To study the mechanical properties of fat crystal networks, rheologic tests are used, which meas ure how the crystallized material responds to applied forces (stress) and deformations (strain) [18].

Foods are edible structures created as a result of the responses of proteins, polysaccharides, and lipids in aqueous media to different processing methods, such as thermal processing, homogenization, and other physical treatments. The processing operations to which foods are subjected affect their structure and microstructure. Most, if not all, of the responses are physical in nature. By definition, rheology is the study of deformation and flow of materials. In foods, measured rheological responses are those at the macroscopic level. However, they are directly affected by the changes and properties at the microscopic level. Fractal dimen sion has been used to characterize food particles in addition to microscopic and size distri bution data. The fractal dimension can be estimated by several techniques such as viscoelastic behavior [25].

3.1 Rheology and texture

Rheology has been defined as the study of the flow and deformation of materials, with spe cial emphasis being usually placed on the former. In flow, elements of the liquid are deform ing resisted by viscosity. Solids when stressed creep, i.e. continue to deform very slowly over a very long time scale. In structured liquids there is a natural rest condition of the mi crostructure that represents a minimum-energy state. When these liquids are deformed, thermodynamic forces immediately begin to operate to restore this rest state. This kind of energy is the origin of elasticity in structured liquids. Alongside these elastic forces are the ever-present viscous forces that produce viscoelastic effects [26].

Rheological methods can be divided into small and large deformation rheology. Small defor mation rheology does not cause structural damage to the sample. They are performed in the linear viscoelastic region (LVR), in which the stress is directly proportional to the strain. Large deformation rheology is based on the deformation of a sample at a constant rate to the point where the force exceeds the structural capacity of the sample, causing it to permanent ly deform and break [18]. Sometimes, oscillatory testing is referred to as small amplitude os cillatory testing because small deformations must be employed to maintain linear viscoelastic behavior. Many processes, such as mastication and swallowing, are only accom plished with very large deformations. Collecting viscoelastic data relevant to this type of problem involves testing in the non-linear range behavior [27].

A frequently used method of measuring linear viscoelastic response is oscillatory testing, i.e. applying an oscillating stress or strain as an input to the liquid and monitoring the resulting oscillatory strain or stress output. Oscillatory tests are performed over a range of frequency. Short times correspond to high frequencies, and long times relate to low frequencies. In a sine-wave-shaped input of either stress or strain the resulting sinusoidal strain or stress out put is separated into solid-like response, which is in phase with the input, and a correspond ing liquid-like response which is π/2 (i.e. 90°) out of phase with the input. The solid-like component at any particular frequency is characterized by the storage modulus, G', and the liquid-like response is described by the complementary loss modulus, G". The behavior nor mally seen for typical viscoelastic liquids is an initial elastic response, thereafter, a delayed elastic response where the deformation rate becomes slower and slower, ending up as a very slow but steady-state deformation at the longest times, i.e. the material is in steady flow. The overall G', G" response of structured liquids is shown in Figure 4 [26].

G' expresses the magnitude of the energy that is stored in the material or recoverable per cycle of deformation, while G" is a measure of the energy which is lost as viscous dissipa tion per cycle of deformation. For a viscoelastic material the resultant stress is also sinusoi dal but shows a phase lag of δ radians when compared with the strain. The phase angle covers the range of 0 to π/2 as the viscous component increases. If G' is much greater than G", the material will behave more like a solid, i.e., the deformations will be essentially elas tic or recoverable. The loss tangent, tan δ, is the ratio of the energy dissipated to that stored per cycle of deformation. When G" is much greater than G', the energy used to deform the material is dissipated viscously and the materials behavior is liquid like [25].

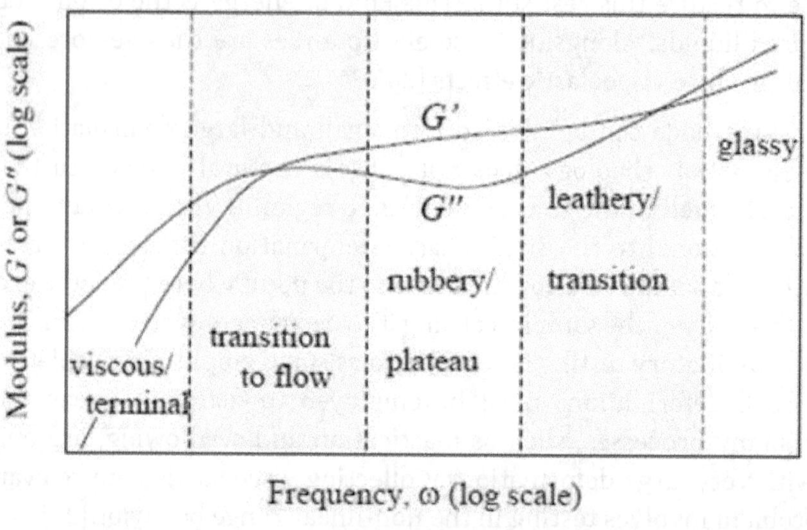

Figure 4. The various regions of an oscillatory test of structured liquids [26].

The penetrometry method and the two-plate compression method are large-deformation tests and are widely used to determine the yield stress or the firmness of a plastic fat. The large-deformation method has been widely used to study the physical properties of fat products, such as the spreadability of shortenings and the hardness of chocolate and milk fat, and the results have been found to correlate well with sensory tests [28].

Texture has been defined as the way in which various constituents and structural ele ments are arranged and combined into a micro- and macrostructure and this structure is externally manifested in terms of flow and deformation [29]. The structural elements of fats consist of solid fat crystals. They are suspended in liquid oil and when present in suf ficient quantity form a three-dimensional network that imparts plastic properties to the fat. The external manifestations of this network structure include a number of physical and mechanical properties such as hardness, softness, spreadability, brittleness, shortening power, and aeration properties. The texture of fats is influenced by a number of factors, including the solids content, the fatty acid and triacylglycerol composition of the solids, the polymorphic behavior of the fat crystals, the size and shape of the crystals, the nature of the crystal network, mechanical treatment, and temperature history. Many of these fac tors are interrelated, making it difficult to establish the effect of each independently [30]. Crystallization usually results in harder materials with higher solid fat contents. In addi tion, microstructural differences must be taken into account when evaluating the function al properties of lipids. The possibility of different polymorphic forms must not be neglected either because it can influence the texture and sensory profile. The texture of plastic fats can be determined by three main methods such as: cone penetrometry, pene tration by a probe, compression between parallel plates. The analyses and the evaluation of food texture are very important in food processing. Some of the attributes such as hardness and adhesiveness can be evaluated by texture analysis.

3.2 Fractal

Once the attraction forces have become larger than the repulsion, and also larger than Brow nian motion, particles can remain together when they collide. The resulting aggregates or flocs have a very complex structure and most of the flocs do not have homogeneous internal structures. The center is usually denser than the outer regions; hence the mass does not change with the third power of the radius as in normal objects with constant density [31].

Many patterns in nature such as the geometry of coastlines, mountains, trees, and vegeta bles, for instance, cannot always be defined adequately by using the familiar straight lines, circles, conic sections, polygons, spheres, quadratic surfaces, etc. Fractal geometry was born out of this lack of geometrical tools. A geometric shape belongs to standard ge ometry when smaller and smaller portions of it become increasingly smooth. For example, a generic curve becomes a straight line, and a generic surface becomes a plane. Fractals are shapes whose roughness and fragmentation neither tend to vanish, nor fluctuate up and down, but remain essentially unchanged as one zooms in continually and examina tion is refined. Hence, the structure of every piece holds the key to the whole structure. Fractals are characterized by two types of symmetries: self-similar and self-affine. In self- similar each part is a linear geometric reduction of the whole, with the same reduction ra tios in all directions. In self-affine, the reductions are still linear but the reduction ratios in different directions are different [32].

A fractal dimension is a powerful means of quantifying the structure of non-Euclidean ob jects by capturing the complexity of a structure's geometry in a single number. The chal lenge, however, is to give physical meaning to the number obtained [24].

The macroscopic rheological properties of the network are influenced by all levels of struc ture defined during the formation of the network, i.e. the structure of the individual triacyl glycerols, the structure of the individual crystalline units formed, or the polymorphic nature of the network, and the microstructural level of structure. The microstructural aggregate or microstructural network present in fat crystal networks scale in a fractal manner in the range between the size of the individual particles composing the sample (microstructural el ements) and the size of the microstructures. For colloidal aggregates and other fractal sys tems (such as fat crystal networks), fractal concept quantifies the way in which the mass of the sample/system increases with its size, according to the fractal dimension [33].

3.3 Fat crystal network

Early nucleation and crystal growth events lead to the formation of submicron primary crys tallites from the melt. These crystallites associate into micron-range particles, which further aggregate into clusters, until a continuous three-dimensional network with voids filled with liquid fat is formed [18].

A structural hierarchy exists within fat crystal networks. Polymorphism has to do with dif ferent molecular packing arrangements of tryacylglicerol molecules, at the nanostructural range, within the primary fat crystals. Once the primary fat crystals are formed, they aggre gate, or grow into each other, to form fat crystal clusters (or aggregate), which in turn cross-link to build up a 3-D fat crystal network. The shape of the fat crystal clusters can be spherulitic, feather-like, blade or needle-shaped. The size of fat crystal clusters can vary from several micrometers to more than 200 µm. Processing conditions can affect the size of the fat crystal clusters [28].

Fats are the main structural constituents of many food products including margarine, chocolate, butter and spreads. The sensory textural characteristics (i.e., spreadability, hard ness, snap) of fat structured foods are dependent on their macroscopic rheological proper ties, which are a consequence of the structure of their underlying fat crystal network. This network arises from the interactions between polycrystalline fat particles, and provides the elastic component, or the solid-like behavior, of a plastic fat. The sensory properties of the fat-structured foods are dependent not only on the amount of solid fat crystals present and their polymorphism, but also their geometry and the spatial distribution of crystalline material [34].

The microstructure of fat crystal networks can be quantified by fractal dimensions, which can describes the combined effects of morphology and spatial distribution patterns of the crystal clusters in fat crystal networks. The usefulness in the quantification of the micro structure of fats using the concept of fractal dimension arises from the possibility of relating structure to physical properties [28]. The fractal model of fat crystal networks explains the scaling behavior of rheological properties of semi-solid fat products to their solid fat content by their microstructure, which can be quantified using fractal dimensions. In general, differ ent microscopy fractal dimensions reflect different aspects of the micro- structure and thus have different physical meanings. An unambiguous agreement between physical fractal di mensions and microscopy fractal dimensions is required [25].

Fat crystal networks are statically self-similar, which means that the microstructure in a fat crystal network looks similar at different magnifications. Fractal structures are created by agglomeration, or clustering, of small particles to form a larger object in a random, iterative fashion under some constraint. In a similar fashion, fat crystal networks are built from clus ters of polycrystalline particles (crystallites) that aggregate in a diffusionally limited, fractal fashion. Fractal mathematics have been used to relate the elastic properties of fat crystal net works to the spatial distribution of the network mass and to link crystallization kinetics and phase behavior to microstructure. The fractal dimension defines the cluster size and has been evaluated by rheology techniques. Rheology is the most common technique for the quantification of microstructure in fat crystal networks and utilizes the relationship of the shear storage modulus (G′) to the volume fraction of network solid mass via the mass fractal dimension of the network [18].

The shape, size, and the strength of the fat crystal flocs making up the fat crystal network are always different. The weakest floc will become a flaw and acts as a stress concentrator. The elastic properties of the network depend on the number of connections between neigh boring structural clusters, rather than on the amount of apparent solids. This implies that the connectivity of the networks increases with an increasing volume fraction of solids. An idealized view of the structure of a fat crystal network showing the one dimensional defor mation of the links between crystal clusters [35]. In this same work, by using a modified fractal model, which describes the increase of G′ with SFC well, show the idea that the stress-carrying mechanism in fat crystal networks is heterogeneous, i.e. since real networks are not fully connected and that connectivity of networks increases with the volume fraction of solids, the load-bearing volume fraction of solids in real networks increases in an expo nential fashion with the apparent volume fraction of solids.

4. FAT FOODS

Fats and oils are the raw materials for liquid oils, shortenings, margarines, and other special ty or tailored products that are functional ingredients in food products prepared by food processors and restaurants and in the home. Humans have used fats and oils for food and a variety of other applications since prehistoric times, as they were easily isolated from their source. Fats and oils found utility because of their unique properties. These ingredients were found to add flavor, lubricity, texture, and satiety to foods. They have also been found to have a major role in human nutrition. Fats and oils are the highest energy source of the three basic foods (carbohydrates, proteins, and fats), and many contain fatty acids essential for health that are not manufactured by the human body [2].

While vegetable fats were used originally as a cheaper substitute for milk fat the ability to specify the properties of vegetable fat has considerable advantages. This ability arises be cause of the science and technology available to the fat processing industry. Some vegetable fats used in foods are not tailor-made but are simply a vegetable fat of known origin and treatment. The commonest example is palm kernel oil (HPKO), which is often used in foods.

4.1 Chocolate products

Chocolate can be described as a suspension consisting of nonfat particles (sugar and cocoa solids and, eventually, milk powder particles) dispersed in cocoa butter as a continuous phase. Molten chocolates represent a dense blend of phospholipid-coated sucrose and co coa particles in liquid fat. Milk chocolate usually contains about 12 g of cocoa mass, 19 g whole milk powder, 48.5 g sugar and, additionally, 20 g added cocoa butter per 100 g chocolate [36].

The characteristic flavor of chocolate has to be developed in several processing steps. Dur ing processing, the components are mixed, refined, and conched to attain desired rheologi cal properties for a final defined product texture and melting characteristics. A conche is a scraped-surface mixer that optimizes flavor development and turns chocolate mass into a flowable liquid. Through shear and longitudinal mixing, acidic flavors and moisture in the cocoa mass are reduced. Upon entering the conche, not all sugar and cocoa particles will be coated with cocoa butter. Fat in the chocolate will be released from the agglomerated choco late mass and spread to cover these particles so that they can flow easily. The final chocolate mass viscosity should be deemed optimal for the ensuing tempering [37].

Cocoa butter, which amounts to 25-36% in finished chocolate, is responsible for the smooth texture, contractability, flavor release, and gloss of the product. The fat phase is the only continuous phase in chocolate, thus responsible for melting behavior and the dis persion of all other constituents. A careful tempering of the chocolate is necessary in or der to obtain the fine crystals in the correct form (β-modification). Cupuassu fat, a similar cocoa butter fat, shows polymorphic behavior like cocoa butter (β form) and needs to be tempered like cocoa butter; at 24-25ºC an α (alpha) form is present. The melting profiles of cocoa butter and cupuassu fat are similar as shown. At all temperatures, cocoa butter has a higher solid fat content than cupuassu fat. This suggests that cupuassu fat would be useful in filled chocolate manufacture as a softer filling fat compatible with cocoa butter. The fatty acid and triacylglycerol compositions of cupuassu fat in comparison with cocoa butter show that palmitic acid in cupuassu fat is present in much smaller amount (7.8%) than in cocoa butter (26.1%); stearic acid is about the same; oleic acid is higher in cupuas su. Particularly notable is the high amount of arachidic acid (20:0) in cupuassu fat. The tri acylglycerol compositions reflect the fatty acid compositions, but give more useful information. Although cupuassu has a higher SOS content than cocoa butter, its contents of POP and POS are much lower reflecting its low level of palmitic acid. Total SOS-type triacylglycerols, i.e. POP+POS+SOS+SOA, is 57% in cupuassu and 83% in cocoa butter. Fractionation, as applied to fats such as shea and sal, would be needed to bring the total SOS-type content to the same as in cocoa butter. Fractionation could be used to modify cupuassu fat to make it more similar to cocoa butter for use as a CBE (cocoa butter equiv alent), with 65% minimum of total SOS-type triacylglycerols [38].

Modified lipids are used in the majority of chocolate and confectionery applications, such as chocolate compounds, filling fats in pralines, aerated products and cold products such as ice cream toppings. Production economics is often related to price, speed of production and equipment requirements, which in turn are related to the raw materials and their ability to crystallize rapidly.

The quality is related to the capacity of the fat to remain stable in terms of appearance, tex ture and taste; and the sensory properties can briefly be described as appearance, smell, taste and the role that fat plays in mouth feel with regard to texture and melt off properties. In chocolate industry fat bloom is still a problem. It modifies (shortens) the shelf life of the end products and makes life difficult for product development. Fat migration is one of the causes of bloom, but it will also soften the products during storage [39].

In chocolate industry, for processing and texture reasons, however, it is not possible to re duce the level too much below 25%. This is insufficient to make a low calorie claim on the product, so two manufacturers have produced fats that melt like cocoa butter but have a lower calorific value. Like lauric fats they are incompatible with cocoa butter and so the products have to be made with cocoa powder [36]. Some fats go into confectionery as a component of other ingredients. The common example is nuts, which contain fats, often of types such as lauric or unsaturated fats. These fats are sometimes the origin of spoilage problems.

Studies correlating chocolate composition and textural or rheological properties are commonly found due to the source of new fat or cocoa butter replacers which strongly affect rheological parameters on chocolate manufacture and final product texture. According to that, adaptations on manufacturing scale have to be done in order to keep the desirable sensory characteristics in the final product. Rheology is a useful feature on setting those issues. Several works have been conducted to study and understand rheological properties of chocolates. The various fats used in chocolate can contain different levels of trisaturated triacyl glycerols. Since these can crystallize out early in the tempering process, they can, in some instances, have an effect on the rheology of the chocolate. Six basic source oils are permitted as non-cocoa vegetable fats (CBE - Cocoa Butter Equivalent) in chocolate throughout European Union - palm oil, shea oil, illipe butter, sal oil, kokum gurgi, and mango kernel oil. Among these six oils, four (palm, shea, sal, and mango kernel) usually have to undergo some form of fractionation process to concentrate the SOS type of triacylglycerol necessary for equivalence to cocoa butter. Palm oil is even more complicated since it contains a significant quantity of trisaturated triacylglycerols which also have to be removed [40].

4.2 Ice cream

Ice cream has been identified as three component foam made up of a network of fat globules and ice crystals dispersed in a high viscosity aqueous phase. The composition of ice cream varies depending on the market requirements and processing conditions. Although the quality of the final product depends largely on the processing and freezing parameters, the ingredients also play an important role. The physical structure of ice cream affects its melting rate and hardness, although the specific relationships have not all been worked out. Structure development in ice cream often is attributed to the macromolecules present in the ice cream mix – milk fat, protein, and complex carbohydrates. Milk fat interacts with other ingredients to develop the texture, mouthfeel, creaminess, and overall sensations of lubricity. Typically, ice cream contains 10 to 16% fat and its type and amount influence the characteristics of the resultant products by affecting their rheological properties. The fat content can influence the size of the ice crystals. Fat globules could mechanically impede the ice crystal growth. Since each type of fat exhibits a specific polymorphism function of its triacyl glycerol composition, the thermal behavior of fats during ice cream processing should influence the physicochemical properties of the intermediate and final products [41].

A typical ice cream formulation has fat (7-15%), lactose (5-7%), other sugars (12-16%), stabilizers, emulsifiers and flavours (0.5%), total solids (28-40%), water (60-72%), milk protein (4-5%). Fat performs several functions in ice cream: it helps to stabilize the foam, it is largely responsible for the creamy texture, it slows down the rate at which ice cream melts and it is necessary to deliver flavour molecules that are soluble in fat but not water. The major sour ces of fat used in industrial ice cream production are butterfat, cream and vegetable fat [42].

Ice creams are metastable systems created from an emulsion o/w employing several unit operations: mixing, heating, cooling, freezing, aerating and packaging. While the ingredients combination is responsible by chemical characteristics, a sophisticated microstructural arrangement constituted by fat globules, ice crystals and air bubbles supported in a highly vis cous matrix dictates mechanical, thermal and sensorial properties.

There are many factors within the microstructure of products, which determine the rheological properties, such as colloidal interactions between disperses components, the junctions between structural elements, the properties of this elements, the interfacial behavior between phases, the rheology, and structure of individual component phase. In order to improve the quality of this very appreciated foodstuff, ingredients research and their impact on formulations are very desirable. In [43] was investigated the potential of a chemically modified polysaccharide (N-succinil chitosan hydrogel) when applied as structuring agent in colloidal systems. It was found that the mixes resulting by combination among chitosan and palm fat presented good characteristics; the enormous structuring power presented by this biomolecule can be very useful to elaborate low-fat formulations with good textural properties. Moreover, taking in account the physiological activity, it can be employed in order to promote best nutritional quality in foods; this biopolymer and their derivatives, can be extensively explored, since appear do not has limitations in its potentialities.

In [41] was found in study that the replacement of hydrogenated vegetable fat by palm fat caused changes in melting ranges of formulations. Higher melting rate was observed by combination between palm fat and fructose syrup. In addition to effects expected on melting behavior and solids content, sugar blends employed in this study affected the air incorporation. There is consense that greater air content increase the melting resistance. However, despite to lower overrun, ice creams made with fructose syrup melted more slowly. Thus, the levels of air added into the products not allow safe conclusions about the influence of this parameter on physical behavior of assessed ice creams in this study. In [44] was evaluated also the influence of the substitution of hydrogenated fat in the manufacture of ice cream formulation with palm fat through rheology analysis, and compare the results obtained with the melting test. The rheological and the melting tests showed a better response from the ageing process, and a better formed structure with the formulation produced with hydrogenated fat. It was suggested that formulations produced with palm fat suffers a poorer partial coalescence by its crystallization profile and less membrane destabilization by the emulsifiers.

In another study was evaluated the influence of the substitution of hydrogenated fat in the manufacture of ice cream formulation with cupuassu fat through rheology analysis, and compare the results obtained with the melting test. The rheological tests showed similar response from the ageing process to both formulations, and the melting tests showed a slower meltdown of the structure with the ice cream produced with cupuassu fat. The results ob tained demonstrated that cupuassu fat is a good substitute for hydrogenated vegetable fat for using in ice cream formulations [45].

4.3 Bakery

The functionality of fats in bakery products can be explained as: development of the structure; lubrication; aeration; heat transfer; moisture retention; improved shelf-life, volume, texture and flavor. In some cases the function of a fat can be either partially or completely replaced by some other ingredient, typically an emulsifier.

Fats shorten the texture of baked products by preventing cohesion of gluten strands during mixing, hence the term shortening. All-purpose shortenings are used primarily for cookies but are also common ingredients in cakes, breads, and icings and are also used for frying applications. The quality of cakes and icings is highly dependent upon aeration; therefore, a variety of very specialized shortenings has been developed over the years to satisfy that de mand. High ratio shortenings (containing mono and diglycerides), designed primarily for cakes, began to appear in the '30s. Fluid cake shortenings were commercialized in the '60s and offer many advantages including pumpability, ease of handling and the option of bulk delivery and storage [46].

Cake is a baked batter made from wheat flour, sugar, eggs, shortening, leavening agents, salt, nonfat dry milk, flavors, and water. Cake batters are essentially a 'foam', that is a system in which air bubbles are trapped and held in an aqueous phase. The main function of fat in cake making is to assist with the incorporation of air into the batter during mixing, and the air bubble size and stability. High-ratio cakes, rich in sugar and fat, are extensively used in the baking industry [47].

Margarine has always had the advantage over butter in that the properties of the product can be tailored to give the best performance in a particular system. For puff pastry, i.e., specialized margarines are easier to work with than butter.

Various bakery margarines are manufactured to meet the technical requirements of particular uses.

The effect of different fats and margarines on the physical properties of cakes was investigated. The low trans fat suggested: greater volume and firmness; resilience comparable to hydrogenated fat; elasticity and chewiness were comparable to other formulations, as well the color parameters of the crumbs [48].

Textural properties are important quality parameters for this type of product. Physical and structure changes during aerated batters processing may alter their performance during baking or the quality of the final product. It is possible to test materials with particles and fiber in suspensions, since flushing effects may reduce sedimentation problems. In [49] was examined the influence of different types of fats (hydrogenated fat, margarine and vegetable oil) in formulation of cake batter, evaluating textural properties by Herschel-Bulkley equation using back extrusion analysis; and it were observed values close in all parameters to samples prepared with margarine and hydrogenated fat. It can be mentioned the break point values could be consider by the industry as an important parameter, pointing the need of less energy in their processes of pumping, i.e.

4.4 Food emulsions

4.4.1 Mayonnaise and salad dressing

Mayonnaise and salad dressing are emulsified, semi-solid fatty foods that by federal regulation must contain not less than 65% and 30% vegetable oil, respectively, and dried whole eggs or egg yolks. Salt, sugar, spices, seasoning, vinegar, lemon juice, and other in gredients complete these products. Pourable and spoonable dressings may be two phase (e.g., vinegar and oil) or the emulsified viscous type (e.g., French). There is a great variety of products available of varying compositions with a wide range in their oil content. Sal ad oils exclusively are used for dressing products; typical choices include soybean, canola and olive oils [46].

Emulsion is a thermodynamically unstable system due to flocculation, creaming, coales cence, phase inversion and Ostwald ripping. Emulsifier is a surfactant which can stabilize the emulsion by absorption at the interface, thereby lowering the interfacial tension. It is usually used to improve the emulsion stability. Proteins and polysaccharides are often applied in emulsion as emulsifier. Proteins are usually used for their surfactant and gelling properties to improve the textural characteristics and stability of emulsion, while polysaccharides are usually added to increase the viscosity or to obtain a gel-like product. It was studied the impact of the use of a biomaterial (N-succinil chitosan hydrogel) in elaboration and structuration of food emulsion, and in substitution of a part of the oil phase. Chitosan showed to be a versatile ingredient since that was capable to modify the rheological proper ties, acting as emulsifying agent, besides its already known antimicrobial and nutritional qualities [50].

4.4.2 Margarine

Margarine and spreads are prepared by blending fats and/or oils with other ingredients such as water and/or milk products, suitable edible proteins, salt, flavoring and coloring ma terials and Vitamins A and D. Margarine must contain at least 80% fat by federal regulation, however, "diet" margarines and spreads may contain 0-80% fat. These products may be for mulated from vegetable oils and/or animal fats, however, the vast preponderance are all vegetable. Non hydrogenated oils typically represent the majority of the fat phase. Lesser amounts of partially hydrogenated fats, that are naturally semisolid at room temperature, and/or hard fractions of certain fats are added to the blend as required to deliver the desired structure and melting properties [46]. At the moment, interesterification technics have been employed to produce tailor fats. Margarine originated as a substitute for butter. The big ad vantage of margarine is that as a manufactured product the properties can be tailored to suit a particular use.

An acceptable margarine must be a soft plastic at room temperature; the ratio of solid or crystalline fat to liquid oil in the mixture must be such that when the fat crystals are of the proper size and well dispersed, the mass will offer some resistance to deformation and sepa ration of solid and liquid fats will be negligible; all the fat crystals must melt completely at body temperature and leave a pasty sensation in the mouth; the fat crystals must not melt abruptly [51].

5. PROSPECTIVE

The prominent brands are currently available globally. The adverse economic effects of a negative reputation on these super brands have compelled major firms to prioritize brand image management. The rise in obesity has emphasized the health implications of foods overall, particularly concerning chocolate and confectionery items. Future fat replacers must satisfy critical requirements, including the effective reduction or replacement of the target fat, affordability relative to the advantages offered, and compliance with safety and regulatory standards, while exhibiting minimal adverse effects. [39, 52].

REFERENCES

1. Chow C.K. Fatty acids in foods and their health implications. Boca Raton: CRC Press; 2008.
2. O'Brien R. D. Fats and oils: formulating and processing for applications. Boca Raton: CRC Press; 2004.

3. Zevenbergen H., De Bree A., Zeelenberg M., Laitinen K., Van Duijn G., Flöter E. Foods with a high fat quality are essential for healthy diets. Annals of Nutrition and Metabolism 2009; 54: 15–24,.

4. Gunstone F.D., Harwood J. L., Dijkstra A.J. The lipid handbook. Boca Raton: CRC: Taylor & Francis; 2007. Gurr M. I. Lipid Biochemistry. Oxford: Blackwell Science; 2002.

5. Eckel R.H., Kris-Etherton P., Lichtenstein A.H., Wylie-Rosett J., Groom A., Stitzel K.F., Yin-Piazza S. Americans' awareness, knowledge, and behaviors regarding fats. Journal of the American Dietetic Association 2009; 109: 288-296.

6. Food And Drug Administration. Trans fat now listed with saturated fat and choles terol on the nutrition facts label. http://www.fda.gov/Food/LabelingNutrition/ ConsumerInformation/ucm110019.htm (Accessed november 2011).

7. Grooten H., Oomen C. Feature ruminant and industrial trans fatty acids: consumption data and health aspects. Is there a difference? Lipid Technology 2009; 2 (21): 36-38.

8. Lumor S. E., Pina-Rodriguez A. M., Shewfelt R. L., Akoh C. C. Physical and sensory attributes of a trans-free spread formulated with a blend containing a structured lip id, palm mid-fraction, and cottonseed oil. Journal of the American Oil Chemistys' Society 2010; 87: 69–74.

9. Wassell P., Young N.W.G. Food applications of trans fatty acid substitutes. International Journal of Food Science and Technology 2007; 42: 503–517.

10. Williams C., Buttriss J. Improving the fat contents of foods. Boca Raton: CRC Press: Woodhead Publishing; 2006.

11. Narine S.S., Marangoni A.G. Relating structure of fat crystal networks to mechanical properties: a review. Food Research International 1999; 32: 227-248.

12. Awad T.S., Rogers M.A., Marangoni A.G. Scaling behavior of the elastic modulus in colloidal networks of fat crystals. Journal of Physical Chemistry B 2004; 108: 171-179.

13. Walstra P. Physical chemistry of foods. New York: Marcel Dekker; 2003.

14. Myerson A.S. Handbook of Industrial Crystallization. Elsevier Science & Technology Books; 2001.

15. Foubert I., Vanrolleghem P. A., Vanhoutte B.; Dewettinck K.. Dynamic mathematical model of the crystallization kinetics of fats. Food Research International 2002; 35: 945–956.

16. Ghotra B.S., Dyal S.D., Narine S.S. Lipid shortenings: a review. Food Research International 2002; 35: 1015–1048.
17. Marangoni A.G. Fat Crystal Network. New York: Marcel Dekker; 2005.
18. Afoakwa E.O., Paterson A., Fowler M., Vieira J. Effects of tempering and fat crystalli sation behaviour on microstructure, mechanical properties and appearance in dark chocolate systems. Journal of Food Engineering 2008; 89:128–136.
19. Kellens M. , Gibon V., Hendrix M., De Greyt W. Palm oil fractionation. European Journal of Lipid Science and Technology 2007; 109: 336–34.
20. Tarrago-Trani M.T., Phillips K. M., Lemar L. E. , Holden J. M. New and existing oils and fats used in products with reduced trans-fatty acid content. Journal of the American Dietetic Association 2006; 106: 867-880.
21. Herrera M.L , Falabella C., Melgarejo M., Añón M.C. Isothermal crystallization of hydrogenated sunflower oil: I — Nucleation. Journal of the American Oil Chemistys' Society 1998; 10: 1273–1280.
22. Jeyarani T., Khan M.I., Khatoon S. Trans-free plastic shortenings from coconut stearin and palm stearin blends. Food Chemistry 2009; 114: 270–275.
23. Marangoni A.G. The nature of fractality in fat crystal networks. Trends in Food Science & Technology 2002; 13: 37–47.
24. Mcclements D.J. Understanding and controlling the microstructure of complex foods. Woodhead publishing; 2007.
25. Barnes H.A. A handbook of elementary rheology. Aberystwyth: University of Wales Institute of Non-Newtonian Fluid; 2000.
26. Steffe J.F. Rheological methods in food process engineering. East Lansing: Freeman Press; 1996.
27. Tang D., Marangoni ,A.G. Microstructure and fractal analysis of fat crystal networks. Journal of the American Oil Chemistys' Society 2006; 5 (83): 377-388.
28. Deman J. M. Principles of food chemistry. 3rd ed. Aspen Publishing: Gaithersburg; 1999.
29. Deman J. M. In: Widlak, N. ed. Physical properties of fats, oils and emulsifiers. Champaign, IL: AOCS Press; 2000.
30. Macosko C.W. Rheology principles, measurements and applications. New York: John Wiley & Sons; 1994.
31. Mandelbrot B. B. Fractal geometry: what is it, and what does it do ? Proceedings of the Royal Society A 1989; 423: 3-16.

32. Narine S.S., Marangoni A.G. Microscopic and rheological studies of fat crystal networks. Journal of Crystal Growth. 1999; 198/199: 1315-1319.

33. Lam R., Rogers M.A., Marangoni A.G. Thermo-mechanical method for the determi nation of the fractal dimension of fat crystal networks. Journal of Thermal Analysis and Calorimetry 2009; 98: 7–12.

34. Marangoni A.G., Tang D. Modeling the rheological properties of fats: a perspective and recent advances. Food Biophysics. 2008; 3: 113–119.

35. Beckett S. T. The science of chocolate. Cambridge: Royal Society of Chemistry Paperbacks; 2009.

36. Afoakwa E. O., Paterson A., Fowler M., Vieira J. Relationship between rheological, textural and melting properties of dark chocolate as influenced by particle size distribution and composition. European Food Research Technology 2008; 227: 1215-1223.

37. Lannes S. C. S.. Cupuassu - A new confectionery fat from Amazonia. Inform-AOCS 2003; 14(1): 40-41.

38. Norberg S. Chocolate and confectionery fats. In: Gunstone F.D. (ed.) Modifying lipids for use in food. Boca Raton:CRC 2006; 488-516.

39. Gonçalves E.V., Lannes S. C. S. Chocolate rheology: a review. Ciência e Tecnologia de Alimentos 2010; 30: 845-85.

40. Silva Junior E. , Lannes S. C. S. Effect of different sweetener blends and fat types on ice cream properties. Ciência e Tecnologia de Alimentos 2011; 31: 217-220.

41. Clarke C. The Science of Ice Cream. Cambridge: RS.C.; 2004.

42. Silva Junior E., Mello K. G. P. C., Polakiewicz B., Lannes S. C. S. Ice cream mixes for mulated with n-succinil chitosan hydrogel characterized by rheo-optic techniques. In: 14th World Congress of Food Science & Technology-IUFoST: proceedings of the 14th World Congress of Food Science & Technology - IUFoST, 2008, Beijing, China. Chinese Institute of Food Science and Technology, 2008.

43. Su F., Lannes S. C. S.. Structural evaluation of ice cream produced with palm fat through rheology. In: International Symposium on Food Rheology and Structure:

44. Abstract book of the 6th International symposium on food rheology and structure, Zurich, Swiss. Lappersdorf: Kerschensteiner Verlag, 2012.

45. Su F., Lannes S. C. S.. Structural evaluation of ice cream produced with cupuassu fat through rheology. In: International Symposium on Food Rheology and Structure: Abstract book of the 6th International symposium on food rheology and structure, Zurich, Swiss. Lappersdorf: Kerschensteiner Verlag, 2012.

46. Institute of Shortening and Edible Oils. Food fats and oil. 2006. http://www.iseo.org/ foodfats.htm (Accessed August 2012)

47. Summu S.G., Sahin S. Food Engineering Aspects of Baking Sweet Goods. New York: CRC Press; 2008.

48. Salas A.G.V., Lannes S. C. S.. The comparison of fat and margarine application on texture and volume of cake. In: XV Semana Farmacêutica de Ciência e Tecnologia: Brazilian Journal of Pharmaceutical Sciences, São Paulo, Brazil. São Paulo:Biblioteca do Conjunto das Químicas, 2010; 46: 82-82.

49. Rios R.V. , Lannes S. C. S. The influence of different fats in cake batter. In: 16th World Congress of Food Science & Technology-IUFoST: proceedings of the 16th World Con gress of Food Science & Technology, 2012, Foz do Iguaçu, Brazil. International Insti tute of Food Science and Technology, 2012.

50. Pinto M. M. M., Dias M.G.C., Lannes S. C. S.. Rheo-optic determination of mayon naise sauce. In: IFT09 Annual meeting + Food Expo: proceedings of the IFT09 Annual meeting, 2009, Anaheim, CA. Chicago: IFT, 2009.

51. Feuge R.O. Tailor made fats for industry. Yearbook of Agriculture 1950-1951. http:// naldc.nal.usda.gov/download/IND43894116/PDF (Accessed August, 2012)

52. The future of fat reduction and replacement in food and drinks. 2011. http:// www.reportlinker.com/p0683180/The-Future-of-Fat-Reduction-and-Replacement-in- Food-and Drinks.html#utm_source=prnewswire&utm_medium=pr&utm_cam paign=Food_Manufacturing (Accessed August 2012)

5 Chapter
MINERAL COMPOSITION OF BLOOD SAUSAGES – A DUAL CASE STUDY

1. INTRODUCTION: AN OVERVIEW

1.1. Significance of evaluating mineral content in food

A balanced diet is crucial for sustaining good health; therefore, the nutritional content of foods is a significant factor of food quality [1]. In this setting, an increasing number of individuals are expressing significant concern regarding the chemical composition of their food. Consequently, the food sector is focused on upholding a high standard of quality in their manufactured products to satisfy the demands of an increasingly discerning consumer. Consequently, a critical concern in the food sector is the assessment of food composition and the implementation of analytical controls. [2].

Food scientists and the food industry have long focused on minerals in food, primarily due to their essential role in human nutrition, including physiological functions, nutritional requirements, and the implications of minerals on safety, particularly concerning mineral toxicity. The human body contains about 60 minerals, although only a select few are deemed essential: iron, calcium, zinc, magnesium, phosphorus, sodium, potassium, manganese, selenium, and copper. These minerals are needed for numerous critical activities, including bone and tooth development, as well as the functioning of neurological, circulatory, renal, and digestive systems; some are also required for the regulation of enzyme systems. [2,3].

Minerals deficiencies in human are common world-wide and there are evidences which suggest that deficiencies may play a main negative role in children's development, pregnancy and elderly health [3]. In this context, Ca, K, Mg and Fe are the most commonly under-consumed minerals in humans' diet [4]. Fe deficiency is the most common and widespread nutritional disorder in the world affecting both developing and industrialized nations [5]. Insufficient intakes of Fe cause anemia, fatigue, poor growth, rickets and impaired cognitive performance in humans [3]. Conversely, the accumulation of undesirable minerals in food can be exacerbated by the continual discharge of toxic chemicals into the environment, primarily resulting from human industrial activities. The pollution of the food supply can lead to increased consumer exposure to harmful metals such as lead, cadmium, arsenic, and mercury, exceeding the tolerated daily consumption levels. [6].

The evaluation of mineral content in food is significant from both nutritional and toxicological perspectives. In recent decades, instrumental analytical techniques utilizing atomic absorption or emission spectrometry, in conjunction with multivariate statistical analysis, have demonstrated efficacy in characterizing food products, distinguishing between quality categories, and verifying food authenticity, including the determination of geographical origin, differentiation of cultivation methods (e.g., organic versus conventional crops), varieties of fruits and vegetables, and food processing practices. [7-10].

The examination of minerals in food is complex due to the considerable variation in concentrations, which can range from parts per billion to percentage levels. The scenario is further exacerbated by inherent seasonal and varietal variations in concentrations within the same food. [11]. The official methods of AOAC provide numerous single-element techniques utilizing colorimetric methods, including UV/Visible spectrophotometry and flame and graphite furnace atomic absorption spectrophotometry. Nevertheless, while no AOAC food procedures presently utilize Inductively Coupled Plasma-Atomic Emission Spectroscopy (ICP-AES), it is a recognized multi-element approach that does not necessitate the use of hazardous solvents from an environmental perspective. [11The technique's excellent specificity, multi-element detection capacity, and favorable detection limits contribute to its application across a wide range of fields. Detection limits generally span from parts per million (ppm) to parts per billion (ppb); however, depending on the element and instrument, detection below ppb may occasionally be attainable. [12]. ICP-AES offers superior reproducibility and an extended quantitative linear range relative to conventional AES, while minimizing molecular interferences due to the elevated temperature (7000-8000 K) of the excitation source (plasma). Conversely, ICP-AES is costlier than traditional AES, and in intricate samples, emission patterns may be challenging to understand. [13].

1.1 Blood sausages, making process and chemical composition

Meat products are generally made from various raw materials (from different origins and suppliers), which are combined at the formulation stage in obedience to criteria of composition, technological factors, sensory characteristics, legal regulations and also economic efficiency and profit [14].

Among meat and meat products, muscle foods are the most commonly consumed. However several edible meat by-products and their derivatives are also importantly consumed in a number of countries, where meat by-products are usually linked to traditional or ethnic foods. Meat by-products are traditionally sold to the lower income market however, by different reasons - one of them could be the increase in tourism - their consumption seems to be increasing and some of the by-products are becoming delicacies in niche markets. Advantageously, meat by-products consumption contributes to increase the edible portion of slaughter animals, Furthermore, meat edible by-products constitute an excellent source of nutrients like essentials amino acids, minerals and vitamins [15,16]. Due to the great variety and specificity of edible meat by-products and their peculiar consumption patterns and their relative low economic value, there is relatively scarce information on their making process and chemical composition.

In some areas of the world, and to different degrees, blood is utilized as an edible meat by- product. For example, for several ethnic groups of Africa and India, blood is the primary source of animal protein, where it holds ritualistic importance. However, in some cultures (Islamic and Jews), blood consumption is seen as a taboo [17,18]. In Europe and Asia, animal blood has been traditionally used in making a variety of foods such as blood sausages, blood pudding, biscuits and bread, as well as blood soups and crackers [19,20].

From the nutritional point of view, blood is a good source of dietary protein, lysine and iron [19,21]. The high iron content of blood (approximately between 400-500 mg of iron per liter), coupled with the high absorption of heme iron compared to non-heme iron, is particularly useful for food based strategies designed to combat iron deficiency anemia. Furthermore, the environmental concern associated with blood disposal at slaughterhouses, together with blood nutritive value, has fostered research and industrial efforts to recover blood or blood components, to be used into a wide range of food products or as dietary supplements [22]. For example, blood or blood proteins (plasma or cellular fractions) are being used in meat products, primarily to increase protein levels and enhance water binding and emulsifying capacity.

Blood sausages are very popular traditional meat products in many parts of the world such as Europe, Latin America or Asia [23-26]. In Europe, blood sausages are normally called morcilla and morcella in Spain and Portugal, black pudding in Great Britain, blutwurst and Thuringer blood sausage in Germany, blodpφlse in Denmark, boudin noir in France, bloed worst in Belgium, blood-tongue sausage and black pudding in Austria, caltabosi cu singe in Hungary, vaerevorst in Estonia, kaszanka in Poland, biroldo in Italy. In Latin American countries, blood sausages are also produced and are named as relleno, prieta, moronga, mocillón in Mexico, Colombia, Peru or Argentina, and Morcela in Brazil; these sausages from Latin America show characteristics similar to those from Europe, especially to those of Iberian Peninsula [25]. In this sense, blood sausages from Latin America can be included into the group of creole meat products, which means that they were originated from the adaptation of former Iberian meat products (brought to America by immigrants) to local condition and circumstances, thus, involving an innovation process at that time.

Nowadays, blood sausages are currently receiving worldwide increasing attention because they have become gourmet products in several countries, thus leading to an increase in their production and potential markets [27]. Furthermore, increasing consumer demand for ethnic specialties has renewed interest in such products, leading to a consequent need to assure safety and longer shelf-lives in an expanding market. Moreover, the Governmental Institutions, e.g., European Union, are getting more involved in the protection of high-quality traditional foods from specific regions or areas, which reflects a policy of supporting the inhabitants of rural areas and promoting regional products [23,27].

Blood sausages are basically made with blood, fat and a variety of vegetable origin food; Moreover the use of meat, pork skin or offal (e.g., liver, intestine) is common, mainly in German blood sausages [28-31]. The vegetable-origin food used is enormously diverse so that, apart from spices and condiments, blood sausages can contain as main ingredients onion, leeks, cereals (rice, oat, flour, bread, etc.), sugar, fruits (apple, plum, etc.), nuts, etc. [32-34]. Other ingredients such as eggs, cream, milk are used in some types of blood sausages in France [32]. Moreover, as any meat product, blood sausages are added with common salt. The NaCl used in blood sausages from Mediterranean Europe tend to be between 1.2 and 1.5 % [35-37], and in blood sausages from Germany [38] and USA [16] tend to be higher, close to 2%. NaCl has a direct effect on the flavour and also increases the shelf-life, decreasing the water activity. Finally, several additives such as curing agents (nitrate and nitrite salts), pH modifiers (such as lactate or acetate salts) or emulsifiers can be also used [30,39].

The making process of blood sausages differs as a result of type, region and manufacturer. However, a common flow chart of the process of most of the blood sausages is depicted in Table 1 [21]. The initial mixture of blood sausages is complex by the number of ingredients used and pre-treatments to which those ingredients have been undergone. For example, meat can be cured previously to the mixture preparation, or pork rind can be cooked and emulsified. Similarly, several ingredients, such as fat, onion or rice, can be cooked before the mixture is prepared. Once prepared, the initial mixture is normally stuffed into natural or artificial casings and the sausage is cooked in hot water until a temperature of 65-75 ºC is reached in the inner part of it [31], and then the sausage is chilled before refrigeration storage. Some varieties of blood sausages are dried and/or smoked after cooking. Once cooked and chilled, most of blood sausages present a dark-red to black colour and a rather firm and sliceable texture [30] due to the formation of a gel structure from the interaction of collagen, starch, blood proteins, etc.; nonetheless, some blood sausages are soft and spreadable.

Table 1. General flow chart of blood sausage making process

Raw matter selection
Preliminary preparation of raw materials (weighting, size reduction, premixing, precooking, curing, etc.)

Mixing
Stuffing
Cooking
Chilling

In general, meat and meat products are generally recognized as good sources of high biological-value proteins, group B vitamins, minerals as well as some other bioactive compounds [15]. The composition of meat products depends on their formulation. Thus, the chemical composition of blood sausages is diverse and would depend on the ingredients and manufacturing process used. As a matter of reference, Table 2 shows the proximate composition of several blood sausages from Europe and Latin America. Moisture is expressed as percentage of fresh weight, and values of protein, fat, available carbohydrate, fibre and ash are expressed as percentage of dry matter. The literature sources for the data are the following (see Table for superscripts correspondence): a[36], b[40], c[38], d[41], e[42], f[35], g[37], h[16], i[43], j[44].

Table 2. Proximate composition of several blood sausages from Europe and America

Location and blood sausage name	Moisture	Protein	Fat	Available carbohydrate	Fibre	Ash
Europe						
De Burgos, Spaina	62.2	13.1	28.7	51.1	1.7	4.3
Asturiana, Spainb	38.5	7.0	69.1	8.0	-	2.9
With onion, Spainb	46.0	20.9	59.4	23.2	0.0	-
Blutwurst, Germanyc	55.9	27.4	65.8	0.0	-	-
Thueringer, Germanyc	66.2	58.9	32.3	0.0	-	-
Verivanukas, Finlandd	61.1	19.3	22.6	43.9	9.8	-
Verimakkara, Finlandd	54.7	28.7	42.0	21.2	6.2	-
Blodpølse, Denmarke	43.7	19.0	36.9	32.0	8.9	3.2
With rice, Portugalf	62.0	28.9	38.9	24.6	-	-
Boudin noir, Franceg	62.0	26.8	58.1	10.8	-	-
America						
Blood Sausage, USAh	47.3	27.7	65.5	2.5	0.0	4.4
Traditional, Chilei	77.8	47.3	38.3	0.0	5.9	8.6
Traditional, Boliviaj	44.5	31.7	57.3	10.8	-	1.8
With tongue, Boliviaj	48.8	41.2	55.5	0.0	-	3.3
Stege, Boliviaj	41.2	31.2	56.8	6.5	-	5.4

Moisture content of blood sausages would depend inversely on the fat content and directly on the amount of moisture evaporated during an eventual drying/smoking stage. As can be seen in Table 2, the ranges of fat, available carbohydrate and protein in dry matter vary from 22.6 to 69.1, 0 to 51.1 and 7 to 58.9, respectively. There are great variations in dry matter composition between sausage types, which can be attributable to differences in the quantities of the main ingredients used, i.e., pork fat, cereals, vegetables, meat or blood. Thus, the presence and levels of fibre are the result of the use of vegetables, namely onion, leek, fruits, etc. Finally, ash content is related to the amount of common salt used in the making process. Regarding to the mineral content of blood sausages, the Fe content is the most reported in literature. Fe content of blood sausages is high due to the use of blood, and amounts reported vary from 6 to 16 mg per 100 g [16,36,42,45].

1.3 Aim of the study

In spite of their popularity and increasing interest, literature on the composition and quality of blood sausages is to our knowledge scarce. The knowledge of the chemical composition of blood sausages presents potential usefulness regarding nutritional, product characterization and quality control aspects. Among the chemical composition, the mineral content of blood sausages seems to be a key point in those aspects.Therefore, the main aim of the present study is to describe and determine, as case studies, the manufacturing process and the chemical composition with particular interest on the mineral content, of two typical blood sausages produced in two different parts of the world: a typical blood sausage with white onion (Allium cepa), from the region of Leon (north-western Spain), known as Morcilla de Leon; and typical blood sausage with white cabbage (Brassica Oleracea var. capitata), from the region of Tumbes (north-western Peru), known as Relleno de Tumbes.

2. MATERIAL AND METHODS

2.1 Making process of the blood sausages

In order to collect information about the making process of the blood sausage Morcilla de Leon, four interviews were conducted with the correspondent production managers at the four main local companies producing this sausage in Leon city. The two-member interview panel asked a set of questions regarding general company characteristics, raw materials used, making process and storage conditions. Moreover, collecting data on the making process of Relleno de Tumbes was carried out by standardized open-ended interviews conducted with 15 homemade manufacturers at the region of Tumbes (Tumbes city and small villages at Zarumilla province). The questions asked were to know information on the raw materials used and the making process followed. In both cases, the interviews were followed by the observation of the sausage making process.

2.2 Chemical analysis

A total of 8 samples of Morcilla de Leon were manufactured by local producers (city of Leon, north-western Spain) and were purchased from local markets. The sample weights were approximately 250 g. Once taken, the sausages were transported under refrigeration (<4 °C) to the laboratory of Department of Food Hygiene and Technology (University of Leon). On the other hand, a total of 12 samples of Relleno de Tumbes were obtained from small local producers and retail stores in Tumbes City (north-western Peru) and small villages around the city. For each sausage sampled, a 300 g sample was packaged individually in a bag and transported in refrigerated containers to the laboratory in Tumbes. Subsequently, samples were frozen at -40° C and were transported to the laboratory at University of Leon where upon arrival at laboratory the samples were kept frozen at -40 °C until the analysis was performed.

Determinations of moisture, fat, protein and ash contents in the sausage samples were performed in duplicate according to methods recommended by the AOAC International [46] – Official Methods nos. 950.46, 991.36, 981.10 and 920.153, respectively. Total dietary fibre was analysed following the AOAC 991.43 standard method [46], using the K-ACHDF 11/06 enzymatic kit (Megazyme, Wicklow, Ireland). Finally, the percentage of available carbohydrates was calculated by difference (100 – the percentage of the rest of components).

The analysis of mineral composition of sausages was performed by ICP-AES on wet digested samples. Duplicate aliquots of approximately 1 g (±0.01) of the previously homogenised samples were digested with 10 ml of concentrated HNO3 in tightly closed screw cap glass tubes for 18 h at room temperature, and then for a further 4 h at 90 °C. For the analysis of sodium, potassium, sulphur and phosphorus, 1 ml of the mineralized solution was added with 8 ml of deionized water and 1 ml of scandium solution as internal standard. In order to determine the levels of calcium, copper, iron, magnesium, manganese and zinc, 3 ml of the digested solution was added with 6 ml of deionized water and 1 ml of Sc solution.

The instrumental analysis was performed with an Optima 2000 DV ICP optical emission spectrometer (PerkinElmer, Waltham, MA, USA). Instrument operating conditions were: radiofrequency power, 1400 W; plasma gas flow, 15.0 l/min; auxiliary gas flow, 0.2 l/min; nebulizer gas flow 0.75 l/min, crossed flow; standard axial torch with 2.0 mm i.d. injector of silica; peristaltic pump flow, 1 ml/min; no. of replicates, 2. The spectrometer was calibrated for Cu, Mn, Zn, Fe, Ca and Mg determinations (at 224.7, 257.61, 213.9, 238.2, 393.4 and 279.6 nm, respectively) with nitric acid/water (1:1, v/v) standard solutions of 2, 5 and 10 ppm of each element, and for Na, P, S and K (at 589.6, 213.6, 182.0 and 766.5, respectively) with nitric acid/water (1:9, v/v) standard solutions of 30, 50 and 100 ppm, respectively.

2.3 Statistical analysis

The software STATISTICA for Windows [47] was used for the statistical treatment of data. Furthermore, a principal component (PC) analysis, unrotated method, using the mineral composition as expressed as non-fat dry matter, was also performed.

3. RESULTS AND DISCUSSION

3.1 Making process

The Morcilla de Leon (Figure 1), typically produced in the region of Leon (north-western Spain), is made from a mixture of chopped onion (used at amounts between 65 and 75 % of total weight), animal fat (lard and/or tallow; 10-20 %), blood (normally from pigs, 10-20 %), rice or breadcrumbs (2-10 %), salt (1-1.5 %), dry powdered paprika (1-2 %; including hot and sweet paprika), garlic and a mixture of spices (usually up to 1 g/kg) composed of several of the following: oregano, cumin, anis, cinnamon or pepper. Normally, onion and rice are precooked with the lard or tallow for 1-2 hours (until the onion becomes soft and tender). At the end of cooking, the condiments, spices and blood (liquid) are added and the mix is stirred from some minutes. Nevertheless, one manufacturer did not precook the onion and fat, and thus all ingredients (raw) were cold-mixed. The mixture, (hot if it was precooked or cold if not precooked) is stuffed in natural pork or beef casings of around 45 mm of diameter, tied or clipped forming 20-cm pieces. After the stuffing of the mix, the sausages are cooked in hot water at 80-90 °C for 20-45 min. After this step, sausages are drained hung at room temperature for a few hours and then chill-stored. This product is usually stored without packaging, and the shelf–life is around 12 days at refrigeration temperatures.

Figure 1. Spanish blood sausage Morcilla de León.

Relleno de Tumbes (Figure 2) is a typical blood sausage from Northern Peru, which consists of a mixture of blood (approximately 30%), pork lard fat (10%), chopped cabbage (40%), chopped red and Chinese onion (5%), chopped fresh paprika (2%; including sweet and hot paprika local varieties), common salt (1.5%) and a number of herbs and spices at low quantities (spearmint, coriander, garlic, cumin, pepper) and a in-situ-prepared annatto oil extract; furthermore, the addition of glutamate is common. The amounts indicated above are roughly estimated because the manufacturers did not use scales and the interviewers did not carry a scale in order to weight the ingredients used in the making process. The blood (liquid or coagulated, sometimes precooked and shredded) is manually mixed with the lard, chopped vegetables and salt. Then, the mix is manually stuffed into natural pork casings (large intestine). The blood sausages are cooked in boiling water for approximately half an hour. After cooking, the blood sausages are cooled and then drained hung.

Figure 2. Peruvian blood sausage Relleno de Tumbes.

3.2 Chemical composition

3.2.1 Proximate composition

The proximate composition of Morcilla de Leon and Relleno de Tumbes are shown in Tables 3 and 4, respectively. Moisture is expressed as percentage of fresh weight, and values of protein, fat, available carbohydrate, fibre and ash as percentage of fresh and dry matter weights. Moisture content variability would mainly depend on fat content and the degree of drying loss during cooling and storage. Furthermore, the presence and variability of protein, fat, available carbohydrates and total dietary fibre would be respectively explained mainly by the amounts of blood, lard or tallow, rice or breadcrumbs, onion or cabbage (plus other vegetal condiments and species) used in the formulation. In fresh weight basis, both types of blood sausages have a similar percentage of moisture. However, Morcilla de Leon shows lower amount of protein and higher of fat, fibre and available carbohydrates, than Relleno de Tumbes, both in fresh and dry weight basis. This is explained by a higher amount of blood and lower of vegetables and fat, being used in the Relleno de Tumbes making process, with respect to those being used for Morcilla de Leon.

Table 3. Proximate composition of the Spanish blood sausage Morcilla de Leon (n = 8).

	Mean ± SD (% of fresh weight)	Mean ± SD (% of dry weight)
Moisture	67.1 ± 5.8	-
Protein	5.2 ± 0.9	16.3 ± 3.6
Fat	14.2 ± 3.9	42.9 ± 7.6
Ash	1.9 ± 0.1	5.9 ± 1.2
Total dietary fibre	3.4 ± 1.5	10.1 ± 3.1
Available carbohydrate	8.2 ± 3.4	25.0 ± 6.3

Table 4. Proximate composition of the Peruvian blood sausage Relleno de Tumbes (n = 12).

	Mean ± SD (% of fresh weight)	Mean ± SD (% of dry weight)
Moisture	71.8 ± 6.9	-
Protein	11.9 ± 2.8	42.4 ± 10.2
Fat	9.4 ± 4.0	33.3 ± 13.9
Ash	2.1 ± 0.9	7.6 ± 3.2
Total dietary fibre	1.1 ± 0.4	3.9 ± 1.3
Available carbohydrate	3.6 ± 1.6	13.8 ± 6.3

3.2.2 Mineral composition

The mineral contents of Morcilla de Leon and Relleno de Tumbes are shown in Tables 5 and 6, respectively. Values (expressed as mg/100 g) are given in all fresh, dry and nonfat dry weight basis. Na is the mineral with the highest concentration, and the mean value seems slightly lower in Morcilla de León than in Relleno de Tumbes, where Na concentration shows a great variability between samples (high standard deviation). In average, Relleno de Tumbes contained higher amounts (approximately twice as much) of Ca and S macroelements and of Fe, Zn and Cu microelements.

The mineral content of blood sausages is the result of the sum of the contributions from all the ingredients used. In order to better ascertain the eventual contribution of ingredients to the mineral content of blood sausages, the mineral composition of the main ingredients used in Morcilla de Leon and/or Relleno de Tumbes is shown Table 7 [16,41,48-50]. From Table 7 and taking into account the quantities of the ingredients used in the making processes of the blood sausages, it can be notice that blood appears to be the main source of Fe and Cu to blood sausages. On the other hand, onion and specially cabbage would be the main sources of K, Ca and Mn. Furthermore, S, P, Mg and Zn are importantly provided by both blood and vegetables, with the the high S content of cabbage being remarkable. Finally, lard seems not to be a good source of minerals and common salt, added at amounts of 1-2% to the sausage mixture, is the major source of Na in sausages (not shown in tables). In this context, the higher content of Fe and Cu in Relleno de Tumbes can be associated to the higher quantity of blood used. Similarly, the high content of Ca and S in cabbage together with the high quantity used in Relleno de Tumbes would account for the higher levels of those minerals with respect to Morcilla de Leon.

Table 5. Essential mineral content (mg/100 g) of the Spanish blood sausage Morcilla de Leon (n = 8).

	Fresh weight	Dry matter	Non-fat dry matter
Macroelements			
Na	623 ± 131	1900 ± 615	3315 ± 1038
K	149 ± 27	452 ± 83	795 ± 121
S	76 ± 9	240 ± 61	402 ± 101
P	45 ± 13	136 ± 21	229 ± 35
Ca	29 ± 8	86 ± 21	146 ± 36
Mg	15 ± 2	48 ± 9	78 ± 16
Microelements			
Fe	10.96 ± 3.30	33.24 ± 10.96	58.71 ± 23.33
Zn	0.37 ± 0.12	1.14 ± 0.35	1.88 ± 0.41
Mn	0.20 ± 0.05	0.59 ± 0.10	1.00 ± 0.18
Cu	0.08 ± 0.02	0.26 ± 0.05	0.42 ± 0.12

From the nutritional point of view, comparing the mineral content of blood sausages (fresh weight basis) with that of pork meat or muscle meat products, such as frankfurters or chorizos [16-51], the blood sausages had considerably higher levels of Ca (more than three times), and Mn, Fe and Cu (more than ten times). On the contrary, amounts of K, P, S and Zn are slightly lower in blood sausages (up to 60% lower than those in meat). The levels of Mg were roughly comparable in meat and blood sausages, and those of Na depends on the quantities of common salt added. Having into account the Reference Labelling Values (RLVs) reported by the Scientific Committee of Food from the European Union [52], which are the following: K (2000 mg), Ca (1000 mg), P (700 mg), Na (600 mg), Mg (375 mg), Fe (14 mg), Zn (10 mg), Mn (2 mg) and Cu (1 mg);

interestingly, a portion of 100 g of blood sausage equals or exceeds the recommended daily intake of Fe and contributes with 10-15% the recommended daily intake of Mn and Cu. Thus, the high iron content of blood, coupled withthe high absorption of heme iron compared to non-heme iron, is particularly useful for food based strategies designed to combat iron deficiency anemia a major global malnutrition problem.

Table 6. Essential mineral content (mg/100 g) of the Peruvian blood sausage Relleno de Tumbes (n = 12)

	Fresh weight	Dry matter	Non-fat dry matter
Macroelements			
Na	706 ± 335	2821 ± 1343	3951 ± 1640
K	142 ± 56	565 ± 223	798 ± 254
S	116 ± 22	411 ± 102	590 ± 108
P	48 ± 19	190 ± 67	271 ± 83
Ca	50 ± 18	180 ± 65	257 ± 83
Mg	14 ± 6	50 ± 21	71 ± 28
Microelements			
Fe	29.01 ± 8.55	101.03 ± 25.39	146.50 ± 36.81
Zn	0.70 ± 0.10	2.44 ± 0.51	3.51 ± 0.52
Mn	0.14 ± 0.07	0.52 ± 0.25	0.74 ± 0.36
Cu	0.13 ± 0.06	0.49 ± 0.25	0.69 ± 0.32

Results of principal component analysis are shown in Figures 3 and 4. Principal component analysis was carried out with the mineral content expressed as mg per 100 g of nonfat dry matter for all the blood sausages analysed in this study. The first principal component (factor 1) accounted for a variance of 42.77% and the second of 19.89%. Figure 3 shows that samples from each type of blood sausage are located in two defined sets of results, which corroborate the differences in mineral contents found between both blood sausages. Figure 4 shows the projection of the variables (mineral contents) on the plane formed by the two principal components. The minerals with higher influence (factor loadings higher than 0.8) on factor 1 are Mn, Zn, and Ca. The mineral with higher influence on factor 2 was K (factor loading > 0.8).

Moreover, in Figure 4 it can be seen that the most correlated mineral contents, as indicated by the highest proximity of points in the plain, were S with Ca and Mn with Zn. The first relation could be explained by the significant contribution of cabbage and onion to the S and Ca content in the sausage mixture. However, the second relation is difficult to explain from the contribution of ingredients. Other remarkable correlation is that of Fe with Cu, with blood being the main source of both of them. This correlation could be not as strong as expected due to the feasible migration of Fe ions to ingredients and sausage mixture from the surfaces of cast iron equipment, i.e., pans, knives, etc. [53], which are frequently present at small homemade sausage producing facilities in small villages. This reason could be responsible for part of the distance between the Fe and Cu points.

Table 7. Essential mineral content (mg/100 g) of the main ingredients used in Morcilla de Leon and/or Relleno de Tumbes

	Blood, pork	Pork fat	Onion	Cabbage
Macroelements				
Na	300	11	3	41
K	50	65	166	161
S	140	-	51	300
P	100	38	35	32
Ca	7	2	22	53
Mg	6	2	11	15
Microelements				
Fe	50	0.2	0.2	0.6
Zn	0.5	0.4	0.2	0.2
Mn	0.0	0.0	0.2	0.2
Cu	0.7	0.0	0.1	0.0

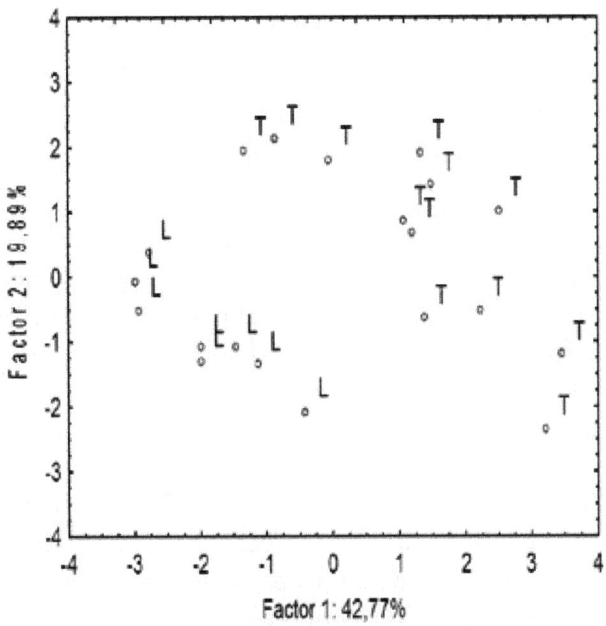

Figure 3. Principal component score plot (two first principal components or factors), considering mineral composition on non-fat dry matter basis, and showing samples according to sausage type: L, Morcilla de Leon; T, Relleno de Tumbes

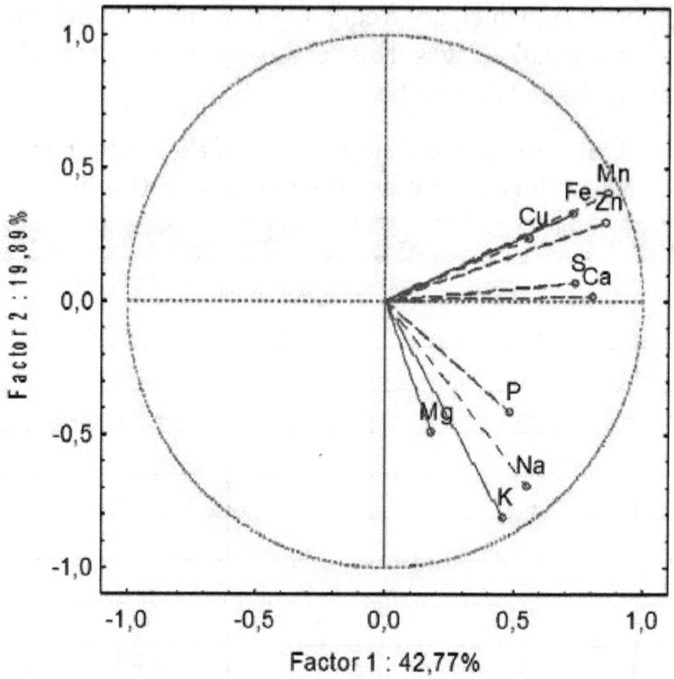

Figure 4. Projection of the normalised factor coordinates of variables (mineral contents) in the 1 x 2 factor plane obtained by the principal component analysis

4. CONCISE SUMMARY

This study delineates the mineral composition of two traditional blood sausages, Morcilla de Leon and Relleno de Tumbes, alongside their proximate composition and general preparation guidelines, thereby contributing to the chemical characterization, dissemination, and preservation of these two traditional meat products.

The diversity and amounts of substances utilized in blood sausage preparation significantly influence their mineral content. Blood supplies significant amounts of Fe, Cu, and Mn to blood sausages from a nutritional perspective. The iron concentration in 100 g of Morcilla de Leon nearly meets the daily requirements for adults, whereas that in Relleno de Tumbes surpasses those requirements.

REFERENCES

1. Millikan M. Nutritional Metals in Foods by AAS. In: Akhyar FM. (ed.) Atomic Absorption Spectroscopy. Rijeka: Intech; 2012. p143-166. Available from http:// www.intechopen.com/books/atomic-absorption-spectroscopy/comparative-assessment-of-the-mineral-content-of-a-latin-american-raw-sausage-made-by-traditional-or (accessed 10 September 2012).
2. Castro FMM, Morgano MA, Nascimiento de Queiroz SC, Mantovani MDB. Relationships of the Minerals and Fatty Acid Contents in Processed Turkey Meat Products. Food Chemistry 2000; 69: 259-265.

3. Cabrera MC, Ramos A, Saadoun A, Brito G. Selenium, Copper, Zinc, Iron and Manganese Content of Seven Meat Cuts from Hereford and Braford Steers Fed Pasture in Uruguay. Meat Science 2000; 84: 518-528.

4. Decker AE, Park Y. Healthier Meat Products as Functional Foods. Meat Science 2010; 86: 49-55.

5. McNeill S, Van Elswik ME. Red Meat in Global Nutrition. Meat Science 2012; 92, 166-173.

6. Nasreddine L, Parent-Massin D. Food Contamination by Metals and Pesticides in the European Union. Should We Worry?. Toxicology Letters 2002; 127, 29-41.

7. Grembecka M, Malinowska E, Szefer P. Differentiation of Market Coffee and its Infusions in View of their Mineral Composition. Science of the Total Environment 2007; 383, 59-69.

8. Kelly SD, Bateman AS. Comparison of Mineral Concentrations in Commercially Grown Organic and Conventional Crops – Tomatoes (Lycopersicon esculentum) and Lettuces (Lactuca sativa). Food Chemistry 2010; 119, 738-745.

9. Luykx DMAM, Van Ruth SM. An Overview of Analytical Methods for Determining the Geographical Origin of Food Products. Food Chemistry 2008; 107, 897-911.

10. Sun S, Guo B, Wei Y, Fan M. Multi-element Analysis for Determining the Geographical Origin of Mutton from Different Regions of China. Food Chemistry 2011; 124, 1151-1156.

11. Barnes KW, Debrah E. Determination of Nutrition Labeling Education Act. Minerals in Foods by Inductively Coupled Plasma-Optical Emission Spectroscopy. Atomic Spectroscopy 1997; 18, 41-54. Available from http://shop.perkinelmer.com.cn/ Content/RelatedMaterials/Articles/atl_barnesfoodatomicspec.pdf (accessed 10 September 2012).

12. Gupta PA, Gupta S. Elemental Profiling: Its Role and Regulations. In: Akhyar FM. (ed.) Atomic Absorption Spectroscopy. Rijeka: Intech; 2012. p37-60. Available from http://www.intechopen.com/books/atomic-absorption-spectroscopy/comparative-assessment-of-the-mineral-content-of-a-latin-american-raw-sausage-made-by-traditional-or (accessed 10 September 2012).

13. Ibáñez E, Cifuentes A. New Analytical Techniques in Food Science. Critical Reviews in Food Science and Nutrition 2001; 41, 413-450.

14. Jimenez-Colmenero F, Pintado T, Cofrades S, Ruiz-Capillas C, Bastida S. Production Variations of Nutritional Composition of Commercial Meat Products. Food Research International 2010; 43, 2378-2384.

15. Toldrá F, Reig M. Innovations for Healthier Processed Meats. Trends in Food Science and Technology 2011; 22, 517-522.
16. USDA. USDA National Nutrient Database for Standard Reference, Release 24, U.S. Department of Agriculture, Agricultural Research Service; 2012 http://ndb.nal.usda.gov/ (accessed 10 September 2012).
17. Davidson A. The Oxford Companion to Food. UK, Oxford University Press; 2011.
18. Guha A, Guha R, Gera S. Study on the Alteration of Bubaline Blood Biochemical Composition Owing to Slaughter. African Journal of Biotechnology 2012; 11, 12134-12137.
19. Liu DC, Ockerman HW. Meat Co-Products. In: Hui YH, Nip W, Rogers RW, Young OA. (eds.) Meat Science and Applications. New York: Marcel Dekker; 2001.
20. Toldrá F, Aristoy MC, Mora L, Reig M. Innovations in Value-Addition of Edible Meat by-Products. Meat Science 2012; 92, 290-296.
21. Mateo J, Cabeza EA, Zumalacárregui JM. Bases de la Tecnología de los Embutidos de Sangre. In: Pérez-Álvarez JA, Fernández-López J, Sayas-Barberá E. (eds.) Industrialización de Productos de Origen Animal (Vol. 1). Orihuela, España: Escuela Politécnica Superior de Orihuela. Universidad Miguel Hernández; 2007, p215-242.
22. Ofori JA, Hsieh YHP. Blood-Derived Products for Human Consumption. Revelation and Science 2011; 1, 14-21.
23. Diez MA, Jaime I, Rovira J. The influence of Different Preservation Methods on Spoilage Bacteria Populations Inoculated in Morcilla de Burgos During Anaerobic Cold Storage. International Journal of Food Microbiology 2009; 132, 91-99.
24. Feiner G. Meat Products Handbook: Practical Science and Technology. Cambridge: Woodhead Publishing Ltd; 2006.
25. Mateo J, Caro I, Figueira AC, Ramos D, Zumalacárregui JM. Meat Processing in Ibero-American Countries: a Historical View. In: Noronha VT, Nijkamp P, Rastoin JL. (eds.) Traditional Food Production and Rural Sustainable. A European challenge. Surrey, UK: Ashgate Publishing Ltd; 2008. p121-134.
26. Choi YS, Choi JH, Han DJ, Kim HY, Lee MA, Kim HW, Lee CH, Paik HD, Kim CJ. Physicochemical and Sensory Characterization of Korean Blood Sausage with Added Rice Bran Fiber. Korean Journal of Food Science and Animal Resources 2009; 29, 260-268.
27. Diez AM, Santos EM, Jaime I, Rovira J. Effectiveness of Combined Preservation Methods to Extend the Shelf Life of Morcilla de Burgos. Meat Science 2009; 81, 171-177.

28. Dehmer NA. La Formación Profesional de los Carniceros y Fabricantes de Embutidos. Guatemala: Instituto Técnico de Capacitación y Productividad - Centro de Capacitación en Tecnología de la Carne; 1995.

29. Reichert JE. Tratamiento Térmico de los Productos Cárnicos. Zaragoza, España: Acribia; 1988.

30. Stiebing A. Tecnología de la Morcilla. Fleischwirtsch. Español 1992; 1, 13-20.

31. Wirth F. Tecnología de los Embutidos Escaldados. Zaragoza, España: Acribia; 1992.

32. Frentz JC, Migaud M. (1976). La Charcuterie Cuite. Généralités et Techniques Actuelles. Vesoul, France: Soussana Editeur; 1976.

33. Luzón-Merino F, Martín-Bejarano S. Fabricación de Productos Cárnicos Tratados por el Calor - Morcillas Cocidas. In: Martín S. (ed.) Enciclopedia de la Carne y los Productos Cárnicos. Plasencia, España: Ediciones Martín & Macias; 2001. p1431-1458.

34. Schiffner E, Oppel K, Lörtzing D. Elaboración Casera de Carne y Embutidos. Zaragoza, España: Acribia; 1996.

35. Cattaneo P, Bonandrini E. "La Morcela de Arroz". Origine - Caratteristiche e Tecnologia di Produzione di un Tipico Insaccato di Sangue Portoghese. Ingegneria Alimentare de le Conserve Animali 1998; 14, 18-27.

36. Santos EM, González C, Jaime I, Rovira J. Physicochemical and Sensory Characterisation of Morcilla de Burgos, a Traditional Spanish Blood Sausage. Meat Science 2003; 65, 893-898.

37. CIC. Le Boudin Noir. Paris : Paris, Francia: Centre d'Information sur les Charcuteries 2003. http://www.infocharcuteries.fr/dmdocuments/dossier_synthese_boudins.pdf (accessed 10 September 2012)

38. Souci SW, Fachmann W, Kraut H. Sausages and Pastries. Food Composition and Nutrition Tables. Stuttgart, Germany: Wissenschaftliche Verlagsgesellschaft mbH; 1989.

39. Frey W. Fabricación Fiable de Embutidos. Zaragoza, España: Acribia; 1985.

40. Martin G, Wert R, Vigili L, Perianes J, Mancilla C, Aza M. Tabla de Composición de Alimentos de la Sociedad Española de Nutrición Básica Aplicada – SENBA; 2005 http://www.5aldia.org/v_5aldia/apartados/apartado.asp?te=179 (accessed 5 March 2008)

41. KTL (2011). Finnish Food Composition Database, Release 14. National Public Health Institute of Finland, Nutrition Unit; 2011. http://www.fineli.fi/foodlist.php?foodname=B%&lang=en (accessed 10 September 2012)

42. Saxholt E, Christensen AT, Møller A, Hartkopp HB, Hess Ygil K. Hels OH. Danish Food Composition Databank, ed. 7.01. Denmark: Department of Nutition, Danish Food Institute – Technical University of Denmark; 2012. http://www.foodcomp.dk/ (accessed 10 September 2012).

43. Bunger A, Alessandri T, Vinagre J, Wittig E, López L. Formulación y Estudio de Embutidos Cárnicos en Envases Flexibles Esterilizables: Morcillas de Sangre. Fleischwirtschaft Español 1992; 2, 15-18.

44. FAO/LATINFOODS. Tabla de Composición de Alimentos de América Latina. Santiago de Chile: FAO, Oficina Regional Para Latinoamérica y el Caribe; 2009. http:// www.rlc.fao.org/es/conozca-fao/que-hace-fao/estadisticas/composicion-alimentos (accessed 10 September 2012)

45. Alcaide E, Gómez R, Carmona, MA, Fernández J. Estudio de los Elementos Minerales en Productos Cárnicos. Alimentaria 1995; 262, 63-67.

46. AOAC. Official Methods 920.153 Ash, 950.46 Moisture, 981.10 Crude Protein, and 991.36 Fat (Crude) Contents in Meat and Meat Products, and 985.29 Total Dietary Fiber in Foods, Enzymatic-Gravimetric Method. In: Cunniff, P. (ed.) Official Methods of Analysis of the AOAC International (Vol. II, 16th ed.). Gaithersburg, MD: AOAC; 1999.

47. StatSoft Inc. STATISTICA (Data Analysis Software System), Version 6. 6. 2001. www.statsoft.com

48. Collazos C, Alvistur E, Vasquez J. Tablas peruanas de composición de alimentos, Séptima edición. Lima, Peru: Centro Nacional de Alimentación y Nutrición, Instituto Nacional de Consumo; 1996.

49. Gorbatov VM. Collection and Utilization of Blood and Blood Proteins for Edible Purposes in the URRS. In: Pearson AM, Dutson TR. (eds.) Edible Meat By-products, Advances in Meat Research, 5. Essex, UK: Elsevier Science Publishers LTD; 1988. p167-196.

50. Rodrigues AS, Fogliano V, Graziani G, Mendes S, Vale AP, Conçalves C. Nutritional Value of Onion Regional Varieties in Northwest Portugal. Electron Journal of Environmental Agricultural Food Chemistry 2003; 2, 519-524.

51. Gonzalez-Tenorio, R, Fernández-Díez A, Caro I, Mateo J. Comparative Assessment of the Mineral Content of a Latin American Raw Sausage Made by Traditional or Non- Traditional Processes. In: Akhyar FM. (ed.) Atomic Absorption Spectroscopy. Rijeka: Intech; 2012. p167-182. Available from http://www.intechopen.com/books/atomic-absorption-spectroscopy/comparative-assessment-of-the-mineral-content-of-a-latin-american-raw-sausage-made-by-traditional-or (accessed 10 September 2012).

52. SCF (Scientific Committee on Food). Opinion of the Scientific Committee on Food on the revision of reference values for nutrition labelling. SCF/CS/NUT/ GEN/18 Final. Bruxelles/Brussels, Belgium: European Commission Health & Consumer Protection Directorate-General; 2003.

53. Quintaes KD, Amaya-Farfan J, Tomazini FM, Morgano MA, Mantovani DMB. Mineral Migration from Stainless Steel, Cast Iron and Soapstone Pans (Steatite) onto Food Simulants. Ciência e Tecnologia de Alimentos 2004; 24, 397-402

SECTION 2
FOOD QUALITY

6 Chapter

OXIDATION AND ANTIOXIDANTS IN AQUATIC AND TERRESTRIAL MEAT FROM PRODUCTION TO CONSUMPTION

1. INTRODUCTION: AN OVERVIEW

Both meat and, particularly, fish are susceptible to significant quality degradation owing to oxidation [1, 2]. Oxidation of lipids in meat and fish products results in rancidity and off-flavors, as well as the formation of various compounds, some of which may have detrimental effects on human health [3]. Oxidation restricts storage duration, thereby impacting the marketing and distribution of fish and meat products. Fish, which is abundant in n-3 polyunsaturated fatty acids (PUFA), is prone to PUFA peroxidation, hence limiting its storage and processing options [4]. Moreover, peroxidative products, especially aldehydes, can interact with certain amino acids to generate carbonyls [5] and protein aggregates [6], resulting in further nutritional deficiencies. Oxidation in red meat and red fish, such as salmon, will not only degrade the lipids but also alter the color, thereby impacting visual consumer appeal. The incorporation of antioxidants is essential to enhance the storage stability, sensory quality, and nutritional value of animal products [9, 10]. Given the beneficial health impacts of long-chain n-3 polyunsaturated fatty acids (PUFA), there is a heightened interest in the production of fish and animal products abundant in n-3 PUFA[11]. Augmenting the concentration of readily oxidized polyunsaturated fatty acids (PUFA) in animal products necessitates a corresponding increase in antioxidants within the final product to safeguard the nutritionally important fatty acids (FA). The significance of a well-balanced amalgamation of polyunsaturated fatty acids (PUFA) and antioxidants, for both product durability and human nutrition, was further underscored by [12]. In addition to the conventional antioxidants utilized in meat and fish, a diverse array of herbs, spices, and fruits is increasingly employed as additives with antioxidative properties [13-17]. In recent years, extensive study has assessed these natural compounds as antioxidative additions in food items, resulting in innovative combinations of antioxidants and the creation of new food products [17-20]. The elevated antioxidant activity of these plant components is primarily attributable to their concentrations of various phenols, anthocyanins, and ascorbic acid, which function as radical scavengers. [21].

In addition to their antioxidative capacity, many of this natural substances have positive effects in the human body and documented health benefits and are therefore highly appreciated food additives [22-27]. So a combination of foods rich in omega 3 PUFA and plant substances rich in phenols and anthocyanins might result in nutritionally very valuable novel food products. These products could play and important role in the prevention of specific chronic-health problems beside dietary supplements where PUFA, probiotics and super- fruits are achieving particular interest in the recent time [23, 28]. Finally nutritionally dense meals may be of interest and importance for people with particularly high nutritional demands, e.g. suffering from malnutrition [29].

For animal foods there are always two possible ways to include antioxidants: Via the feed or post mortem during the processing. Depending on the type of antioxidant, the one or the other way will be more effective. In general fat soluble antioxidants like tocopherol are more effective when present in the feed, while water soluble ones like vitamin C are more effective when added during processing [30, 31]. In addition there are synergistic effects between different antioxidants as for example shown for tocopherol and ascorbic acid [32] so a good combination of all available tools might be able to boost antioxidative protection for certain products.

This chapter will provide a review of the primary utilized and evaluated antioxidants, their synergistic effects, and the potential enhancement of nutritional value. The impact of feeding, along with a range of processing and preservation methods for animal products—spanning both ancient and contemporary techniques—will be discussed, and their implications on oxidative stability will be clarified.

2. GENERAL EFFECTS OF LIPID OXIDATION IN MEAT AND FISH

Lipid oxidation is omnipresent in meat and fish and their products. Especially in products with a high amount of unsaturated FA, oxidation leads to rancidity, off-flavour and taste and to the formulation of toxic substances [2, 33, 34]. In the food industry a great deal of research and attention is spend on the on-going oxidative processes. The main aim is always to protect the raw material and the products as good as possible from oxidation through the whole process and during storage.

2.1 Short introduction to lipids

In order to get a whole picture about lipid oxidation, it is important to know some basics about lipids and FA. FA consist of carbon chains with a methyl (CH3) group at one end and a carboxyl (COOH) group at the other. The C atoms in the chain can either be saturated or unsaturated meaning they form double bonds between each other. The FA which do not have double bonds are called saturated FA (SFA), those having one double bond are called monounsaturated FA (MUFA) and those with two or more double bonds are called polyunsaturated FA (PUFA) (Fig. 1). The FA are generally named in the scheme X:Y n-z where X is the number of carbon atoms in the chain, Y the number of double bonds and z the number of the last carbon atom with a double bond counted from the methyl end (see Fig. 1).

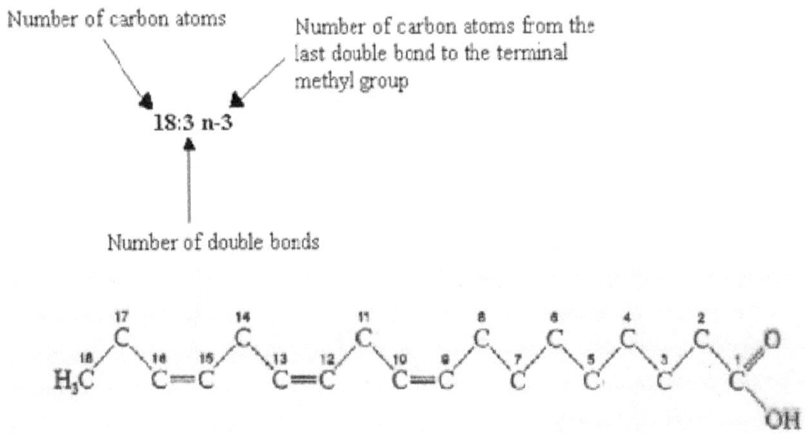

Figure 1. Linolenic acid, 18:3 n-3

The n stands in spoken language for omega so a FA with the last double bond at the third carbon atom from the methyl end is an omega 3 FA while the one with the last double bond at the sixth carbon atom from the methyl group is an omega 6 FA and so on. A very good in depth review about the classification and chemistry of FA and also about their biological functions has been done by [35, 36].

2.2 Reactivity of lipids to oxidation

The reactivity of unsaturated FA increases with their chain length and number of double bounds [37, 38]. Beside the number of double bonds also the placing of the double bonds and the form of the FA determine their oxidative reactivity. In general the n-3 FA are more prone to oxidation than the n-6 and those are more prone to oxidation than the n-9 FA [38].

In animal tissues the lipids are usually divided into two main classes: polar lipids (PL) and neutral lipids (NL). NL consist mainly of triacylglycerols (TAG) which are three FA bound to a glycerol molecule, and minor amounts of mono- and diacylglycerols, whereas PL include mainly phospholipids which are diacylglycerols including a phosphatic acid derivate [39]. TAG serve mainly as an energy source, whereas phospholipids are mainly constituents of the cell and organelle membranes being essential for their functionality and fluidity [39-41]. Phospholipids are in general more unsaturated due to their functionality and therefore also more prone to oxidation. In addition free FA (FFA) can occur in raw or processed tissues due to enzymatic breakdown of acylglycerols or phospholipids. The reactivity to oxidation is in general TAG>phospholipids>FFA.

The complicated thing about oxidation is that once it started a cascade of reactions will occur with each new molecule increasing the reaction speed and variability (Fig. 2). The kinetics of oxidation in meat and meat products are described by [38] and [42]. Oxidation leads to the formation of lipid radicals (L.) that react further to lipid peroxides (LOO.) and hydroperoxides (LOOH). Auto oxidation in meat and fish can be initiated by light, heat, presence of metal ions and radicals. Very low concentrations of radicals are needed to start the reaction. Once initiated, oxidation propagates in a chain reaction (steps 2-6). In the termination reactions, lipid peroxides (LOO.) will react freely, forming a wide range of more stable products including aldehydes, alkanes and conjugated diens.

Initiation LH + Initiator →
L

(1)

Propagation L· + O_2 →
LOO

(2)

. + LH → LOOH +
L·

(3)

Termination LOO+ LOO·non-radical
products (4)

. + LO·non-radical
products

(5)

LOO· + L·
products

(6)

In meat and muscle there are different possibilities to measure the degree of oxidation. The most used ones are listed very briefly here to facilitate the understanding of oxidation parameters used in this chapter:

- The peroxide value: determines the amount of hydroperoxides, which are among the primary products. However, as the peroxides are not stable and react further the results have to be evaluated carefully as, with on-going oxidation the peroxides first increase and reach a maximum but after a while the reaction speed towards secondary oxidation products is faster and the peroxide value decreases again [43].
- TBARS: Another very frequently method is the measurement of thiobarbituric reactive substances (TBARS). Thiobarbituricacid (TBA) reacts with malondialdehyd a secondary oxidation product from PUFA with 3 or more double bonds to a pink complex that can be measured at 532 nm. However the problem with that method is, that other substances also form coloured complexes with TBA and might result in wrong estimation of the oxidation status [44].
- Iodine value: A very traditional method which is still used sometimes to measure the iodine value as a number for the amount of lipid double bonds and the decrease of that number over time as a sign for oxidation.

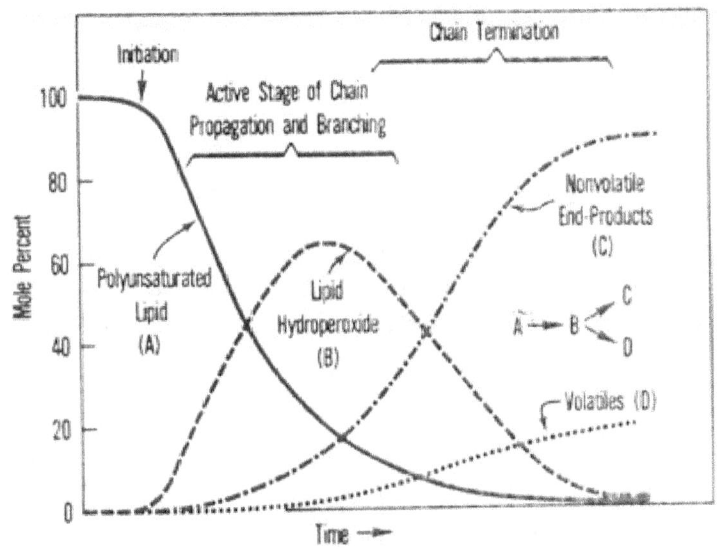

Figure 2. Hypothetical autoxidation of a polyunsaturated lipid as a function of time [47]

- Volatile lipid oxidation products by Headspace GC-MS: During the last decade also more advanced methods have been used more and more for evaluation of oxidation. Content of Hexanal and other volatiles has been shown to give a quite good picture of oxidation status and mechanisms [45, 46]. However as these measurement are quite time consuming and expensive they are still not used routinely.

- Free fatty acids: The amount of free FA (FFA) is actually a value for lipolysis. But as the FFA are oxidised faster than bound FA, they can be regarded as a measurement for increased oxidative reactivity of the muscle or product.

3. ANTIOXIDANTS IN FEEDS

Antioxidants can be introduced into the muscle by different means. Coming first in the natural chain from farm to fork would be to add the antioxidants via the feed. Also in the feed antioxidants are needed, to stabilize the lipids in the feed during storage, especially true is that for fish feed with high contents of PUFA.

The main used antioxidant in feeds is the fat soluble Vitamin E, normally added in the form of tocopherol acetate. Vitamin E is a generic name for all substances that have the biological function of α-tocopherol. These include the tocopherols with a saturated phytyl side-chain, (Fig. 3) and tocotrienols with an unsaturated isoprenoid side-chain, substituted to a chroman head. The different forms of tocopherols and tocotrienols are specified by the use of the Greek letters α, β, γ and δ, to denote the number and position of methyl groups linked to the chroman head [48].

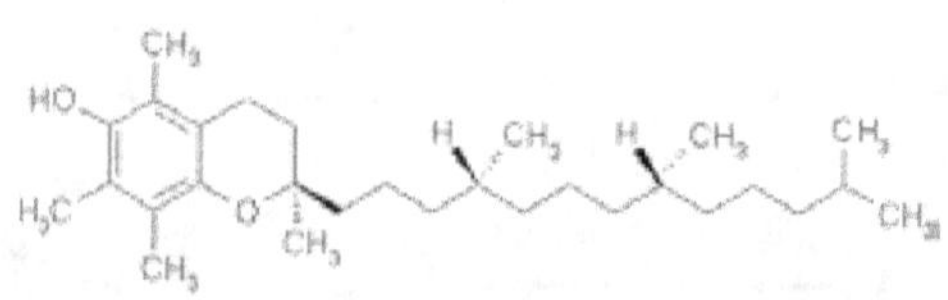

Figure 3. Structure of α-tocopherol

The water soluble vitamin C, ascorbic acid is another antioxidant used in feeds (Fig. 4). However studies have shown that in the live animal tocopherol shows a greater effect, while ascorbic acid works better added post mortem [30, 31].

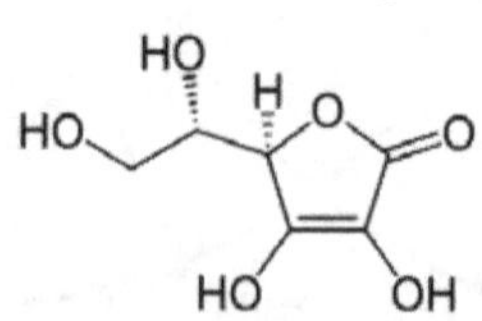

Figure 4. Structure of ascorbic acid

A third group of natural occurring antioxidants are the also fat soluble carotenoids, the precursors of retinol (vitamin A). They are for example found in corn. Carotenoids are hydrocarbons built from eight isoprene bodies (40 C atoms) (Fig. 5). Due to their structure and the conjugated double bonds, both vitamin E and the carotenoids, are radical scavengers that can build relatively stable radicals. In addition, carotenoids, tocopherols and tocotrienols are quenchers for singlet oxygen.

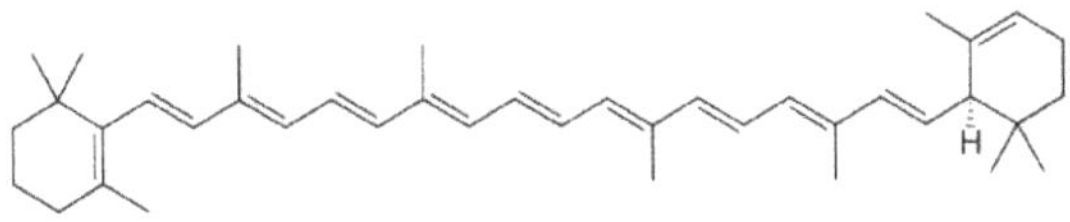

Figure 5. Structure of α-carotene

Carnosine (Fig. 6), a dipeptide, occurring in skeletal muscle which has also been tested as potential antioxidant, however added post mortem [49]. For example [30] suggested using a combination of feed additives and post mortem added antioxidants as for example a feed supplementation with a-tocopheryl acetate and post mortem applied carnosine.

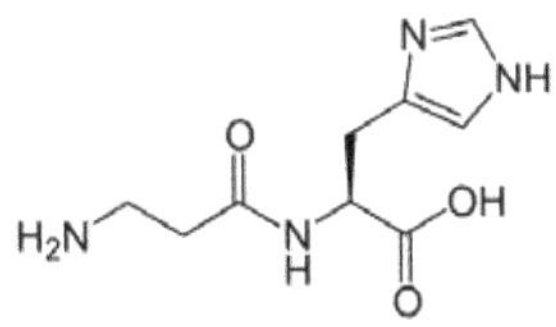

Figure 6. Structure of carnosine

Squalene is triterpene (30 C atoms) (Fig.7) that is present in plants and animal tissues as a key intermediate in the biosynthetic pathway to steroids. It has similar to the carotenoids conjugated double bonds and can hence build stable radicals and has been investigated as possible antioxidant [50]. Significant amounts of squalene in plant sources are detected in e.g. olive oil, wheat germ oil, bran oil and yeast [51] as well as in Amaranthus grain and Ecchium plants [52].

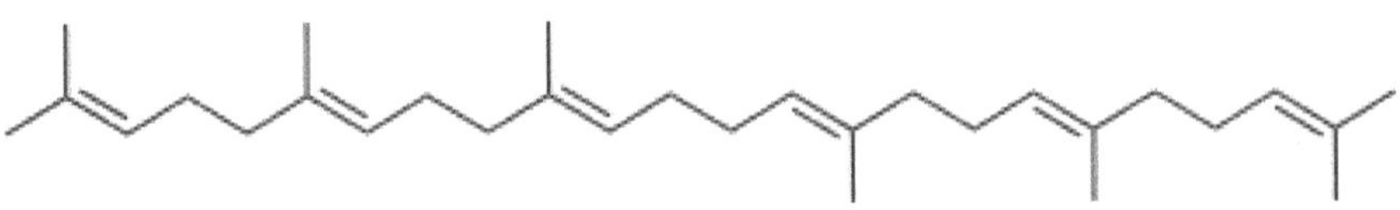

Figure 7. Structure of squalene

As mentioned before, there is also a growing interest to use novel sources of natural antioxidants for feeds, as for example from diverse vegetables [53, 54] or spices [55] or from more exotic sources as algae and lichen [56, 57]. [56] showed for example that feeding chicken with microalgae did not only increase muscle content of the long chain omega 3 FA docosahexaenoic acid, but also increased content of carotenoids and squalene. [58] supplemented pigs with cranberry powder and found both pro- and anti-oxidative effects, most possibly depending on muscle origin and later processing.

Moreover, there are also always interactions between different nutrients [59-61], which have to be taken into consideration when planning how to achieve antioxidative protection of animal foods. For example did high dietary lipid increase also muscle astaxanthin accumulation in salmon (Salmo salar) [60]. Astaxanthin is a carotenoid that gives the pink colour to salmon muscle but can also act as an antioxidant. In another study [48] showed that dietary intake of sesamin increased uptake of α-tocopherol in rats, which suggests that it is possible to increase the bioavailability of antioxidants through feed composition. However, if this mechanism is also valid for fish and other mammals, remains to be investigated.

Concerning the oxidation occurring in the feeds during storage [62] showed that ascorbic acid could protect vitamin E from oxidation in the diet for hybrid tilapia.

Also for the nutritional status of the animals the dietary added antioxidants are of importance. [63] was able to increase the survival of juvenile angelfish (Pterophylum scalare) with a combination of supplemented tocopherol and ascorbic acid in comparison to only tocopherol in the feed. Low dietary vitamin C content has shown to increase requirement of vitamin E in juvenile salmon [64], suggesting that the deficiency of one antioxidant will lead to the increased use of the available ones. However, bioavailability, efficiency and interactions with other substances might vary between different species as summarized by [63].

4. ANTIOXIDANTS ADDED DURING PROCESSING/EFFECT OF PROCESSING TECHNIQUES

4.1 Oxidation factors in muscle and product

4.1.1 Metals

Oils and animal foods always contain a small amount of metals which are too difficult to remove, as for example iron from myoglobin, hemoglobin and the iron storage protein ferritin, or copper, zinc and heavy metals that are present in enzymes and metalloproteins [40, 65]. Another source of metals in animal food products are the machines used during processing, from which minor amounts of iron can get into the products either by abrasion or due to acidic dissolving of metals from the surface. A third source can be migration of metals from the packaging. These metals are present in so low amounts that they do not have a physiological effect; however, they can have pro oxidative effects [66].

4.1.2 Salt

Salt is used for the preservation of meat and fish. Due to its water activity lowering effect and the withdrawl of free water, salt decreases the solubility of oxygen as well as the activity of enzymes and bacteria. In addition chloride ions are also toxic to certain microorganisms. However it is also a pro-oxidant [67].

4.1.3 Oxygen, light and temperature

The more and longer a product is exposed to light and oxygen, the higher is the risk and speed of oxidation. When fish or meat is cut into pieces or minced, the surface is substantially increased and thereby the accessibility for oxygen. As light and increased temperature enhance oxidation [40, 68], during processing temperature and the processing time should be kept as low and short as possible respectively.

4.2 Different preservation and processing techniques:

Various processes including cooling, salting, drying, smoking and heating have been used for a long time to preserve meat and fish and to obtain a variety of products with characteristic organoleptic characteristic [2, 69, 70] Processing is a primarily way to preserve meat, but also adds to its value. However, different processing steps can also negatively affect meat quality, and change for example lipid quality traits. Heating of meat and meat products e.g. hot smoking, can disrupt the cell membranes and promote lipid oxidation [71], which affects the nutritional and sensory properties of the meat product. Use of antioxidants during processing or alternative more gentle processing methods can reduce these negative effects.

4.2.1 Chilling or cooling

Fresh meat is sold chilled at a temperature of about +4 °C. Preservation of meat quality is an important criterion for its shelf life, since raw, chilled meat has traditionally been a perishable product [1, 72]. In order to prolong the chilled storage time advanced packing techniques or various additives are used in addition, which will be described in more detail in the following.

Fresh fish is usually transported and sold on flaked ice, keeping the temperature slightly above 0°C; more recently also ice slurries have been used [2, 73]. To make these ice slurries even more effective different additives to fish as well as to the ice slurry have been used. Examples are natural antioxidants, ozone or organic acid mixtures. [73] evaluated the effect of organic acids mixed into the ice slurry on lipid oxidation and found slightly decreased oxidation. On the other hand [74] showed a significant decrease of lipid oxidation when fish was stored in ice made with water extracts from rosemary or oregano. [2] gives a good review on different additives to slurry ice and summarizes among others that addition of ozone declined microbial spoilage and did not increase oxidation. Addition of antioxidants directly to the fish will be discussed further on in this chapter.

4.2.2 Super chilling or deep chilling

Super chilling or deep chilling means in general to chill the products to a temperature close to or just below the initial freezing point, which is for the most food products between -0.5 and – 2.8 °C (reviewed by [75]). In regard to lipid oxidation it is important that there are no ice crystals formed, as they can destroy organelle membranes and thereby release enzymes and enhance oxidation potential [76]. This technique is used for example for deer meat exported from New Zealand to Europe. The international trading demands new techniques to provide longer storage times. Due to long transport distances and high export quantities, deer meat from New Zealand is stored in vacuum packages and deep chilled to -1.5 °C, and can be considered as fresh meat up to 14 weeks after slaughter [77]. The low temperatures, combined with vacuum, retard bacterial growth, lipid oxidation and color deterioration. There is not done much work on lipid oxidation during deep chilling, however [75] suggested that the improved shelf life and quality reported from deep chilled foods is also indirect resulted by a reduction of lipid deterioration.

Also in fish and seafood deep chilling has been applied successfully and shown to retard microbial growth and extend shelf life of for example prawn (Penaeus japonicas) [78] salmon and cod (Gadus morhua) [79, 80], however without investigating the effects on lipid oxidation.

4.2.3 Freezing

Frozen storage has since long been a method considered sufficient to preserve meat over longer time periods [81], however freezing can also negatively influence structural and chemical properties of meat, e.g. increase content of FFA and lipid oxidation products [82, 83]. [83] reviews some aspects related to lipid oxidation during and after freezing and describes the amount of unfrozen water as one important factor for primary oxidation. The amount of unfrozen water depends on the freezing temperature and in line with that an optimum freezing temperature for meats of -40°C has been suggested by [84]. At this temperature only a minor fraction of the water is unfrozen. In agreement with that [85] showed a significant increase of lipid oxidation products in pork stored at -18°C. Besides the temperature, the formation of ice crystals during freezing is a critical point [76] and the larger ice crystals are formed the higher is the risk of membrane disruption and increased oxidation.

An important element to avoid increased oxidation after thawing should therefore be the formation of small ice crystals during freezing. The faster and more homogeneous the freezing happens, the smaller and more uniform the formed ice crystals will be [86]. Some recent developed fast freezing techniques suitable for muscle foods are high pressure freezing, pressure shift freezing, cryogenic freezing and the already since longer time used air-blast freezing [86, 87]. [87] showed that substantial smaller ice crystals were formed in Norway Lobster (Nephrops norvegicus) when pressure shift freezing was used compared to air blast freezing. However, most of the papers, studying effects of freezing, evaluate only texture, drip loss and sensory, therefore not much is known about the effects of different techniques on lipid oxidation.

Once frozen, it is important for the maintenance of the small ice crystals that a stable temperature is kept, as thawing and refreezing as well as temperature fluctuations lead to formation of bigger ice crystals [86 1895]. Beside different techniques of the freezing itself, the injection or dipping of antifreeze proteins for both meat and fish has shown some success to force the formation of preferably small ice crystals (reviewed by [86]). Addition of antifreeze proteins has also shown to inhibit recrystallization of small ice crystals into bigger ones (reviewed by [86]).

4.2.4 Salting and curing

Meat curing and salting of fish are among the oldest preservation techniques man has used [88]. As described above salt has pro-oxidative effects. A demonstration of the pro-oxidative effect of salt in muscle foods can be found in [89]. During the process of meat curing with salt, nitrite is usually added to keep the nice pink color of the meat. The nitrite exchanges the oxygen ligand in the oxymyoglobin (Fig. 8), which is responsible for the bright red colour of fresh meat, and builds another very stable pink colored complex, the nitrosylmyoglobin [88]. [90] showed in addition an antioxidative effect of nitrite in meat and discussed different possible reaction mechanisms.

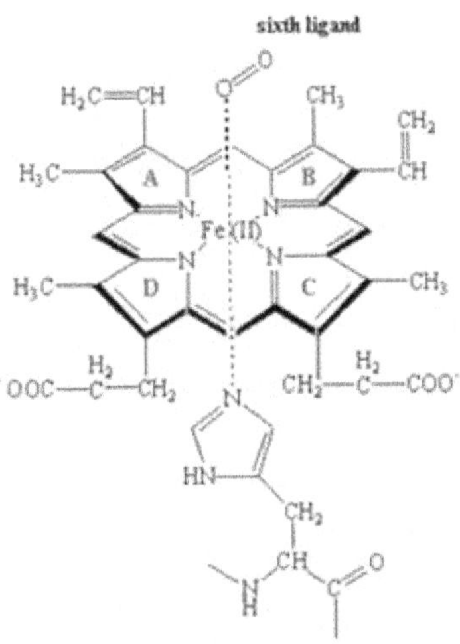

Figure 8. Structure of oxymyoglobine

Salting of fish is very commonly used traditional preservation process [91]. In many cases as for example in the traditional salted herring, salting and ripening takes a quite long ripening time. However [91] showed also that only a modest increase of peroxide values occurred during the ripening time. However as the FFA increased substantially in the brine and as peroxide values show only primary oxidation products substantial oxidation might have happened undetected. An indicator for oxidative stress during the ripening process is the drop of α- tocopherol to approximately 50% after 371 days in that study. However as also pointed out in the section about drying in this chapter, part of the oxidation products might be part of the desired characteristic organoleptic properties[66] investigated the effect of trace metals in the used salt during the salting of cod, showing a significant increase of TBARS with increasing copper concentration and during the salting time. Various attempts have been made to study oxidation during various salting processes and to find ways to inhibit or decrease lipid oxidation. [92] showed that salting initially protected chub mackerel (Scomber japonicus) from oxidation, however after 12 days of storage TBARS values were significantly higher in salted non-smoked fish compared to unsalted non-smoked fish. However, [93] showed that partial replacement of NaCl with KCl decreased lipid oxidation in salted mackerel as well as addition of ascorbic acid to brine solution. [94] found that EDTA prevented copper induced oxidation in salt brined cod, while added citrate enhanced oxidation and ascorbate had no effect in that study.

Besides the use of salting for preservation, fish is sometimes mildly salted to improve sensory characteristics and water holding capacity, where contents of only 0.1-0.3% can give significantly improved water holding capacity [17] but also resulted in increased oxidation levels in herring (Clupea harengus).

4.2.5 Drying

Dry curing and drying of meat also involve pro-oxidative factors, as there are: long exposure to air, dehydration and absence of nitrite [95]. Enzymatic activity can lead to high amounts of FFA, which are more prone to oxidation than TAG [34, 88]. In products such as dry-cured ham or dry-cured salami, a certain amount of volatiles, lipid oxidation compounds and lipolysis products is desired since they are responsible for the particular taste of these products [4, 96]. In line with this [97] showed that the traditional drying process of reindeer meat led to significantly increased oxidation parameters compared to the raw meat or smoked reindeer meat. On the other hand, excessive amounts of oxidation products result in off-flavors and rancid taste [98] and should hence be avoided.

[70] reviewed that in dry cured hams the lipases stay active for several month and hence can produce high amounts of FFA in the tissue. [99] confirmed this in their experiment showing a constant increase of FFA during 24 month of aging of dry cured hams. However TBA values did not increase significantly during that time. Evaluating how to avoid excessive oxidation in dry cured Parma hams, [100] showed that dietary tocopherol could decrease oxidation even in hams with an increased proportion of unsaturated FA.

Also various dried fish products exist, however most work done on dried fish products deals with microbial spoilage or sensory aspects as for example in cod [101, 102] and only few works also evaluate lipid oxidation. But [103] compared different drying methods for dried milkfish (Chanos chanos), a traditional Taiwanese product. In their study cold air drying resulted in significant lower TBA values than the traditional sun drying or hot air drying. They conclude that both light and temperature were important factors which induce increased oxidation. These results agreed with [104] who evaluated different drying methods on dried yellow corvenia (Pseudosciaena manchurica) and found higer oxidation in products made by the traditional sun drying process. Other works found on dried fish [66, 94] investigated heavily salted cod, and are hence discussed in the previous section about salt.

4.2.6 Smoking

Smoking is another traditional method to preserve meat and fish and create new products. [88] described the antioxidative activity of some of the smokes components. The various techniques and the types of wood used lead to the characteristic taste of the final product [105]. However as hot or warm smoking also includes increased temperature over a longer period and the meat parts are usually salted before smoking, also always some oxidation is initiated. In line with that effects are in general more complex, considering the various pro- and antioxidative aspects of this way of processing. For example [92] showed that smoking initially increased oxidation in chub mackerel (Scomber japonicus) but that it had lipid oxidation decreasing effects during storage, leading to lower TBARS values in the smoked fish compared to the unsalted non-smoked fish after 6 days.

In addition the smoke contains also substances that have been associated adverse health effects [106]. Therefore different processing methods as for example the use of liquid smoke have been investigated. [107] showed that a combination of liquid and traditional smoke were more effective inhibiting lipid oxidation in bacon than traditional smoke alone. These results were ascribed to a possible higher content of phenols in the samples processed with the combined smoking procedure. Contradictory [108] showed that traditional smoke resulted in lower TBA values compared to the use of liquid smoke in smoked beef tongue after 5-30 days storage. [109] compared traditional cold smoking and electrostatic smoking of salmon and concluded that electrostatically smoked fillets had a higher loss of lipids, but were less oxidized than traditional smoked fillets.

4.2.7 Packaging

From an oxidation point of view, packing should be tight and compact so that the surface and oxygen access are minimized. However this will not always meet the customers' expectations of product presentation, so naturally compromises have to be made. Packaging systems and technologies have developed rapidly during the last decades [1]. Both in meat and fish the principal function is to limit bacterial spoilage and growth. In red meats also the preservation of a bright red color is important, which is an indicator of freshness for the consumers [7]. This will be reached for example by keeping a high percentage of oxygen in a modified atmosphere package (MAP), while most bacteria are inhibited by an increased concentration of CO2 at the same time. For the different types of meat and fish the perfect gas mixture differs. A good overview is given by [110]. In fat fish due to the high oxidation risk a gas mixture without oxygen is generally recommended. [49] tried to use various antioxidants combined with modified atmosphere and showed increased lipid and color stability when a combination of rosemary and ascorbic acid was used in MAP.

Examples for the application of vacuum packing technique are given by [111] for fish burgers and by [80] for salmon fillets. Unfortunately these studies have not investigated oxidation in normal versus vacuum packing. However [112] investigated the effect of different storage conditions on oxidation in burgers made from rabbit meat and found decreased oxidation when vacuum packing was used.

4.2.8 Other preservation methods (irradiation)

Low dose irradiation is a very effective method to kill many bacteria including Salmonella and Escherichia coli, but it is also known to generate hydroxyl radicals and could hence lead to increased oxidation in meat and fish products [113]. [114] evaluated the effect of low dose irradiation up to 9.43 kGy on different meats (pork, beef, lamb and turkey) and found only low dependency between lipid oxidation values and the irradiation dose. However slightly higher values of malondialdehyd were found in turkey breast with the highest dose compared to the other meats at the same dose. [115] found increased oxidation values in pacu (Piaractus mesopotamicus) fish after irradiation. Nevertheless, in the same experiment the researchers showed addition of antioxidants α-tocopherol, BHT or rosemary extract could inhibit the oxidation accelerated by irradiation.

4.2.9 Canned meat and fish products

In canned fish the major part of oxidation seemed to occur due to the heating step before and during sterilization [116, 117]. [117] reviewed that also the storage conditions (time and temperature) before the actually canning do have a significant influence on the final content of oxidation products. The longer the storage time and the higher the storage temperature the more oxidation and lipolysis will take place and the higher the content of easily oxidable FFA will be. Beside these factors also the filling media seemed to have a significant impact [117]. [116] showed a significant increase of TBA values in silver carp canned with brine, sunflower oil and soybean oil while olive oil seemed not to enhance oxidation. On the other hand [117] evaluated the effect of natural antioxidants from the canning oil on canned tuna and found protective effects against lipid oxidation from extra virgin olive oil rich in phenols and also partly from soybean oil rich in tocopherols. Highest oxidation was found in tuna canned in brine in that experiment. [117] ascribed that to a possible accumulation of the PUFA at the oil-water surface. In general the results show that even added antioxidants like spices or other plant antioxidants could have a positive effect against oxidation in canned fish products. However, to our knowledge the effect of the addition of antioxidants or the effect of spices present in the brine has not been investigated yet.

In canned meat products the situation is expected to be similar as in fish, but not as much research as on fish products concerning oxidation has been executed. This might be due to the fact that there are more canned fish products on the market and that fish is known to have higher susceptibility to oxidation due to its higher content of PUFA. However, in one of the few more recent studies [118] investigated the importance of the raw product composition and found lowest oxidation in the product with lowest fat content.

4.3 Ready to eat and fast food products

There is a wide variety of ready to eat products from meat and fish available on the market, such as sausages, meat- or fish balls, paté's and many more. As these products often include minced or grinded meat and several other ingredients beside the raw muscle as well as they require several processing steps, all of these will have an influence on the oxidation behavior. On the other hand this creates a great chance to add antioxidants or to optimize processing techniques and packaging towards the lowest possible oxidation status of the final product. In general it can be said that also in this case the fish products will be the ones which are more prone to oxidation due to their more unsaturated FA composition. However also other aspects play a role as for example [119] found comparable cholesterol oxide values in one of three pork paté's as in a cod paté, while two other pork patés and a tested salmon and anchovy paté had lower values.

Antioxidants additives in fast food products are for example rosemary extract showing an antioxidative effect in mackerel burgers [111] and or as a more novel ingredient yerba mate extracts, which enhanced lipid stability in beef hamburgers [120]. [121] showed protective effects of oregano and thyme oil in ready to eat squid rings and [122] showed antioxidative effects of various herbs in pork patties.

But also processing methods or packaging can be used to increase oxidative stability. For instance [111] used vacuum packing in addition with the applied antioxidants in the mackerel burgers, while [123] evaluated a combination of irradiation and different packing environments to increase shelf life in pork patties.

Sausages are very favorite and omnipresent meat products around the world. A wide variety of categories such as raw, cooked, dry fermented, cooked smoked, raw smoked or precooked sausages exist. Through the addition of especially spices the oxidation in these products can efficiently be decreased [124] for example used Spanish paprika and garlic or a mixture of nitrite, nitrate and ascorbic acid in chorizo type sausages. They concluded that paprika showed a potent antioxidant capacity in this type of product and that a mixture of 3% paprika and 1% garlic had similar antioxidative effects as a traditional used curing mixture of nitrite, nitrate and ascorbic acid. [125] tested rosemary as spice and natural antioxidant in fermented goat sausages and found lower oxidation and increased values of overall sensory. Adding Palatase M. (from Rhizomucor miehei) to dry fermented sausages in order to improve sensory aspects, resulted in increased FFA content, however no correlation with higher TBARS could be found by [126]. On the other hand, the authors found an increased amount of volatile compounds which could indicate an increased oxidation due to the added enzyme.

5. ANTIOXIDANTS IN MEAT AND FISH PRODUCTS

Antioxidants delay or inhibit the process of oxidation, even when present in low concentrations [127]. Some antioxidants function as radical scavengers or peroxide decomposers, while others quench singlet oxygen, remove catalytic metal ions or oxygen, or inhibit enzymes. The cellular antioxidants can be classed as low molecular substances and enzymes that are either water-soluble or fat-soluble.

5.1 Spices and herbs

It is well known that phenolic compounds from spices and herbs have an antioxidative potential due to their possibility to act as a radical scavengers [128, 129]. A short review and a list of some polyphenols with their respective antioxidant activity can be found in [129]. Various spices have hence been tested in a wide range of products from sausages over meatballs to fish fillets and fish oil [124, 125, 130, 131]. [16] showed that 1.5% sage added to meatballs decreased oxidation and limited undesirable changes in the composition.

The advantage in the use of spices and herbs is that they are natural and in case of various products often are anyway included in the spicing or that they blend in to the desired taste of the final product. Consumers appreciate having natural antioxidants in their products over synthetic ones. However a problem might be if the taste of the used spice/herb does not fit with the product or gives a too strong side taste. For example [111] found that addition of 0.4 % rosemary gave improved shelf life of fish burgers with an acceptable taste for the consumers, while 0.8% of rosemary gave a too intense taste [132] compared the antioxidative activities of 22 commonly used herbs and spices added in different amounts to pork meat and found highest antioxidant capacity in sansho, ginger and sage. Furthermore also addition of rosemary, thyme, oregano and allspice resulted in up to 64% inhibition of lipid oxidation.

5.2 Fruits and berries

The high antioxidant capacity of berries is particularly due to their content of different phenols, anthocyanins and ascorbic acid [21]. Besides health benefits related to their natural antioxidants, colour attributes of berries are also of interest in food processing, as colour plays a vital role to the acceptability of foods. A wide range of various fruits and berries has shown antioxidant capacity as for example cranberries, elderberries, black currant and many more [133, 134].

For example, grape seed extracts were used to inhibit lipid oxidation in muscle from chicken, beef and pork [14] as well as in turkey meat [135] and polyphenols extracted from grape pomance inhibited lipid oxidation in fish muscle [136].

Even more unusual combinations have been tested successfully, as for example the antioxidative effect of various berry concentrates as marinades for herring fillets [17]. Other applications are cranberry juice powder as antioxidant rich feed for pigs [58], cranberry extract as additive to separated turkey and ground pork meat [137], grape antioxidant dietary fibre in minced fish [138] or tomatoes in beef patties.

5.3 Antioxidants from other sources

Beside spices, herbs and fruits also teas and other possible sources for natural antioxidants have been evaluated.

Among others, tea catechins have been tested and used as antioxidants in various food products. Extracted tea catechins from green tea showed significant potential to inhibit lipid oxidation in red meat, poultry and in fish muscle [139]. Instant green tea has shown to slow down oxidation in frozen mackerel [140].

Chitosan is the deacetylated form of chitin and has been shown to have antibacterial and antifungal properties and has therefore reached some attention as food additive [141]. In its original form it is ineffective as antioxidant, however [141] have shown that as a glucose complex chitosan exhibited both antimicrobial and antioxidative effects in pork salami.

Besides research is constantly searching for new sources of antioxidants as for example tomato seed oil from tomato pomance (industrial tomato waste) [142] or industrial onion waste [54]. These antioxidant rich waste products could be added to the animals feed as successfully shown by [53] where eggs from chicken fed tomato byproducts contained higher amounts of lycopene compared to normal eggs. Or on the other hand extracts from these byproducts could be used as additives directly to the food products as suggested by [143]. Similarly byproducts from wine and olive oil byproducts inhibited oxidation in minced fish and frozen mackerel fillets respectively [138, 144]. As a more exotic possible additive [145] investigated antioxidant properties of Indian red seaweeds.

5.4 Synergistic effects and interactions

As mentioned before, in addition to the antioxidative effect a substance has alone, there are as well always interactions that can influence the bioavailability, the antioxidative effect and mechanisms between the various nutrients.

For example vitamine C and vitamine E have been found to interact as antioxidants, tocopheroxy radicals are reduced back to tocopherols by ascorbic acid [32]. As in meat and fish products this mechanisms takes place at the border between lipid and water phase, the radical is removed from the lipid phase and the lipid oxidation process due to that radical is terminated. [146]

Describes the function of carotenoids in what he calls antioxidant networks, were carotenoids act together with other antioxidants at interfaces as for example xanthophylls and carotenoids in egg yolks and fish. Similar to the synergistic action of tocopherols and ascorbic acid, the more hydrophilic (iso) flavonoids and their glycosides regenerate the lipophilic carotenoids which are active as radical scavengers in the lipid phase. In another mechanism the more hydrophilic xanthophylls act via the membranes between water/lipid interfaces in synergism with more lipophilic carotenoids. [146] defines concluding two types of conditions how carotenoids function: (i) in "equilibrium" with other antioxidants in thermodynamically controlled networks serving as color indicators of good antioxidant status and (ii) as antioxidants active through radical scavenging in networks with kinetically controlled regeneration. Furthermore carotenoids also showed to enhance antioxidant activity of vitamin E [147]. Moreover [109] reported that astaxanthin and tocopherol act via different mechanisms in salmon and hence improve stability against oxidation at different stages of oxidation.

Squalene has been found to protect α-tocopherol in oxidation processes [148], probably in a similar way of action than those described for the carotenoid networks.

The use of combined added antioxidant and other preserving techniques has also shown effects as earlier presented in the case of [49] where modified atmosphere packaging was used in combination with antioxidants and where also the combination of two different antioxidants, namely rosemary and ascorbic acid gave the best results. Furthermore [93] showed that a combination of various preservation techniques gave the best results against lipid oxidation in salted mackerel. Combined frozen storage at -18°C in a vacuum package and added ascorbic acid at the same time as 50% of the NaCl was replaced by KCl gave the best results in that study.

On the other hand [94] showed that ascorbate might have pro-oxidative effects due to concentration and depending on the presence or absence of other oxidants. For instance did ascorbate concentrations below 50ppm in combination with 5 ppm copper in the brine prevented formation of TBARS while concentrations above 500ppm in absence of copper had pro-oxidative effects.

Effects of other nutrients on antioxidant uptake and accumulation were shown by [109] who showed a positive correlation between fat content and tocopherol accumulation and a negative correlation between fat content and content of ascorbic acid in salmon. [149] showed negative effects of high dietary astaxanthin on α-tocopherol deposition in rainbow trout *(Oncorhynchus mykiss)*

There is still much room for novel combinations that might give improved oxidative stability to varying products and hence more research in that field is strongly needed.

6. ADDITIONAL VALUE IN ANTIOXIDANT RICH FOODS

Generally, antioxidants maintain product quality by improving shelf-life, nutritional quality and other aspects related to quality. Meat and fish products have been successfully enhanced with different spices and new food ingredients, to prevent oxidation and increase thereby both nutritional value, storage stability and sensory.

The positive effects of tocopherol and ascorbic acid on human health in their property as vitamins are obvious. But beyond that compounds like polyphenols, carotenoids and catechins have shown to influence human health thanks to various other properties beside their antioxidative capacity. For instance [150] gave a valuable review on antioxidant and antimicrobial effects of various berries and their impact on human health and [151] reviewed the anti-inflammatory properties, effects on cancer, diabetes, the immune system, and ocular health of asthaxanthin. [152] reviews the anti-inflammatory, anti-allergic, antimicrobial and cancer-preventive effects of polyphenols, which are mainly due to their antioxidant activity; and describes that polyphenols furthermore can directly bind with signaling molecules involved in inflammatory mechanisms and carcinogenesis and thereby regulate cell activity. In line with that review, [153] gave an overview about the various positive effects of tea catechins on human health, as for example protection against bacterial induced dental caries and antiviral properties.

Examples for the health effects associated with different berries and fruits are numerous: Cranberries are known for their prevention of urinary tract infections [154]. [155] reviews the antioxidative and cardio-protective actions of Chilean blackberries and [156] reviewed the positive effect of grape juice, berries and walnuts on age related diseases. The authors discuss that beside the antioxidative and anti-inflammatory effects, polyphenols as for example antocyanins and proanthocyanidins enhance neuronal communication and neuronal signaling and decreased oxidative and inflammatory stress occurring due to aging.

As already mentioned before, besides inhibiting oxidation, plant substances can also have protective effect against microbial growth. [141] described the combined antioxidative and antimicrobial effects of chitosan as well as [157, 158] showed that oregano and cranberry inhibit Heliobacter pylori and Listeria monocytogenes in fish and meat beside their antioxidative capacity. Hence some plants with antioxidant capacity are also protecting from food poisoning.

7. PERSPECTIVE ON NOVEL FOODS

The diverse qualities of natural chemicals such as polyphenols, carotenoids, terpenes, and catechins present untapped opportunities for the creation of innovative food products. The significance of a balanced amalgamation of polyunsaturated fatty acids (PUFA) and antioxidants, for both product stability and human nutrition, was delineated by [12]. Increasing the quantity of polyunsaturated fatty acids (PUFAs), particularly n-3 fatty acids, necessitates a corresponding increase in antioxidants to maintain optimal storage stability of fish and meat products and to avert oxidation [12]. Therefore, the amalgamation of fish or meat with diverse plant items such as berries or spices may be intriguing from nutritional, sensory, and technological perspectives.

In addition to these nutritionally dense meals rich in PUFA, antioxidants, and other nutrients, such foods may be particularly crucial for individuals with elevated nutritional requirements, such as elderly individuals experiencing malnutrition [29]. The array of innovative dishes that may be created is extensive; a recent study highlights fish dishes enriched with polyphenols derived from berries [17]. Additionally, [159] delineates the idea of FOSHU (foods for specified health use) and enumerates nine innovative meat products that have received approval in Japan, asserting beneficial impacts on several facets of human health.

REFERENCES

1. Jeremiah, L.E., Packaging alternatives to deliver fresh meats using short- or long- term distribution. Food Research International, 2001. 34(9): p. 749-772.
2. Medina, I., J.M. Gallardo, and S.P. Aubourg, Quality preservation in chilled and frozen fish products by employment of slurry ice and natural antioxidants. International Journal of Food Science and Technology, 2009. 44(8): p. 1467-1479.
3. Ames, B.N., M.K. Shigenaga, and T.M. Hagen, Oxidants, Antioxidants, And The Degenerative Diseases Of Aging. Proceedings of the National Academy of Sciences of the United States of America, 1993. 90(17): p. 7915-7922.
4. Gray, J.I., E.A. Gomaa, and D.J. Buckley, Oxidative Quality and Shelf Life of Meats. Meat Science, 1996. 43(1): p. S111-S123.

5. Uchida, K. and E.R. Stadtman, Covalent Attachment Of 4-Hydroxynonenal To Glyceraldehyde-3-Phosphate Dehydrogenase - A Possible Involvement Of Intramolecular And Intermolecular Cross-Linking Reaction. Journal of biological Chemistry, 1993. 268(9): p. 6388-6393.

6. Buttkus, H., Preparation And Properties Of Trout Myosin. Journal of the Fisheries Research Board of Canada, 1966. 23(4): p. 563-573.

7. Faustman, C. and R.G. Cassens, The biochemical basis for discoloration in fresh meat: a review. Journal of Muscle Foods, 1990. 1: p. 217-243.

8. Scaife, J.R., et al., Influence of alpha-tocopherol acetate on the short- and long-term storage properties of fillets from Atlantic salmon Salmo salar fed a high lipid diet. Aquaculture Nutrition, 2000. 6(1): p. 65-71.

9. Kazimierczak, R., et al., Antioxidant content in black currants from organic and conventional cultivation. Electronic Journal of Polish Agricultural Universities, 2008. 11(2): p. art 28.

10. Ladikos, D. and V. Lougovois, Lipid Oxidation in Muscle Foods - a Review. Food Chemistry, 1990. 35(4): p. 295-314.

11. Wood, J.D., et al., Effects of fatty acids on meat quality: a review. Meat Science, 2003. 66(1): p. 21-32.

12. Kamal-Eldin, A. and J. Pickova, Balance between polyunsaturated fatty acids and antioxidants in nutrition. Lipid Technology: 2008. 20(4) 80-83.

13. Bhale, S.D., et al., Oregano and rosemary extracts inhibit oxidation of long-chain n-3 fatty acids in menhaden oil. Journal of Food Science, 2007. 72(9): p. C504-C508.

14. Brannan, R.G. and E. Mah, Grape seed extract inhibits lipid oxidation in muscle from different species during refrigerated and frozen storage and oxidation catalyzed by peroxynitrite and iron/ascorbate in a pyrogallol red model system. Meat Science, 2007. 77(4): p. 540-546.

15. Haak, L., K. Raes, and S. De Smet, Effect of plant phenolics, tocopherol and ascorbic acid on oxidative stability of pork patties. Journal of the Science of Food and Agriculture, 2009. 89(8): p. 1360-1365.

16. Karpinska, M., J. Borowski, and M. Danowska-Oziewicz, The use of natural antioxidants in ready-to-serve food. Food Chemistry, 2001. 72(1): p. 5-9.

17. Sampels, S., et al., Berry Marinades Enhance Oxidative Stability of Herring Fillets. Journal of Agricultural and Food Chemistry, 2010. 58(23): p. 12230-12237.

18. Coma, V., et al., Edible antimicrobial films based on chitosan matrix. Journal of Food Science, 2002. 67(3): p. 1162-1169.

19. Perumalla, A.V.S. and N.S. Hettiarachchy, Green tea and grape seed extracts - Potential applications in food safety and quality. Food Research International, 2011. 44(4): p. 827-839.

20. Girones-Vilaplana, A., et al., A novel beverage rich in antioxidant phenolics: Maqui berry (Aristotelia chilensis) and lemon juice. Lwt-Food Science and Technology, 2012. 47(2): p. 279-286.

21. Pantelidis, G.E., et al., Antioxidant capacity, phenol, anthocyanin and ascorbic acid contents in raspberries, blackberries, red currants, gooseberries and Cornelian cherries. Food Chemistry, 2007. 102(3): p. 777-783.

22. Aron, P.M. and J.A. Kennedy, Flavan-3-ols: Nature, occurrence and biological activity. Molecular Nutrition & Food Research, 2008. 52(1): p. 79-104.

23. Giusti, M.M. and R.E. Wrolstad, Acylated anthocyanins from edible sources and their applications in food systems. Biochemical Engineering Journal, 2003. 14(3): p. 217-225.

24. Hsu, C.L. and G.C. Yen, Phenolic compounds: Evidence for inhibitory effects against obesity and their underlying molecular signaling mechanisms. Molecular Nutrition & Food Research, 2008. 52(5): p. 624-625.

25. Karaaslan, M., et al., Phenolic fortification of yogurt using grape and callus extracts. Lwt-Food Science and Technology, 2011. 44(4): p. 1065-1072.

26. Blando, F., C. Gerardi, and I. Nicoletti, Sour cherry (Prunus cerasus L) anthocyanins as ingredients for functional foods. Journal of Biomedicine and Biotechnology, 2004(5): p. 253-258.

27. Ghosh, D., Potential role of polyphenol-fortified foods and beverages on vascular health. Agro Food Industry Hi-Tech, 2009. 20(6): p. 25-26.

28. Runestad, T. Functional Ingredients market overview. 2007 http://www.functionalingredientsmag.com/article/Business-Strategies/-em-functional-ingredients-em-market-overview.aspx [accessed 3.11.2009].

29. Wilson, M.-M.G. and J.E. Morley, Invited Review: Aging and energy balance. J Appl Physiol, 2003. 95(4): p. 1728-1736.

30. Morrissey, P.A., et al., Lipid stability in meat and meat products. Meat Science, 1998. 49: p. S73-S86.

31. Bou, R., et al., Influence of dietary fat source, alpha-tocopherol, and ascorbic acid supplementation on sensory quality of dark chicken meat. Poultry Science, 2001. 80(6): p. 800-807.

32. Packer, J.E., T.F. Slater, and R.L. Willson, Direct observations of a free redical interaction between vitamin E and vitamin C. Nature, 1979. 278: p. 737-739.

33. Dobarganes, C. and G. Marquez-Ruiz, Oxidized fats in foods. Current opinion in clinical nutrition and metabolic care, 2003. 6(2): p. 157-163.

34. Enser, M., What is lipid oxidation? Food Science and Technology-Lebensmittel-Wis senschaft & Technologie, 1987. 1: p. 151-153.

35. Tvrzicka, E., et al., Fatty Acids As Biocompounds: Their Role In Human Metabolism, Health And Disease - A Review. Part 1: Classification, Dietary Sources And Biological Functions. Biomedical Papers-Olomouc, 2011. 155(2): p. 117-130.

36. Kremmyda, L.S., et al., Fatty Acids As Biocompounds: Their Role In Human Metabolism, Health And Disease - A Review. Part 2: Fatty Acid Physiological Roles And Applications In Human Health And Disease. Biomedical Papers-Olomouc, 2011. 155(3): p. 195-218.

37. Kubow, S., Routes of formation and toxic consequences of lipid oxidation products in foods. Free Radical Biology and Medicine, 1992. 12(1): p. 63-81.

38. Cosgrove, J.P., D.F. Church, and W.A. Pryor, The kinetics of the autoxidation of poly-unsaturated fatty-acids. Lipids, 1987. 22 (5): p. 299-304.

39. Henderson, R.J. and D.R. Tocher, The lipid composition and biochemistry ofd freshwater fish. Progress in Lipid Research, 1987. 26: p. 281-347.

40. Jacobsen, C., et al., Preventing lipid oxidation in seafood, in Improving Seafood Products for the Consumer, T. Borresen, Editor. 2008, Woodhead Publishing. p. 426-460.

41. Scollan, N.D., et al., Manipulating the fatty acid composition of muscle and adipose tissue in beef cattle. British Journal of Nutrition, 2001. 85(1): p. 115-124.

42. Kanner, J., Oxidative Processes in Meat and Meat-Products - Quality Implications. Meat Science, 1994. 36(1-2): p. 169-189.

43. Khayat, A. and D. Schwall, LIPID OXIDATION IN SEAFOOD. Food Technology, 1983. 37(7): p. 130-140.

44. Sun, Q., et al., Aldehyde reactivity with 2-thiobarbituric acid and TBARS in freeze- dried beef during accelerated storage. Meat Science, 2001. 57(1): p. 55-60.

45. Olsen, E., et al., Analysis of early lipid oxidation in smoked, comminuted pork or poultry sausages with spices. Journal of Agricultural and Food Chemistry, 2005. 53(19): p. 7448-7457.

46. Varlet, V., C. Prost, and T. Serot, Volatile aldehydes in smoked fish: Analysis methods, occurence and mechanisms of formation. Food Chemistry, 2007. 105(4): p. 1536-1556.

47. Gardner, H.W., in Xenobiotics in Foods and Feeds, J.M. Finley and D.E. Schwass, Editors. 1983, Am. Chem. Soc.: Washington D.C. p. 63-84.
48. Frank, J., Dietary phenolic compounds and vitamin e bioavailability : model studies in rats and humans., in Acta Universitatis Agriculturae Suecia, Agraria, 446,. 2004, Swedish Univerisity of Agriculture, Department of Food Science: Uppsala, Sweden.
49. Sanchez-Escalante, A., et al., The effects of ascorbic acid, taurine, carnosine and rosemary powder on colour and lipid stability of beef patties packaged in modified atmosphere. Meat Science, 2001. 58(4): p. 421-429.
50. Amarowicz, R., Squalene: A natural antioxidant? European Journal of Lipid Science and Technology, 2009. 111(5): p. 411-412.
51. Bosku, D., Olive oil, in Mediterranean diets, A.P. Simopoulos and F. Visioli, Editors. 2000, Karger: Basel, Freiburg, Paris, London, New York, New Delhi, Bangkok, Singapore, Tokyo, Sydney. p. 56-77.
52. He, H.P., et al., Extraction and purification of squalene from Amaranthus grain. Journal of Agricultural and Food Chemistry, 2002. 50(2): p. 368-372.
53. Knoblich, M., B. Anderson, and D. Latshaw, Analyses of tomato peel and seed byproducts and their use as a source of carotenoids. Journal of the Science of Food and Agriculture, 2005. 85(7): p. 1166-1170.
54. Benítez, V., et al., Characterization of Industrial Onion Wastes (Allium cepa L.): Dietary Fibre and Bioactive Compounds. Plant Foods for Human Nutrition, 2011. 66(1): p. 48-57.
55. O'Grady, M.N., et al., An assessment of dietary supplementation with tea catechins and rosemary extract on the quality of fresh beef. Meat Science, 2006. 73(1): p. 132-143.
56. Kalogeropoulos, N., et al., Nutritional evaluation and bioactive microconstituents (carotenoids, tocopherols, sterols and squalene) of raw and roasted chicken fed on DHA-rich microalgae. Food Research International, 2010. 43(8): p. 2006-2013.
57. Gülçin, I., et al., Determination of antioxidant activity of lichen Cetraria islandica (L) Ach. Journal of Ethnopharmacology, 2002. 79(3): p. 325-329.
58. Larrain, R.E., et al., Color changes and lipid oxidation in pork products made from pigs fed with cranberry juice powder. Journal of Muscle Foods, 2008. 19(1): p. 17-33.
59. Hilton, J.W., THE Interaction of vitamins, minerals and diet composition in the diet of fish. Aquaculture, 1989. 79(1-4): p. 223-244.

60. Hamre, K., et al., Antioxidant vitamins, minerals and lipid levels in diets for Atlantic salmon (Salmo salar, L.): effects on growth performance and fillet quality. Aquaculture Nutrition, 2004. 10(2): p. 113-123.

61. Palace, V.P. and J. Werner, Vitamins A and E in the maternal diet influence egg quality and early life stage development in fish: a review. 2006: p. 41-57.

62. Shiau, S.-Y. and C.-Y. Hsu, Vitamin E sparing effect by dietary vitamin C in juvenile hybrid tilapia, Oreochromis niloticus◊O. aureus. Aquaculture, 2002. 210(1-4): p. 335-342.

63. Norouzitallab, P., et al., Comparing the efficacy of dietary alpha-tocopherol with that of dl-alpha-tocopheryl acetate, both either alone or in combination with ascorbic acid, on growth and stress resistance of angelfish, Pterophylum scalare, juveniles. Aquaculture International, 2009. 17(3): p. 207-216.

64. Hamre, K., et al., Vitamins C and E interact in juvenile Atlantic salmon (Salmo salar, L). Free Radical Biology and Medicine, 1997. 22(1-2): p. 137-149.

65. Rulíšek, L.r. and J. Vondrášek, Coordination geometries of selected transition metal ions (Co2+, Ni2+, Cu2+, Zn2+, Cd2+, and Hg2+) in metalloproteins. Journal of Inorganic Biochemistry, 1998. 71(3–4): p. 115-127.

66. Lauritzsen, K., G. Martinsen, and R.L. Olsen, Copper induced lipid oxidation during salting of COD (Gadus morhua L.). Journal of Food Lipids, 1999. 6(4): p. 299-315.

67. Rhee, K.S. and Y.A. Ziprin, Pro-oxidative effects of NaCl in microbial growth-controlled and uncontrolled beef and chicken. Meat Science, 2001. 57(1): p. 105-112.

68. Nawar, W.W., Chemical changes in lipids produced by thermal processing. Journal of Chemical Education, 1984. 61(4): p. 299-302.

69. Pearson, A.M. and T.A. Gillett, Processed Meats. 1996, New York: Chapman & Hall.

70. Chizzolini, R., E. Novelli, and E. Zanardi, Oxidation in traditional Mediterranean meat products. Meat Science, 1998. 49: p. S87-S99.

71. Gray, J.I. and A.M. Pearson, Rancidity and warmed-over flavor. Advanced Meat Research, 1987. 3: p. 221-269.

72. Gill, C.O., Extending the storage life of raw chilled meats. Meat Science, 1996. 43(Supplement 1): p. 99-109.

73. Garcia-Soto, B., et al., Preservative effect of an organic acid-icing system on chilled fish lipids. European Journal of Lipid Science and Technology, 2011. 113(4): p. 487-496.

74. Quitral, V., et al., Chemical changes during the chilled storage of Chilean jack mackerel (Trachurus murphyi): Effect of a plant-extract icing system. Lwt-Food Science and Technology, 2009. 42(8): p. 1450-1454.
75. Kaale, L.D., et al., Superchilling of food: A review. Journal of Food Engineering, 2011. 107(2): p. 141-146.
76. Rahelić, S., S. Puač, and A.H. Gawwad, Structure of beef Longissimus dorsi muscle frozen at various temperatures: Part 1—histological changes in muscle frozen at -10, -22, -33, -78, -115 and -196°C. Meat Science, 1985. 14(2): p. 63-72.
77. Deer Industry New Zealand. Venison, Shelf Life and Storage. 2012; Available from: http://www.nzgib.org.nz/n95.html.
78. Ando, M., et al., Effect of Super Chilling Storage on Maintenance of Freshness of Kuruma Prawn. Food Science and Technology Research, 2004. 10(1): p. 25-31.
79. Duun, A.S. and T. Rustad, Quality changes during superchilled storage of cod (Gadus morhua) fillets. Food Chemistry, 2007. 105(3): p. 1067-1075.
80. Duun, A.S. and T. Rustad, Quality of superchilled vacuum packed Atlantic salmon (Salmo salar) fillets stored at -1.4 and -3.6°C. Food Chemistry, 2008. 106(1): p. 122-131.
81. Jeremiah, L.E., Effect of frozen storage and protective wrap upon the cooking losses, palability, and rancidity of fresh and cured pork cuts. Journal of Food Science, 1980. 45(1-2): p. 187-196.
82. Miller, A.J., S.A. Ackerman, and S.A. Paumbo, Effects of frozen storage on functionality of meat for processing. Journal of Food Science, 1980. 45(5-6): p. 1466-1471.
83. Leygonie, C., T.J. Britz, and L.C. Hoffman, Impact of freezing and thawing on the quality of meat: Review. Meat Science, 2012. 91(2): p. 93-98.
84. Estevez, M., Protein carbonyls in meat systems: A review. Meat Science, 2011. 89(3): p. 259-279.
85. Hernandez, P., J.L. Navarro, and F. Toldra, Effect of frozen storage on lipids and lipolytic activities in the longissimus dorsi muscle of the pig. Zeitschrift für Lebensmitteluntersuchung und -Forschung A, 1999. 208(2): p. 110-115.
86. Li, B. and D.-W. Sun, Novel methods for rapid freezing and thawing of foods – a review. Journal of Food Engineering, 2002. 54(3): p. 175-182.
87. Chevalier, D., et al., Comparison Of Air-blast And Pressure Shift Freezing On Norway Lobster Quality. Journal of Food Science, 2000. 65(2): p. 329-333.

88. Gray, J.I. and A.M. Pearson, Cured Meat Flavor, in Advances in Food Research, C.O. Chichester, E.M. Mrak, and B.S. Schweigert, Editors. 1984, Academic Press, INC. p. 1-86.

89. Kanner, J., S. Harel, and R. Jaffe, LIPID-PEROXIDATION OF MUSCLE FOOD AS AFFECTED BY NACL. Journal of Agricultural and Food Chemistry, 1991. 39(6): p. 1017-1021.

90. Morrissey, P.A. and J.Z. Tichivangana, The antioxidant activities of nitrite and nitro sylmyoglobin in cooked meats. Meat Science, 1985. 14: p. 175-190.

91. Andersen, E., M.L. Andersen, and C.P. Baron, Characterization of oxidative changes in salted herring (Clupea harengus) during ripening. Journal of Agricultural and Food Chemistry, 2007. 55(23): p. 9545-9553.

92. Goulas, A.E. and M.G. Kontominas, Effect of salting and smoking-method on the keeping quality of chub mackerel (Scomber japonicus): biochemical and sensory attributes. Food Chemistry, 2005. 93(3): p. 511-520.

93. Park, J.N., et al., Partial replacement of NaCl by KCl in salted mackerel (Scomber japonicus) fillet products: effect on sensory acceptance and lipid oxidation. International Journal of Food Science and Technology, 2009. 44(8): p. 1572-1578.

94. Lauritzsen, K. and R.L. Olsen, Effects of antioxidants on copper induced lipid oxidation during salting of cod (Gadus morhua L.). Journal of Food Lipids, 2004. 11(2): p. 105-122.

95. Vestergaard, C.S. and G. Parolari, Lipid and cholesterol oxidation products in dry- cured ham. Meat Sci, 1999. 52(4): p. 397-401.

96. Pastorelli, G., et al., Influence of dietary fat, on fatty acid composition and sensory properties of dry-cured Parma ham. Meat Science, 2003. 65(1): p. 571-580.

97. Sampels, S., J. Pickova, and E. Wiklund, Fatty acids, antioxidants and oxidation stability of processed reindeer meat. Meat Science, 2004. 67(3): p. 523-532.

98. Mottram, D.S., Flavour formation in meat and meat products: a review. Food Chemistry, 1998. 62(4): p. 415-424.

99. Coutron-Gambotti, C. and G. Gandemer, Lipolysis and oxidation in subcutaneous adipose tissue during dry-cured ham processing. Food Chemistry, 1999. 64(1): p. 95-101.

100. Bosi, P., et al., Effects of dietary high-oleic acid sunflower oil, copper and vitamin E levels on the fatty acid composition and the quality of dry cured Parma ham. Meat Science, 2000. 54(2): p. 119-126.

101. Jonsdottir, R., et al., Flavor and Quality Characteristics of Salted and Desalted Cod (Gadus morhua) Produced by Different Salting Methods. Journal of Agricultural and Food Chemistry, 2011. 59(8): p. 3893-3904.

102. Lauritzsen, K., et al., Effects of calcium, magnesium and pH during salt curing of cod (Gadus morhua L). Journal of the Science of Food and Agriculture, 2004. 84(7): p. 683-692.

103. Hwang, C.-C., et al., Effect of salt concentrations and drying methods on the quality and formation of histamine in dried milkfish (Chanos chanos). Food Chemistry, 2012. 135(2): p. 839-844.

104. Gwak, H.J. and J.B. Eun, Changes in the chemical characteristics of Gulbi, salted and dried yellow corvenia, during drying at different temperatures. Journal of Aquatic Food Product Technology, 2010. 19(3-4): p. 274-283.

105. Jónsdóttir, R., et al., Volatile compounds suitable for rapid detection as quality indicators of cold smoked salmon (Salmo salar). Food Chemistry, 2008. 109(1): p. 184-195.

106. Forsberg, N.D., et al., Effect of native American fish smoking methods on dietary exposure to polycyclic aromatic hydrocarbons and possible risks to human health. Journal of Agricultural and Food Chemistry, 2012. 60(27): p. 6899-6906.

107. Coronado, S.A., et al., Effect of dietary vitamin E, fishmeal and wood and liquid smoke on the oxidative stability of bacon during 16 weeks' frozen storage. Meat Science, 2002. 62(1): p. 51-60.

108. Gonulalan, Z., A. Kose, and H. Yetim, Effects of liquid smoke on quality characteristics of Turkish standard smoked beef tongue. Meat Science, 2004. 66(1): p. 165-170.

109. Espe, M., et al., Atlantic salmon (Salmo salar, L) as raw material for the smoking industry. II: Effect of different smoking methods on losses of nutrients and on the oxidation of lipids. Food Chemistry, 2002. 77(1): p. 41-46.

110. Torri, L., B. Baroni, and M.R. Baroni, Modified atmosphere. 2009, Food Packages Free Press.

111. Ucak, I., Y. Ozogul, and M. Durmus, The effects of rosemary extract combination with vacuum packing on the quality changes of Atlantic mackerel fish burgers. International Journal of Food Science and Technology, 2011. 46(6): p. 1157-1163.

112. Fernandez-Espla, M.D. and E. Oneill, Lipid Oxidation in Rabbit Meat under Different Storage- Conditions. Journal of Food Science, 1993. 58(6): p. 1262-1264.

113. Jo, C., J.I. Lee, and D.U. Ahn, Lipid oxidation, color changes and volatiles production in irradiated pork sausage with different fat content and packaging during storage. Meat Science, 1999. 51(4): p. 355-361.

114. Hampson, J.W., et al., Effect of low dose gamma radiation on lipids in five different meats. Meat Science, 1996. 42(3): p. 271-276.

115. Sant'Ana, L.S. and J. Mancini-Filho, Influence of the addition of antioxidants in vivo on the fatty acid composition of fish fillets. Food Chemistry, 2000. 68(2): p. 175-178.

116. Naseri, M., et al., Effects of different filling media on the oxidation and lipid quality of canned silver carp (Hypophthalmichthys molitrix). International Journal of Food Science and Technology, 2011. 46(6): p. 1149-1156.

117. Medina, I., et al., Effect of Packing Media on the Oxidation of Canned Tuna Lipids. Antioxidant Effectiveness of Extra Virgin Olive Oil. Journal of Agricultural and Food Chemistry, 1998. 46(3): p. 1150-1157.

118. Abdullah, B.M., Properties of five canned luncheon meat formulations as affected by quality of raw materials. International Journal of Food Science and Technology, 2007. 42(1): p. 30-35.

119. Echarte, M., et al., Evaluation of the nutritional aspects and cholesterol oxidation products of pork liver and fish pates. Food Chemistry, 2004. 86(1): p. 47-53.

120. Ferreira, E.L., et al., Natural Antioxidant from Yerba Mate (Ilex paraguariensis St. Hil.) Prevents Hamburger Peroxidation. Brazilian Archives of Biology and Technology, 2011. 54(4): p. 803-809.

121. Sanjuas-Rey, M., et al., Effect of oregano and thyme essential oils on the microbiological and chemical quality of refrigerated (4 degrees C) ready-to-eat squid rings. International Journal of Food Science and Technology, 2012. 47(7): p. 1439-1447.

122. El-Alim, S.S.L.A., et al., Culinary herbs inhibit lipid oxidation in raw and cooked minced meat patties during storage. Journal of the Science of Food and Agriculture, 1999. 79(2): p. 277-285.

123. Ahn, D.U., et al., Packaging and irradiation effects on lipid oxidation and volatiles in pork patties. J Food Sci, 1998. 63(1): p. 15-9.

124. Aguirrezabal, M.M., et al., The effect of paprika, garlic and salt on rancidity in dry sausages. Meat Science, 2000. 54(1): p. 77-81.

125. Nassu, R.T., et al., Oxidative stability of fermented goat meat sausage with different levels of natural antioxidant. Meat Science, 2003. 63(1): p. 43-49.

126. Ansorena, D., et al., Addition of Palatase M (Lipase from Rhizomucor miehei) to dry fermented sausages: Effect over lipolysis and study of the further oxidation process by GC-MS. Journal of Agricultural and Food Chemistry, 1998. 46(8): p. 3244-3248.

127. Halliwell, B. and J.M.C. Cuttteridge, Free radicals in biology and medicine. 2nd edition ed. 1988, Oxford, UK: Claredon Press.

128. Robards, K., et al., Phenolic compounds and their role in oxidative processes in fruits. Food Chemistry, 1999. 66(4): p. 401-436.

129. Rice-Evans, C., N. Miller, and G. Paganga, Antioxidant properties of phenolic compounds. Trends in Plant Science, 1997. 2(4): p. 152-159.

130. Bhale, S.D., et al., Oregano and rosemary extracts inhibit oxidation of long-chain n-3 fatty acids in menhaden oil. Journal of Food Science, 2007. 72: p. C504-C508.

131. Afonso, M.d.S. and L.S. Sant'ana, Effects of pretreatment with rosemary (Rosmarinus officinalis L.) in the prevention of lipid oxidation in salted tilapia fillets. Journal of Food Quality, 2008. 31(5): p. 586-595.

132. Tanabe, H., M. Yoshida, and N. Tomita, Comparison of the antioxidant activities of 22 commonly used culinary herbs and spices on the lipid oxidation of pork meat. Animal Science Journal, 2002. 73: p. 389-393.

133. Abuja, P.M., M. Murkovic, and W. Pfannhauser, Antioxidant and prooxidant activities of elderberry (Sambucus nigra) extract in low-density lipoprotein oxidation. Journal of Agricultural and Food Chemistry, 1998. 46(10): p. 4091-4096.

134. Hakkinen, S.H., et al., Content of the flavonols quercetin, myricetin, and kaempferol in 25 edible berries. Journal of Agricultural and Food Chemistry, 1999. 47(6): p. 2274-2279.

135. Mielnik, M.B., et al., Grape seed extract as antioxidant in cooked, cold stored turkey meat. Lwt-Food Science and Technology, 2006. 39(3): p. 191-198.

136. Pazos, M., et al., Activity of grape polyphenols as inhibitors of the oxidation of fish lipids and frozen fish muscle. Food Chemistry, 2005. 92(3): p. 547-557.

137. Lee, C.H., J.D. Reed, and M.P. Richards, Ability of various polyphenolic classes from cranberry to inhibit lipid oxidation in mechanically separated turkey and cooked ground pork. Journal of Muscle Foods, 2006. 17(3): p. 248-266.

138. Sánchez-Alonso, I., et al., Effect of grape antioxidant dietary fibre on the prevention of lipid oxidation in minced fish: Evaluation by different methodologies. Food Chemistry, 2007. 101(1): p. 372-378.

139. Tang, S., et al., Anti-oxidant activity of added tea catechins on lipid oxidation of raw minced red meat, poultry and fish muscle. International Journal of Food Science & Technology, 2001. 36(6): p. 685-692.

140. Alghazeer, R., S. Saeed, and N.K. Howell, Aldehyde formation in frozen mackerel (Scomber scombrus) in the presence and absence of instant green tea. Food Chemistry, 2008. 108: p. 801-810.

141. Kanatt, S.R., R. Chander, and A. Sharma, Chitosan glucose complex - A novel food preservative. Food Chemistry, 2008. 106(2): p. 521-528.

142. Shao, D., et al., Study of Optimal Extraction Conditions for Achieving High Yield and Antioxidant Activity of Tomato Seed Oil. Journal of Food Science, 2012. 77(8): p. E202-E208.

143. Kalogeropoulos, N., et al., Bioactive phytochemicals in industrial tomatoes and their processing byproducts. LWT - Food Science and Technology, 2012. 49(2): p. 213–216.

144. Pazos, M., et al., Physicochemical properties of natural phenolics from grapes and olive oil byproducts and their antioxidant activity in frozen horse mackerel fillets. Journal of Agricultural and Food Chemistry, 2006. 54(2): p. 366-373.

145. Ganesan, P., C.S. Kumar, and N. Bhaskar, Antioxidant properties of methanol extract and its solvent fractions obtained from selected Indian red seaweeds. Bioresource Technology, 2008. 99(8): p. 2717-2723.

146. Skibsted, L.H., Carotenoids in Antioxidant Networks. Colorants or Radical Scavengers. Journal of Agricultural and Food Chemistry, 2012. 60(10): p. 2409-2417.

147. Bohm, F., et al., Carotenoids enhance vitamin E antioxidant efficiency. Journal of the American Chemical Society, 1997. 119(3): p. 621-622.

148. Psomiadou, E. and M. Tsimidou, Stability of Virgin Olive Oil. 2. Photo-oxidation Studies. Journal of Agricultur and Food Chemistry, 2002. 50(4): p. 722-727.

149. Jensen, C., et al., Effect of dietary levels of fat, alpha-tocopherol and astaxanthin on colour and lipid oxidation during storage of frozen rainbow trout (Oncorhynchus mykiss) and during chill storage of smoked trout. Zeitschrift Fur Lebensmittel-Untersuchung Und-Forschung a-Food Research and Technology, 1998. 207(3): p. 189-196.

150. Heinonen, M.I., Antioxidant activity and antimicrobial effect of berry phenolics - a Finnish perspective. Molecular Nutrition & Food Research, 2007. 51(6): p. 684-691.

151. Hussein, G., et al., Astaxanthin, a carotenoid with potential in human health and nutrition. Journal of Natural Products, 2006. 69(3): p. 443-449.

152. Chirumbolo, S., Plant phytochemicals as new potential drugs for immune disorders and cancer therapy: really a promising path? Journal of the Science of Food and Agriculture, 2012. 92(8): p. 1573-1577.

153. Narotzki, B., et al., Green tea: A promising natural product in oral health. Archives of Oral Biology, 2012. 57(5): p. 429-435.

154. Howell, A.B., Bioactive compounds in cranberries and their role in prevention of urinary tract infections. Molecular Nutrition & Food Research, 2007. 51(6): p. 732-737.

155. Céspedes, C.L., et al., Antioxidant and cardioprotective activities of phenolic extracts from fruits of Chilean blackberry Aristotelia chilensis (Elaeocarpaceae), Maqui. Food Chemistry, 2008. 107(2): p. 820-829.

156. Joseph, J.A., B. Shukitt-Hale, and L.M. Willis, Grape Juice, Berries, and Walnuts Affect Brain Aging and Behavior. Journal of Nutrition, 2009. 139(9): p. 1813S-1817S.

157. Lin, Y.T., et al., Inhibition of Helicobacter pylori and Associated Urease by Oregano and Cranberry Phytochemical Synergies. Appl. Environ. Microbiol., 2005. 71(12): p. 8558-8564.

158. Lin, Y.T., R.G. Labbe, and K. Shetty, Inhibition of Listeria monocytogenes in fish and meat systems by use of oregano and cranberry phytochemical synergies. Applied and Environmental Microbiology, 2004. 70(9): p. 5672-5678.

159. Arihara, K., Strategies for designing novel functional meat products. Meat Science, 2006. 74(1): p. 219-229

7 Chapter ENHANCED FRUIT PROCESSING FOR HUMAN HEALTH

1. INTRODUCTION: AN OVERVIEW

Fruits are a fundamental component of the human diet. A burgeoning interest exists in the correlation between fruit and vegetable diet and enhanced health outcomes. Research has demonstrated that biologically active constituents in plant-based foods, particularly phytochemicals such as polyphenols and carotenoids, play a significant role in mitigating the risks of chronic diseases, including cancer, cardiovascular disease, diabetes, and Alzheimer's disease, among others. The initial section of the chapter offers a concise overview of the connections between fruit-derived antioxidants, other biologically active chemicals, and their potential health advantages.

Global fruit output is on the rise. Notwithstanding the rise in worldwide fruit production, a considerable quantity of fruit is lost or wasted owing to inadequate post-harvest management. The second section of the chapter presents data on the current state of post-harvest losses in specific fruits and strategies to mitigate these losses. Consequently, transforming fruits into value-added goods is a strategy to mitigate post-harvest losses and enhance fruit consumption.

Fresh-cut fruits, or minimally processed fruits, are items that are partially prepared, retain a fresh-like quality, and are ready for consumption. Recently, fresh-cut fruits have gained popularity due to their alignment with consumer desire for quick, ready-to-eat foods that exhibit fresh-like quality. Nonetheless, fresh-cut fruits exhibit more perishability than entire fruits. The third section of the chapter discusses recently developed methodologies for enhancing the value of fresh-cut fruits, focusing on the application of natural antimicrobials, anti-browning agents, edible coatings, modified atmosphere packaging (MAP), 1-methylcyclopropene (1-MCP), and vacuum impregnation (VI).

2. FRUITS AND HUMAN HEALTH

The consumption of fruits and vegetables is rising due to substantial evidence linking their nutritional intake to numerous health benefits (Kaur & Kapoor 2001, Rupasinghe et al. 2012). Epidemiological studies indicate that the consumption of fruits and vegetables may prevent various chronic diseases, including cardiovascular disease (Weichselbaum 2010, Al-Dosari et al. 2011; Thilakarathna and Rupasinghe

2012), type II diabetes (Johnston et al. 2002, Yu et al. 2012b), and certain cancers (De Mejía & Prisecaru 2005, Lala et al. 2006, Sun & Liu 2008, Lippi & Targher 2011). The disease prevention effects of fruits may be attributed to the availability of health-promoting phytochemicals, including carotenoids (Chichili et al. 2006), flavonoids (Yu et al. 2012a), various phenolic compounds (Masibo & He 2008), and vitamins (Lippi & Targher 2011, Gutierrez 2008). Moreover, the health-protective effects may be mostly attributed to complex combinations of interacting natural compounds rather than a singular component in these plant-derived diets (Lila 2007). Table 1 provides a review of specific fruit-derived antioxidants and other health-promoting substances for disease prevention.

3. FRUIT PRODUCTION AND POST-HARVEST LOSS

3.1 Fruit production

Fruit production is increasing dramatically worldwide. According to the FAO, the total world fruit production in 2008 was 572.4 million tons, and the number climbed to 609.2 million tons in 2010 (FAO 2010). Among these fruits, thirty percent of which were tropical fruits, with water melon occupied of 59.2%, mongo and guavas of 20.5% and pineapple of 11.4% (Rawson et al. 2011).

Despite the increasing food production at the global level, about one-third of the food produced in the world is lost or wasted (Prusky 2011), among which, post-harvest stage losses and marketing stage losses are major losses.

3.2 Post-harvest loss of fruits

Despite of food production is increasing globally, a significant amount of the food for human consumption is lost or wasted, especially perishable foods such as fruits and vegetables (Prusky 2011). The amount of food lost each year is equivalent to more than half of the world's annual cereals production (2.3 billion tonnes in 2009/2010) (Gustavsson et al. 2011).

It is hard to give precise information on the amount of fruit losses generated globally, because fruit losses vary greatly among varieties, countries, and climatic regions, and there is no universally applied method for measuring losses. As a consequence, the food loss data during post-harvest are mostly estimated and the variations are from 10% to 40% (Prusky 2011). Table 2 lists some examples of post-harvest losses of selected fruits in India, Egypt and United States.

Table 1. Fruit-based health promoting compounds and postulated disease prevention

Source	Active component	Prevention mechanism	Disease	References
Grape	Anthocyanins	Anti-proliferative	Cancer	Lala et al. 2006
	Flavonoids	Inhibition of HNR-adduct formation	Macular degeneration and cataract	Yu et al. 2012a
	Resveratrol	Antioxidant, anti- inflammatory, activation of SIRT1	Alzheimer's	Sun et al. 2010
	Resveratrol	Normalize iron and Ca2+, increase SOD activity	Cardiotoxicity	Mokni et al. 2012
	Resveratrol	Enhance insulin secretion	Diabetes	Yu et al. 2012b
Apple	Polyphenols	Antioxidant, cell cycle modulation	Cancer	Sun & Liu 2008
	Polyphenols	Antioxidant, multiple mechanisms	Cardiovascular	Weichselbaum 2010
	Phloridzin	Anti-inflammatory, bone resorption	Bone protection	Puel et al. 2005
	Polyphenols	Reduce amyloid-β formation	Alzheimer's	Chan & Shea 2009
	Phloretin-2'-O-Glucoside	Delay glucose absorption	Diabetes	Johnston et al. 2002
Banana	Lectins (Bioactive protein)	Cell cycle arrest, apoptosis	Cancer	De Mejía & Prisecaru 2005
	Polyphenols	Antioxidant, reduce LDL modification	cardiovascular	Yin et al. 2008
	Polyphenols	Antioxidant	Alzheimer’s	Heo et al. 2008
Pineapple	Bromelain	Proteolytic enzyme regulation	Anti-inflammatory	Hale et al. 2010
Mango	Phenolic compounds	Antioxidant, multiple mechanisms	Degenerative diseases	Masibo & He 2008

Table 2. Post-harvest losses in selected fruits

Source	Post-harvest loss %				References
	Farm	Wholesale	Retail	Total	
Grape	15.1	6.9	6.0	28.0	Kader 2010
Grape	7.3	4.2	2.9	14.4	Murthy et al. 2009
Grape	N/A	N/A	7.6	N/A	Buzby et al. 2009
Mango	15.6	8.9	5.3	29.7	Murthy et al. 2009
Mango	N/A	N/A	14.5	N/A	Buzby et al. 2009
Banana	5.5	6.7	16.7	28.8	Murthy et al. 2009
Banana	N/A	N/A	8.0	N/A	Buzby et al. 2009
Papaya	N/A	N/A	54.9	N/A	Buzby et al. 2009
Pineapple	N/A	N/A	14.6	N/A	Buzby et al. 2009
Kiwi	N/A	N/A	12.7	N/A	Buzby et al. 2009
Apple	N/A	N/A	8.6	N/A	Buzby et al. 2009
Avocado	N/A	N/A	9.3	N/A	Buzby et al. 2009
Tomato	9.0	17.9	16.3	43.2	Kader 2010

3.3 Prevention and reduction of post-harvest loss

Methods of preventing losses of fruits and vegetables could be found from papers and fact sheet written by Singh and Goswami (2006), Sonkar et al. (2008), Prusky (2011) and DeEII and Murr (2009). These methods include selection of new cultivars with firm fruits and longer postharvest life, minimizing physical damage during harvesting and postharvest handling, control and monitoring of temperature and relative humidity, use of controlled or modified atmosphere storage, use of pre- and post-harvest fungicides (hydrogen peroxide) before and after harvest and use of physical treatment such as ozonation technology. Table 3 gives examples of use of controlled atmosphere storage of selected fruits.

Table 3. Controlled atmosphere storage conditions of selected fruits

Source	Temp (ºC)	RH (%)	O2 (%)	CO2 (%)	Storage life	References
Grape	0-5	90-95	5-10	15-20	> two weeks	Singh and Goswami 2006
Mango	10-15	90	3-7	5-8	> two weeks	Singh & Goswami 2006

Banana	12-16	90	2-5	2-5	> two weeks	Singh & Goswami 2006
Papaya	10-15	90	2-5	5-8	> two weeks	Singh & Goswami 2006
Pineapple	8-13	90	2-5	5-10	> two weeks	Singh & Goswami 2006
Kiwi	0-5	90	1-2	3-5	> two weeks	Singh & Goswami 2006
Avocado	5-13	90	2-5	3-10	> two weeks	Singh & Goswami 2006
Apple (Empire)	1-2	N/A	1.5-2.5	1.5-2.0	5-8 months	DeEll & Murr 2009
Apple (Gala)	0	N/A	1.5-2.5	1.5-2.5	5-8 months	DeEll & Murr 2009
Apple (Golden Delicious)	0	N/A	1.5-2.5	1.5-2.5	5-8 months	DeEll & Murr 2009
Apple (McIntosh)	3	N/A	1.0-2.5	0.5-2.5	5-8 months	DeEll & Murr 2009
* Temp: temperature; RH: Relative humidity						

4. FRUIT PROCESSING AND PRESERVATION

Processed fruit products generally include minimally processed fruit products such as fresh- cut fruit, fermented fruit products such as cider, wine and vinegar, traditional thermally processed fruit products such as jam, jelly, juice and beverage, novel non-thermal processed fruit products such as juice and beverage, etc. A comprehensive review has been given by the same authors on novel non-thermal processed fruit product preservation including juices and beverages (Rupasinghe & Yu 2012). At the same time, fresh-cut fruit stands out to be a promising food that meets the demand of consumers for convenient and ready-to-eat fruits with a fresh-like quality. In this case, this part of the chapter would give emphasis on fresh-cut fruit processing and preservation.

4.1 Fresh-cut fruit processing

The sales of fresh-cut produce have grown from approximately $5 billion in 1994 to $10–12 billion in 2005, which is about 10% of total produce sales (Rupasinghe et al. 2005). Fresh-cut fruits and vegetables are products that are partially prepared, maintain a fresh-like state and no additional preparation is necessary for use and eating (Watada & Qi 1999). Figure 1 shows the flowchart of fresh-cut fruit processing. It generally includes washing, and/or peeling, cutting, and/or slicing or wedging and packaging. Dipping solutions or edible coating materials could be applied during dipping or coating process.

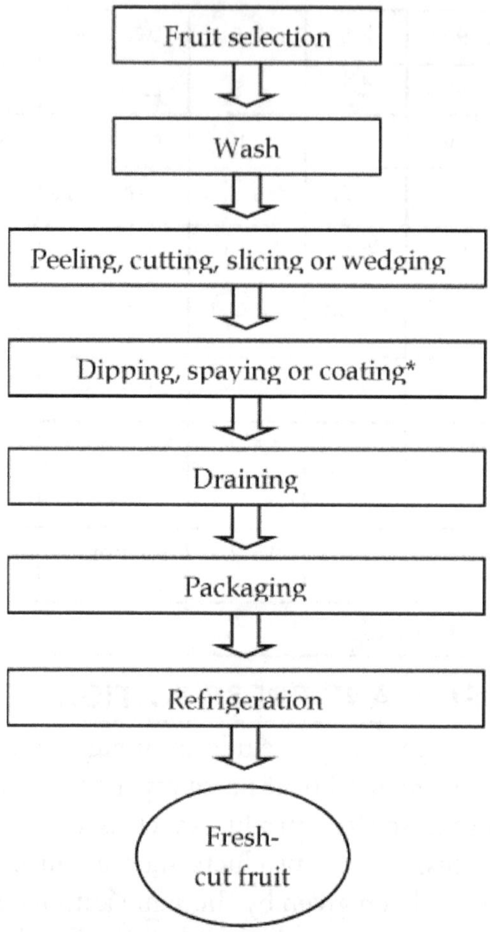

Figure 1. Major steps for fresh-cut fruit processing (revised from Corbo et al. 2010) * During this process, natural preservatives or edible coating materials could be applied

4.2 Fresh-cut fruit preservation

Fresh-cut fruits are more perishable than whole fruits, because the tissue integrity of fruits is more easily altered during processing. Post-cut quality of fresh-cut fruits suffers from wound induced biochemical and physiological changes such as water loss, accelerated respiration and cut-surface browning as well as microbiological spoilage (Kader 2002, Chiabrando & Giacalone 2012). Therefore, preservation of fresh-cut fruits needs combinative efforts of antimicrobial agents, anti-browning substances as well as packaging strategies. A detailed review was given by Oms-Oliu et al. (2010) about recent approaches for preserving quality of fresh-cut fruits.

4.2.1 Antimicrobial agent

During the preparatory steps of fresh-cut fruit processing, the natural protection of fruit is removed and chances of contamination may increase. Damage of tissues allows the growth and fermentation of some species of yeasts such as Saccharomyces cerevisiae and the attack by pathogenic microorganisms such as Listeria monocytogenes, Salmonella spp., Staphylococcus aureus and Escherichia coli O157:H7 (Martin-Belloso et al. 2006). Therefore, the searching for methods to retard microbial growth is of great interest to researchers and fresh-cut industry.

Traditionally, the most commonly used antimicrobials are potassium sorbate and sodium benzoate. However, consumer demand for natural origin, safe and environmental friendly food preservatives is increasing. Natural antimicrobials such as organic acids, herb leaves extracts and oils, chitozan and bacteriocins have shown feasibility for use in some food products including fresh-cut fruits (Gould 2001, Corbo et al. 2009). Some of them have been considered as Generally Recognized As Safe (GRAS) additives in foods. Selected natural antimicrobials and their status for GRAS additives are listed in Table 4.

Cinnamon as an antimicrobial agent has been investigated in fresh-cut apple slices (Muthuswamy et al. 2008). Ethanol extract of cinnamon bark (1% to 2% w/v) and cinnamic aldehyde (2 mM) could reduce E. coli O157:H7 and L. innocua in vitro. Ethanol extract of cinnamon bark (1% w/v) reduced significantly the aerobic growth of bacteria inoculated in fresh-cut apples during storage at 6°C up to 12 days. It was also found that cinnamic aldehyde has greater antimicrobial activity than potassium sorbate (Muthuswamy et al. 2008).

Carvacrol and cinnamic acid could delay microbial spoilage of fresh-cut melon and kiwifruit. Dipping of fresh-cut kiwifruit in carvacrol solutions at up to 15 mM reduced total viable counts from 6.6 to less than 2 log CFU/g for 21 days of storage at 4°C. Also, treatment with 1 mM of carvacrol or cinnamic acid reduced viable counts on kiwifruit by 4 and 1.5 log CFU/g for 5 days of storage at 4°C and 8°C, respectively (Roller & Seedhar 2002).

Vanillin was also proved to be a practical preservative for processing fresh-cut mango and apples under refrigerated conditions. Fresh-cut mango slices were dipped for 1 min in solutions containing 80 mM vanillin before being packaged. Results indicated that treatment with 80 mM vanillin significantly delayed ($P < 0.05$) the development of total aerobic bacteria and yeast and mold populations of fresh-cut mangoes stored at 5 and 10 °C for up to 14 and 7 d, respectively (Ngarmsak et al. 2006). Also, a dip of 12 mM vanillin incorporated with a commercial anti-browning dipping solution (calcium ascorbate, NatureSeal™) inhibited the total aerobic microbial growth by 37% and 66% in fresh-cut 'Empire' and 'Crispin' apples, respectively, during storage at 4 °C for 19 days. Furthermore, vanillin (12 mM) did not influence the control of enzymatic browning and softening by NatureSeal (Rupasinghe et al. 2006).

Table 4. Selected natural antimicrobial agents and their status for GRAS additives*

Name	Origin	GRAS status
Rosemary	Plant	Yes
Cinnamon	Plant	Yes
Cinnamic acid	Plant	Yes
Clove	Plant	Yes
Lactoperoxidase	Animal	No
Lemon (peel, balm, grass)	Plant	Yes
Lime	Plant	Yes
Nisin	Microorganism	Yes
Chitozan	Animal	No
Carvacrol	Plant	Yes
Citric acid	Plant	Yes
Ascorbic acid	Plant	Yes
Vanillin	Plant	Yes

* Revised from USFDA (2006): Food Additive Status List

4.2.2 Anti-browning agents

Enzymatic browning is also a major concern on the extension of shelf-life of fresh-cut fruit (Oms-Oliu et al. 2010). It is caused by the enzymatic oxidation of phenols to quinones by enzymes, typically polyphenoloxidases, in the presence of oxygen. Quinones are then subjected to further reactions, leading to the formation of browning pigments (Ozoglu & Bayindirli 2002, Jeon & Zhao 2005). Traditionally, sulfites have been used for browning prevention. However, their use on fresh-cut fruit and vegetables was banned in 1986 by the FDA owing to their potential hazards to health (Buta et al. 1999). Therefore, various alternative substances, such as honey, citric acid, ascorbic acid, calcium chloride, calcium lactate and calcium ascorbate, among others, have being used to retard browning in fresh-cut fruit (Jeon & Zhao 2005, Oms-Oliu et al. 2010). These anti-browning products are not often used alone because it is difficult to achieve efficient browning inhibition, and combination of them would give preferable results. Table 5 gives examples of anti-browning treatment on fresh-cut Apples.

Examples of anti-browning treatment on other fresh-cut fruits including banana, kiwifruits, mango, among others could be found in Table 3 in Oms-Oliu et al. (2010)'s paper.

Table 5. Examples of anti-browning treatments assessed on fresh-cut fruits

Cultivar of apple	Anti-browning agent	Storage conditions	References
Gala	10% honey solution with vacuum impregnating	3°C for 14 days	Jeon & Zhao (2005)
Granny Smith	1% w/v of citric acid/CaCl2 and 1% of ascorbic acid/CaCl2	4°C for 5 days	Chiabrando & Giacalone (2012)
Granny Smith	0.05% w/v of sodium chlorite and 1% of calcium propionate	10°C for 14 days	Guan & Fan (2010)
Golden Delicious	1% w/v of citric acid/CaCl2 and 1% of ascorbic acid/CaCl2	4°C for 5 days	Chiabrando & Giacalone (2012)
Golden Delicious	80 mg/L acidic electrolyzed water (AEW) followed by 5%calcium ascorbate	4°C for 11 days	Wang et al. (2007)
Granny Smith	1% w/v of citric acid/CaCl2 and 1% of ascorbic acid/CaCl2	4°C for 5 days	Chiabrando & Giacalone (2012)
Red Delicious	300 mg /L sodium chlorite (SC) and 300 mg /L citric acid	5°C for 14 days	Luo et al. (2011)
Scarlet Spur	1% w/v of citric acid/CaCl2 and 1% of ascorbic acid/CaCl2	4°C for 5 days	Chiabrando & Giacalone (2012)

4.2.3 Edible coating

The incorporation of antimicrobial and anti-browning agents to fresh-cut fruits could be done by dipping, spaying or edible coating treatment. Dipping or spraying aqueous solutions to fruit pieces containing antimicrobial agents, antioxidants, calcium salts or functional ingredients such as minerals and vitamins are widely used to improve quality of fresh-cut fruit. However, the effectiveness of these compounds could be better improved with their incorporation into edible coatings. The application of edible coatings to deliver active ingredients is one of the recent progresses made for shelf-life extension of fresh-cut fruits. Detailed information on edible coating for fresh-cut fruits could be found in review papers from Vargas et al. (2008), Rojas-Graü et al. (2009) and Valencia-Chamorro et al. (2011).

Edible coatings may be defined as a thin layer of material that covers the surface of the food and can be eaten as a part of the whole product. Therefore, the composition of edible coatings has to be food grade or GRAS. Furthermore, the coating materials need to be transparent, odourless, permeable for water vapour and selectively permeable to gases and volatile compounds (Kester & Fennema 1986).

Ingredients that can be used to form edible coatings include polysaccharides such as cellulose, starch, alginate, chitosan, pectin, carrageenan, gum Arabic, guar gum and xanthan gum, proteins such as zein, gluten, soy, whey protein, lipids such as beeswax, lecithin, cocoa butter and fatty acids (Vargas et al. 2008). Examples of edible coating treatment on fresh-cut apples are listed in Table 6.

Table 6. Examples of edible coating treatment on fresh-cut apples

Cultivar of apple	Functional ingredients	Concentration (%)	Coating materials	References
Gala	N/A	N/A	Cassava starch, glycerol, carnauba wax, stearic acid	Chiumarelli & Hubinger 2012
Gala	N/A	N/A	Chitosan	Wu et al. 2005
Fuji	Oregano oil	0.1 – 0.5 (v/v)	Apple puree, alginate	Rojas-Graü et al. 2007
Fuji	Lemongrass	1.0 – 1.5 (v/v)	Apple puree, alginate	Rojas-Graü et al. 2007
Fuji	Vanillin	0.3 – 0.6 (v/v)	Apple puree, alginate	Rojas-Graü et al. 2007
Fuji	Cinnamon	0.7 (v/v)	Alginate	Raybaudi-Massilia et al. 2008
Fuji	Clove	0.7 (v/v)	Alginate	Raybaudi-Massilia et al. 2008
Fuji	Lemongrass	0.7 (v/v)	Alginate	Raybaudi-Massilia et al. 2008
Fuji	Cinnamaldehyde	0.5 (v/v)	Alginate	Raybaudi-Massilia et al. 2008
Fuji	Citral	0.5 (v/v)	Alginate	Raybaudi-Massilia et al. 2008
Fuji	Ascorbic acid, $CaCl_2$	1.0 (w/v)	Carrageenan	Lee et al. 2003
Fuji	Ascorbic acid, $CaCl_2$	1.0 (w/v)	Whey protein concentrate	Lee et al. 2003
Fuji	Ascorbic acid, $CaCl_2$	1.0 (w/v)	Whey protein concentrate	Lee et al. 2003
Golden Delicious	Ascorbic acid	0.5-1.0 (w/v)	Whey protein concentrate, beeswax	Perez-Gago et al. 2006
Golden Delicious	Cysteine	0.1-0.5 (w/v)	Whey protein concentrate, beeswax	Perez-Gago et al. 2006
Granny Smith	Ascorbic acid, citric acid	0.5 (w/v)	Pectin, apple purée	McHugh & Senesi 2000

4.2.4 Modified atmosphere packaging (MAP) and 1-methylcyclopropene (1-MCP)

The respiration rate of fresh-cut fruits is greater than that of intact fruits (Kader 1986). The increased respiration rate can induce the ethylene synthesis, increase enzymatic activity, promote oxidation of phenolic compounds and microbial growth, and therefore contributes to quality losses such as color and firmness. In this case, the control of respiration is essential for maintaining quality and prolonging the shelf life of fresh-cut fruits (Rocha & Morais 2003).

Modified atmosphere packaging (MAP) is a technology which offers the optimum gas conditions around the product by adjusting the barrier properties of the packaging film (Simpson and Carevi 2004). Various approaches to prolong the shelf life of fresh-cut products, such as edible coatings and refrigeration could be applied in combination with MAP (Rupasinghe 2005).

1-Methylcyclopropene (1-MCP) may retard or inhibit the generation of ethylene, the natural ripening hormone which is undesirable in terms of storage of certain fruits. Therefore, 1- MCP is becoming a commercial tool (SmartFresh, AgroFresh Inc., Philadelphia) for extending the shelf-life and quality of certain fruits and plant products (Rupasinghe et al. 2005). 1- MCP can be applied immediately after harvest (Aguayo et al. 2006; Mao & Fei 2007), just before fresh-cut processing or at both steps (Calderón-López et al. 2005; Vilas-Boas & Kader 2007). However, treatment of intact fruit with 1-MCP before fresh-cut processing is easier and more convenient than after processing. Moreover, the increase in ethylene production promoted by peeling, slicing or wedging could be prevented by the pre-use of 1-MCP (Rupasinghe et al., 2005).

4.2.5 Vacuum impregnation

Osmotic treatments have been traditionally used as a pre-treatment step in freezing, canning and frying to improve the quality of the final produce (Alzamora et al., 2000). Among developments in osmotic treatments of fruit products, vacuum impregnation (VI) may be the latest (Zhao & Xie 2004). The VI technique is performed by applying a vacuum pressure in a tank or oven containing the immersed product for a short time and then restoring the atmospheric pressure with the product remains immersed (Martínez-Monzó et al., 1998). The process of VI is a hydrodynamic mass transfer process based on an exchange between internal gas or liquid and an external liquid phase (Zhao & Xie, 2004). VI technique could be used to develop novel minimally processed fruit products with value-addition since nutritional and bioactive ingredients could be incorporated into the fruit based products during VI process (Xie & Zhao, 2003; Guillemin et al., 2008, Rößle 2011) and which gives a bright future for VI application in fresh-cut fruits. Table 7 gives examples of VI treatment on fresh- cut fruits.

Table 7. Examples of VI treatment conditions on fresh-cut fruits or value-added products

Type of fruit	VI treatment conditions				References
	VI solution	VI pressure (mmHg)	VI time (min)	Restoration Time (min)	
Apple (Gala)	20% (w/w) of HFCS, Ca, Zn	50	15	30	Xie & Zhao 2003
Apple (Gala)	10% (w/w) of honey	75	15	30	Jeon & Zhao 2005
Strawberry	8ºBrix of glucose solution	37.5	5	5	Castelló et al. 2006
Apple (Empire)	15ºBrix of grape juice, 1.6% of $CaCl_2$ (w/v), 0.05% of NaCl (w/v), 0.1% of vitamin E (v/v)	152.4	10	22	Joshi et al. 2010
Apple (Empire)	20-40 % (v/v) of maple syrup, 1.6% of $CaCl_2$ (w/v), 0.05% of NaCl (w/v)	152	10	22	Joshi et al. 2011
Apple (Granny Smith)	50% (v/v) of honey	525	10	10	Rößle et al. 2011

HFCS: High fructose corn syrup

5. CONCISE SUMMARY

Fruits serve not only as a staple diet but also offer significant health benefits beyond their fundamental nutritional value. Nonetheless, the quantitative and qualitative losses of fruits are substantial during post-harvest, marketing, processing, and storage. The prevention of these losses in post-harvest management can be achieved by many approaches, including controlled or modified environment packing and the application of ozonation technology.

Conversely, the promotion of minimally processed fruit products, such as fresh-cut fruit, on the commercial market is a practical, financial, and environmentally sustainable alternative to traditional processing methods. Fresh-cut fruits exhibit greater perishability than whole fruits due to biochemical and physiological alterations, including water loss, enhanced respiration, cut-surface browning, and microbial spoiling. Consequently, the preservation of fresh-cut fruits necessitates a synergistic application of antimicrobial agents, anti-browning compounds, and packing techniques.

Natural or Generally Recognized as Safe (GRAS) additives are commonly utilized as antibacterial agents, anti-browning agents, or bioactive components. The application of antimicrobial and anti-browning chemicals to fresh-cut fruits may be achieved through dipping, spraying, or edible coating techniques. The utilization of edible coatings to administer active substances represents a recent advancement in prolonging the shelf life of fresh-cut fruits. It may be utilized in conjunction with modified atmosphere packing (MAP), 1-methylcyclopropene (1-MCP), and refrigeration for enhanced outcomes.

Moreover, vacuum impregnation (VI) may serve as an effective method for integrating health-promoting natural substances into fresh-cut fruits, alongside tasty coatings. The VI approach may be employed to create innovative minimally processed fruit products that enhance value by integrating nutritional and bioactive components.

REFERENCES

1. Aguayo, E., Jansasithorn, R. & Kader, A. A. (2006). Combined effects of 1-Methylcyclopropene, calcium chloride dip, and/or atmospheric modification on quality changes in fresh-cut strawberries, Postharvest Biology and Technology 40(3): 269-278.
2. Alzamora SM, Tapia MS, Leunda A, Guerrero SN, Parada-Arias E. 2000. Minimal Presser vation of fruits: a cited project, in: Lozano, J.E., Anon, C., Parada-Arias, E. & Borabosa-Canovas GV (ed.), Trend in Food Engineering, Technomic Publishing Company,Pennsylvania, USA, pp. 205-225.
3. Buta, J. G., Moline, H. E., Spaulding, D. W. & Wang, C. Y. (1999). Extending storage life of fresh-cut apples using natural products and their derivatives, Journal of Agricultural and Food Chemistry 47(1): 1-6.
4. Buzby, J. C.,Wells, H. F., Axtman, B. & Mickey, J. (2009). Supermarket loss estimates for fresh fruit, vegetables, meat, poultry, and seafood and their use in the ERS loss- adjusted food availability data, United States Department of Agriculture, Economic Research Service, Economic Information Bulletin, 44: 20.
5. Calderon-Lopez, B., Bartsch, J. A., Lee, C. Y. & Watkins, C. B. (2005). Cultivar effects on quality of fresh cut apple slices from 1-methylcyclopropene (1-MCP)-treated apple fruit, Journal of Food Science 70(3): S221-227.
6. Castelló, M. L., Fito, P. J. & Chiralt, A. (2006). Effect of osmotic dehydration and vacuum impregnation on respiration rate of cut strawberries, LWT - Food Science and Technology 39(10): 1171-1179.
7. Chan, A. & Shea, T. B. (2009). Dietary supplementation with apple juice decreases endo- genous amyloid-β levels in murine brain, Journal of Alzheimer's Disease 16(1): 167- 171.
8. Chiabrando, V. & Giacalone, G. (2012). Effect of anti-browning agents on color and related enzymes in fresh-cut apples during cold storage, Journal of Food Processing and Preservation 36(2): 133-140.

9. Chiumarelli, M. & Hubinger, M. D. (2012). Stability, solubility, mechanical and barrier properties of cassava starch - carnauba wax edible coatings to preserve fresh-cut apples, Food Hydrocolloids 28(1): 59-67.
10. Corbo, M. R., Bevilacqua, A., Campaniello, D., D'Amato, D., Speranza, B. & Sinigaglia, M. (2009). Prolonging microbial shelf life of foods through the use of natural compounds and non-thermal approaches - A review, International Journal of Food Science and Technology 44(2): 223-241.
11. Corbo, M.R., Speranza, B., Campaniello, D., D'Amato, D. & Sinigaglia, M. (2010). Fresh-cut fruits preservation: current status and emerging technologies, in Mendez- Vilas, A (Ed.), Current research, Technology and Education Topics in Applied Microbiology and Microbial Biotechnology, Formatex Research Center, Badajoz, Spain, pp. 1143-1154.
12. DeEII J. R. & Murr, D. P. (2009). CA storage guidelines and recommendations for apples, http://www.omafra.gov.on.ca/english/crops/facts/03-073.htm
13. De Mejía EG & Prisecaru VI. (2005). Lectins as bioactive plant proteins: a potential in cancer Treatment, Critical Reviews in Food Science and Nutrition 45 (6): 425-445.
14. FAO (2010). World fruit production in metric tonnes. http://faostat.fao.org/default.aspx.
15. Gould, G. W. (2001). Symposium on 'nutritional effects of new processing technologies' – new processing technologies: An overview, Proceedings of the Nutrition Society 60(4): 463-474.
16. Guan, W. & Fan, X. (2010). Combination of sodium chlorite and calcium propionate reduces enzymatic browning and microbial population of fresh-cut "granny smith" apples, Journal of Food Science 75920, M72-M77
17. Guillemin, A., Guillon, F., Degraeve, P. et al. (2008). Firming of fruit tissues by vacuum- infusion of pectin methylesterase: visualisation of enzyme reaction, Food Chemistry 109, 368–378.
18. Gustavsson, J., Cederberg, C., Sonesson, U., van Otterdijk, R. & Meybeck, A. (2011). Global food losses and food waste: extent causes and prevention, Food and Agriculture Organization of the United Nations. International congress Save Food!, International packaging industry fair Interpack 2011, Dusseldorf, Germany.
19. Hale, L. P., Chichlowski, M., Trinh, C. T. & Greer, P. K. (2010). Dietary supplementation with fresh pineapple juice decreases inflammation and colonic neoplasia in IL-10- deficient mice with colitis, Inflammatory Bowel Diseases 16(12): 2012-2021.

20. Heo, H.J., Choi, S.J., Choi, S-G, Lee. J.M. & Lee, C.Y. (2008). Effects of banana, orange, and apple on oxidative stress-induced neurotoxicity in PC12 cells, Journal of Food Science 73(2), H28-H32.

21. Jeon, M. & Zhao, Y. (2005). Honey in combination with vacuum impregnation to prevent enzymatic browning of fresh-cut apples, International Journal of Food Sciences and Nutrition 56(3): 165-176.

22. Johnston, K. L., Clifford, M. N. & Morgan, L. M. (2002). Possible role for apple juice phenolic compounds in the acute modification of glucose tolerance and gastrointestinal hormone secretion in humans, Journal of the Science of Food and Agriculture 82(15): 1800-1805.

23. Joshi, A. P. K., Rupasinghe, H. P. V. & Pitts, N. L. (2010). Sensory and nutritional quality of the apple snacks prepared by vacuum impregnation process, Journal of Food Quality 33(6): 758-767.

24. Joshi, A. P. K., Rupasinghe, H. P. V. & Pitts, N. L. (2011). Comparison of non-fried apple snacks with commercially available fried snacks, Food Science and Technology International 17(3): 249-255.

25. Kader, A.A., (1986). Biochemical and physiological basis for effects of controlled and Modified atmospheres on fruits and vegetables, Food Technol. 40: 99-104.

26. Kader, A. (2002). Quality parameters of fresh-cut fruit and vegetable products, in Lamikanra, O. (ed.), Fresh-Cut Fruits and Vegetables, Science, Technology and Market, CRC Press, Boca Raton, Florida, USA, pp. 11–19.

27. Kader, A. A. (2010). Handling of horticultural perishables in developing vs. developed countries, Acta Horticulturae 877: 121–126.

28. Kaur, C. & Kapoor, H. C. (2001). Antioxidants in fruits and vegetables - the millennium's health, International Journal of Food Science and Technology 36(7): 703-725.

29. Kester, J.J. & Fennema O. R. (1986). Edible films and coatings: A review, Food Technol. 40: 47– 49.

30. Lala, G., Malik, M., Zhao, C., He, J., Kwon, Y., Giusti, M. M. & Magnuson, B. A. (2006). Anthocyanin-rich extracts inhibit multiple biomarkers of colon cancer in rats, Nutrition and Cancer 54(1): 84-93.

31. Lee, J. Y., Park, H. J., Lee, C. Y. & Choi, W. Y. (2003). Extending shelf-life of minimally processed apples with edible coatings and anti-browning agents, Lebensmittel Wissenschaft und Technologie 36: 323-329.

32. Lila, M. A. (2007). From beans to berries and beyond: Teamwork between plant chemicals for protection of optimal human health, Annals of the New York Academy of Sciences 1114: 372-380.

33. Luo, Y., Lu, S., Zhou, B. & Feng, H. (2011). Dual effectiveness of sodium chlorite for enzymatic browning inhibition and microbial inactivation on fresh-cut apples, LWT- Food Science and Technology 44(7): 1621-1625.

34. Mao, L. & Fei, G. (2007). Application of 1-methylciclopropene prior to cutting reduces wound responses and maintains quality in cut kiwifruit, J. Food Eng. 78: 361– 365.

35. Martin-Belloso, O., Soliva-Fortuny, R. & Oms-Oliu, G. (2006). Fresh-cut fruits, in: Hui, Y.H. (Ed.), Handbook of Fruits and Fruit Processing, Blackwell Publishing, Oxford, pp. 129– 144.

36. Martínez-Monzó, J., Martínez-Navarrete, N., Chiralt, A. & Fito, P. (1998). Mechanical and structural changes in apple (var. granny smith) due to vacuum impregnation with cryoprotectants, Journal of Food Science 63(3): 499-503.

37. Masibo, M. & He, Q (2008). Major mango polyphenols and their potential significance to human health, Comprehensive Reviews in Food Science and Food Safety 7(4): 309-319.

38. McHugh, T. H. & Senesi, E. (2000). Apple wraps: a novel method to improve the quality and extend the shelf life of fresh-cut apples, Journal of Food Science 65: 480-485.

39. Mokni, M., Hamlaoui-Guesmi, S., Amri, M., Marzouki, L., Limam, F. & Aouani, E. (2012). Grape seed and skin extract protects against acute chemotherapy toxicity induced by doxorubicin in rat heart, Cardiovascular Toxicology 1-8.

40. Murthy, D. S., Gajanana, T. M., Sudha, M. & Dakshinamoorthy, V. (2009). Marketing and post-harvest losses in fruits: Its implications on availability and economy, Indian Journal of Agricultural Economics 64(2): 259-275.

41. Muthuswamy, S., Rupasinghe, H. P. V. & Stratton, G. W. (2008). Antimicrobial effect of cinnamon bark extract on Escherichia coli O157:H7, Listeria innocua and fresh-cut apple slices, Journal of Food Safety 28(4): 534-549.

42. Ngarmsak, M., Delaquis, P., Toivonen, P., Ngarmsak, T., Ooraikul, B. & Mazza, G., (2006). Antimicrobial activity of vanillin against spoilage microorganisms in stored fresh- cut mangoes, J. Food Prot. 69: 1724–1727.

43. Oms-Oliu, G., Rojas-Graü, M. A., González, L. A., Varela, P., Soliva-Fortuny, R., Hernando, M. I. H., . . . Martín-Belloso, O. (2010). Recent approaches using chemical treatments to preserve quality of fresh-cut fruit: A review, Postharvest Biology and Technology 57(3): 139-148.

44. Ozoglu, H. & Bayindirli, A. (2002). Inhibition of enzymatic browning in cloudy apple juice with selected anti-browning agents, Food Control 13: 213-221.

45. Perez-Gago, M. B., Serra, M. & del Rio, M. A. (2006). Color change of fresh-cut apples coated with whey protein concentrate-based edible coatings, Postharvest Biology and Technology 39: 84-92.

46. Prusky, D. (2011). Reduction of the incidence of postharvest quality losses, and future prospects, Food Security 3(4): 463-474.

47. Puel, C., Quintin, A., Mathey, J., Obled, C., Davicco, M. J., Lebecque, PCoxam, V. (2005). Prevention of bone loss by phloridzin, an apple polyphenol, in ovariectomized rats under inflammation conditions, Calcified Tissue International 7(5): 311-318.

48. Rawson, A., Patras, A., Tiwari, B. K., Noci, F., Koutchma, T. & Brunton, N. (2011). Effect of thermal and non-thermal processing technologies on the bioactive content of exotic fruits and their products: Review of recent advances, Food Research International, 4(7): 1875-1887.

49. Raybaudi-Massilia, R. M., Rojas-Grau, M. A., Mosqueda-Melgar, J. & Martín -Belloso, O. (2008). Comparative study on essential oils incorporated into an alginate-based edible coating to assure the safety and quality of fresh-cut Fuji apples, J. Food Prot. 71: 1150–1161.

50. Rößle, C., Brunton, N., Gormley, T. R. & Butler, F. (2011). Quality and antioxidant capacity of fresh-cut apple wedges enriched with honey by vacuum impregnation, International Journal of Food Science and Technology 46(3): 626-634.

51. Rocha, A. M. C. N. & Morais, A. M. M. B. (2003). Shelf life of minimally processed apple (cv. Jon gold) determined by colour changes, Food Control 14(1): 13-20.

52. Rojas-Graü, M. A., Raybaudi-Massilia, R. M., Soliva-Fortuny, R. C., Avena-Bustillos, R. J., McHugh, T. H. & Martín-Belloso, O. (2007). Apple puree-alginate edible coating as carrier of antimicrobial agents to prolong shelf-life of fresh-cut apples, Postharvest Biol. Technol. 45: 254–264.

53. Rojas-Graü, M. A., Soliva-Fortuny, R. & Martín-Belloso, O. (2009). Edible coatings to incorporate active ingredients to fresh-cut fruits: A review, Trends in Food Science and Technology 20(10): 438-447.

54. Roller, S & Seedhar, P. (2002). Carvacrol and cinnamic acid inhibit microbial growth in fresh- cut melon and kiwifruit at 4° and 8°C, Letters in Applied Microbiology 35: 390– 394.

55. Rupasinghe, H.P.V., Murr, D.P., DeEll, J.R. & Odumeru, J. A. (2005). Influence of 1- methylcyclopropene (1-MCP) and NatureSealTM on the quality of fresh-cut 'Empire' and 'Crispin' apples, J. Food Qual. 28: 289-307.

56. Rupasinghe, H.P.V., Boulter-Bitzer, J, Ahn, T. & Odumeru, J.A. (2006). Vanillin inhibits pathogenic and spoilage microorganisms in vitro and aerobic microbial growth on fresh-cut apples, Food Res. Intern. 39: 575-580.

57. Rupasinghe, H. and Yu, L. J. 2012. Emerging preservation methods for fruit juices and beverages, in El-Samragy Y. (Eds.), Food Additives, InTech - Open Access Publisher, Rijeka, Croatia. ISBN: 978-953-51-0067-6. pp: 65-82.

58. Rupasinghe, H.P.V., Thilakarathna, S. & Nair, S. (2012). Polyphenols of Apples and Their Potential Health Benefits, in: J. Sun, K. N. Prasad, A. Ismail, B. Yang, X. You and L. Li (Eds.), Polyphenols: Chemistry, Dietary Sources and Health Benefits, Nova Science Publishers, Inc. Hauppauge, NY, USA, ISBN: 978-1-62081-868-8.

59. Simpson, R. & Carevic, E. (2004). Designing a modified atmosphere packaging system for foodservice portions on non-respiring foods: optimal gas mixture and food/ headspace ratio, Foodservice Research International 14: 257-272.

60. Singh, A. K. & Goswami, T. K. (2006). Controlled atmosphere storage of fruits and vegetables: A review, Journal of Food Science and Technology 43(1): 1-7.

61. Sonkar, R. K., Sarnaik, D. A., Dikshit, S. N., Saroj, P. L. & Huchche, A. D. (2008). Post- harvest management of citrus fruits: A review, Journal of Food Science and Technology 45(3): 199-208.

62. Sun, J. & Liu, R. H. (2008). Apple phytochemical extracts inhibit proliferation of estrogen-dependent and estrogen-independent human breast cancer cells through cell cycle modulation, Journal of Agricultural and Food Chemistry 56(24): 11661-11667.

63. Sun, A. Y., Wang, Q., Simonyi, A. & Sun, G. Y. (2010). Resveratrol as a therapeutic agent for neurodegenerative diseases, Molecular Neurobiology 41(2-3): 375-383.

64. Thilakarathna, S.H. & H.P.V. Rupasinghe. 2012. Anti-atherosclerotic effects of fruit bioactive compounds: A review of current scientific evidence, Can J. Plant Sc. 92: 407-419.

65. U.S. Food and Drug Administration. (2006). Food Additive Status List. http://www.cfsan.fda.gov/~dms/opa-appa.html

66. Valencia-Chamorro, S. A., Palou, L., Delŕio, M. A. & Pérez-Gago, M. B. (2011). Antimicrobial edible films and coatings for fresh and minimally processed fruits and vegetables: A review, Critical Reviews in Food Science and Nutrition 51(9): 872-900.

67. Vargas, M., Pastor, C., Chiralt, A., McClements, D. J. & González-Martínez, C. (2008). Recent advances in edible coatings for fresh and minimally processed fruits, Critical Reviews in Food Science and Nutrition 48(6): 496-511.

68. Vilas-Boas, E. & Kader, A. (2007). Effect of 1-MCP on softening of fresh-cut kiwifruit mango and persimmon slices, Post. Biol. Technol. 43: 238–244.

69. Wang, H., Feng, H. & Luo, Y. (2007). Control of browning and microbial growth on fresh-cut apples by sequential treatment of sanitizers and calcium ascorbate, Journal of Food Science 72(1), M1-M7.

70. Watada, A.E. & Qi, L. (1999). Quality of fresh-cut produce, Postharvest Biol. Technol. 15: 201–205.

71. Weichselbaum, E., Wyness, L. & Stanner, S. (2010). Apple polyphenols and cardio vascular disease - a review of the evidence, Nutrition Bulletin 35(2): 92-101.

72. Wu, T., Zivanovic, S., Draughon, F.A., Conway, W. S. & Sams, C. E. (2005). Physicochemical properties and bioactivity of fungal chitin and chitosan, J. Agric. Food Chem. 53: 3888–3894.

73. Xie, J. & Zhao, Y. (2003). Nutritional enrichments of fresh apple (Royal Gala) by vacuum impregnation, International Journal of Food Sciences and Nutrition 54: 387–398.

74. Yin, X., Quan. J. & Kanazawa T. (2008). Banana prevents plasma oxidative stress in healthy individuals, Plant Foods in Human Nutrition 63 (2): 71-76.

75. Yu, C., Nandrot, E. F., Dun, Y. & Finnemann, S. C. (2012a). Dietary antioxidants prevent age-related retinal pigment epithelium actin damage and blindness in mice lacking $\alpha v\beta 5$ integrin, Free Radical Biology and Medicine 52(3): 660-670.

76. Yu, W., Fu, Y. -. & Wang, W. (2012b). Cellular and molecular effects of resveratrol in health and disease, Journal of Cellular Biochemistry 113(3): 752-759.

77. Zhao, Y. & Xie, J. (2004). Practical applications of vacuum impregnation in fruit and vegetable processing, Trends in Food Science & Technology 15; 434–451.

8 Chapter DIFFERENTIATED FOODS FOR CONSUMERS WITH NEW DEMANDS

1. INTRODUCTION: AN OVERVIEW

In recent decades, the food business has addressed the increasing customer demand for products that offer benefits beyond just nutritional value, resulting in billions of dollars in the global market. The lifestyle, convenience and rapidity of food preparation, together with alterations in dietary habits among the populace, all indicate the rising prevalence of chronic diseases attributable to the consumption of high-calorie foods and insufficient physical activity.

Progress in food science has elucidated the functions and mechanisms of bioactive compounds, thereby endorsing the incorporation of ingredients and the formulation of foods that promote a healthy diet linked to a healthy lifestyle. While functional foods ought to be ingested in their natural form rather than as supplements or capsules, the incorporation of bioactive ingredients or components into the formulation and processes of these supplements can serve as a catalyst for industry innovation and enhance the capacity to provide products of superior quality.

Dairy products have historically been linked to health benefits, a perception that persists to some extent; consequently, innovations in this sector typically involve the utilization of lactic acid bacteria (LAB), probiotic microorganisms, or the incorporation of functional ingredients and bioactive metabolites. Diverse methods, including encapsulation, may be employed to safeguard and preserve the viability of microorganisms in food products.

There is a trend towards utilizing inexpensive and sustainable novel materials that provide qualities compatible with controlled release of ingredients.

The notion of functional starter cultures, which may not qualify as probiotics yet could enhance product quality or yield physiological benefits for consumers, warrants more investigation. Besides probiotic properties, alternatives encompass in situ cultures that impede pathogenic contaminants through antimicrobial action; degrade or eliminate toxic compounds; synthesize vitamins or exopolysaccharides (EPSs); enhance viscosity, body, or texture; and promote adherence to specific sites within the host.

The binding of EPS mucoid bacteria to the protein matrix enhances viscous behavior, and certain EPSs generated by LAB confer health benefits through their prebiotic and hypocholesterolemic properties, immunomodulatory capacity, or anticancer effects. Confirming these observations, some authors have reported that the production of exopolysaccharides by specific bifidobacteria can enhance the viscosity of fermented foods, thereby influencing their rheological properties, and can thus be regarded as natural additives favored by consumers, capable of substituting plant or animal stabilizers.

The utilization of the unique properties of LAB to enhance their efficacy in foods or food items for vegetarians and anyone with dietary or religious constraints offers an alternative to specialized goods. This category encompasses meals that are lactose-free, enriched with fiber, devoid of animal products, and have elevated levels of antioxidant bioactive components (e.g., isoflavones, aglycones, oligosaccharides). Fruits and vegetables possess elevated concentrations of advantageous compounds (e.g., antioxidants, vitamins, fiber, and minerals), and the incorporation of lactic acid bacteria (LAB) and probiotics can enhance these properties further. An understanding of their behavior in fruit and vegetable matrices as carriers for probiotics or bioactive compounds is essential and remains largely unexamined in both literature and industrial practices.

Nonetheless, it is imperative that the rising pressure or process aligns with sustainable practices across the entire production chain concerning economic, environmental, and social challenges. Every stage of the process that enhances a product's value or mitigates waste or effluent generation will align with the principles of clean production.

This chapter will concentrate on the recovery of by-products and innovative applications of plant materials, as well as the enhancement of resources and advantageous effects of food combinations to produce value-added functional products, providing alternatives for consumers seeking to enhance their health through specialty foods.

2. ANTIOXIDANTS FROM PLANT SOURCES

In recent years, natural compounds have generated great interest due to the correlation between carcinogenic effects and the ingestion of synthetic compounds. Natural compounds such as phenolics, carotenoids and organic acids are widely found in plants and vegetables and have been the target of numerous studies because they exhibit strong antioxidant activity in addition to the ability to reduce the incidence of cancer in humans [1-4].

There is an equilibrium between the antioxidant defence system and the pro-oxidants in the human body, which are mainly reactive oxygen species (ROS) and reactive nitrogen species (RNS). The majority of reactive species (RS) originate in endogenous metabolic processes, whereas exogenous sources may include excess iron or copper in the diet, smoking, exposure to environmental pollutants, inflammation, bacterial infections, radiation, prolonged emotional stress and unbalanced intestinal microflora. The abnormal formation of RS may occur in vivo and cause damage to lipids, proteins, nucleic acids or carbohydrates in cells or tissues, and an imbalance with regard to pro-oxidants gives rise to oxidative stress (OS) [5].

Antioxidants impede or delay the in vivo oxidation reactions of lipids and other molecules or foods, inhibiting or retarding the chain propagation of free radicals generated by oxidation, such as hydroxyl radicals (•OH). In general, antioxidants are aromatic compounds that possess at least one free hydroxyl; they may be synthetic, such as BHA (butylhydroxyanisole), or natural, such as terpenes and phenolic compounds [2, 6-9].

Many studies have demonstrated that the consumption of antioxidants in food reduces the effects of the oxidative processes that naturally occur within the organism, aiding the natural endogenous protection mechanisms, such as the activities of superoxide dismutase, catalase and peroxidase, which together with vitamins E, C and A; enzymes; and other antioxidants and reduced glutathione (GSH) constitute the integrated antioxidant defence system (IADS) of the human body [4, 5, 7-8].

Flavonoids belong to the polyphenol group, which can be further divided into 11 smaller classes, including isoflavones, anthocyanins, flavans and flavanones. Their basic structure (Figure 1) comprises a flavone nucleus with 2 benzene rings (A and B) bonded to a heterocyclic pyran ring (C) [10,11].

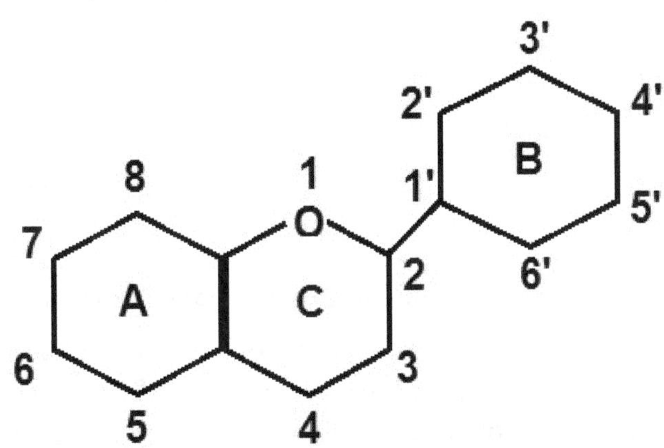

Figure 1. General structure of a flavonoid [10]

Isoflavones are phenolic compounds found mainly in beans and soybean derivatives and vary in concentration from 0.1 to 5 mg/g. They are distinguished by the substituents on the benzene ring, which are classified into 4 distinct forms: β-glycosides (daidzin, genistin and glycitin), acetyl-glycosides (acetyldaidzin, acetylgenistin and acetylglycitin), malonyl-glycosides (malonyldaidzin, malonylgenistin and malonylglycitin) and aglycones (daidzein, genistein and glycitein). As a result, there are a total of 12 different forms, with the β-glycoside forms bonded at position 7 of the benzene ring to a glucose molecule (Figure 2). The consumption of isoflavones is related to the prevention of several diseases, such as breast cancer, colon cancer and cardiovascular problems. In a study performed by Silva, Carrão- Panizzi and Prudêncio (2009) comparing different varieties of soybean, the authors found a prevalence of glycosidic and malonyl-glycosidic isoflavones in the beans, with higher levels of the aglycone forms in the BRS 267 soybean variety with cooking [10,13-15].

According to Arora, Nair and Strasburg et al. (1998), all isoflavone forms display antioxidant action, which varies widely according to the structure. In addition, the genistein form, with hydroxyl groups at positions 5, 7 and 4, has a greater antioxidant strength, which is evident by its structure, as shown in figure 2.

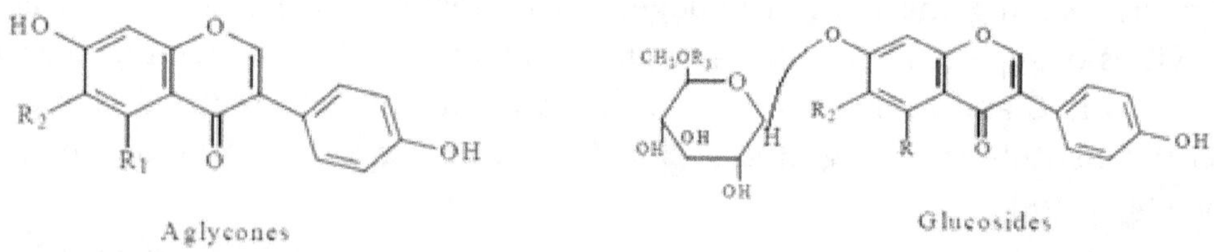

R1	R2	R3	Compounds
H	H	-	daidzein
OH	H	-	genistein
H	OCH_3	-	glycitein
H	H	H	daidzin
OH	H	H	genistin
H	OCH_3	H	glycitin
H	H	$COCH_3$	6"-O-Acetyldaidzin
OH	H	$COCH_3$	6"-O-Acetylgenistin
H	OCH_3	$COCH_3$	6"-O-Acetylglycitin
H	H	$COCH_2COOH$	6"-O-Malonyldaidzin
OH	H	$COCH_2COOH$	6"-O-Malonylgenistin
H	OCH_3	$COCH_2COOH$	6"-O-Malonylglycitin

Figure 2. Chemical structures of the 12 isoflavones found in soybean [10]

Chaiyasut et al. (2010) evaluated the effect of the time of Aspergillus oryzae fermentation in soybean on the isoflavone profile and the antioxidant capacity through ABTS cation (2,2'azinobis-[3-ethylbenzthiazoline-6-sulfonic acid]) and iron reduction (FRAP) assays. According to the authors, there was a significant increase in aglycone isoflavones (daidzein and genistein) and a reduction in glycosilades (daidzin and genistin) with a longer exposure time to the fermentation process.

This trend was reflected in the antioxidant activity, with the greater antioxidant capacity displayed by samples with a longer fermentation time due to an increase in the aglycone forms. These results were similar to those found by Barbosa et al. (2006), who evaluated the isoflavone profile and the amount of phenolic compounds in different soybean-based products and the influence of these products on the antioxidant capacity. Their results showed that the antioxidant capacity is related not only to the amount of total phenolic compounds but also to the amount and forms of the aglycones and the types of conjugation.

However, anthocyanins are considered to be natural pigments, as they exhibit colours that are visible to the human eye and may be found in flowers, fruits and vegetables. Anthocyanins belong to the flavonoid group and are not synthesised by the human body; when ingested, they help the immune system by decreasing the action of radicals formed during respiration, and they are naturally found in several plants [19,20].

Anthocyanins are glycosides of anthocyanidins (Figure 3) and may have different sugars bonded to their ringed structure. They are classified as mono, di or triglycosides, and the diglycoside and triglycoside forms are more stable than the monoglycoside forms. They display colour variations according to their structural forms, pH value, number of hydroxyls and methoxyls and temperature [21-23].

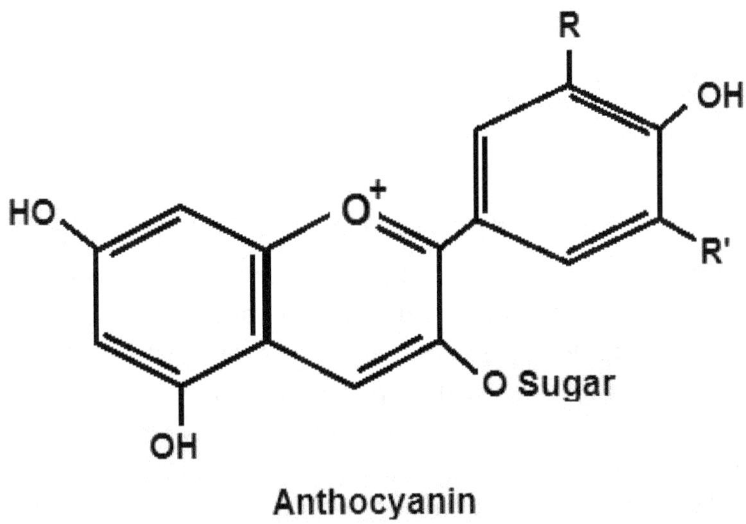

Figure 3. General structure of an anthocyanin [23]

According to Levi et al. (2004), there may be 4 structures in an aqueous medium depending on the pH value: the flavylium cation, the quinoidal base, carbinol and chalcone (Figure 4).

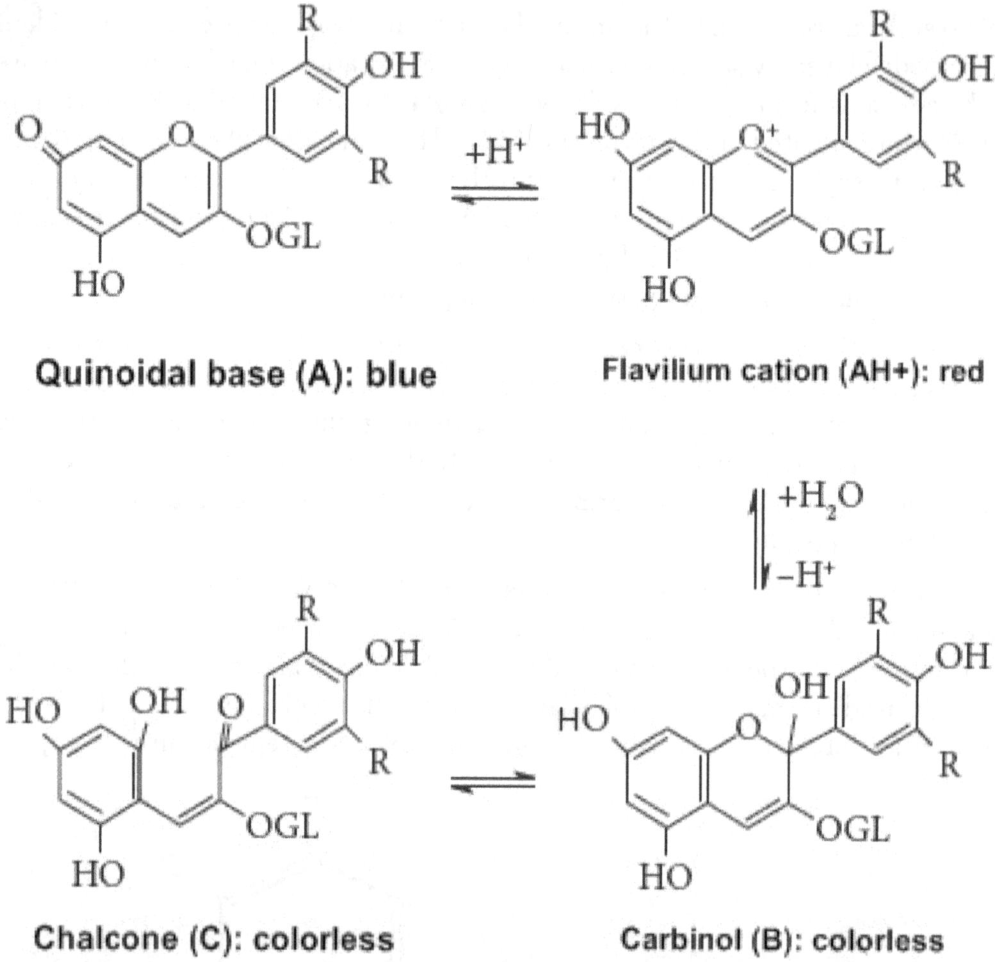

Figure 4. Molecular structures found at different pH values [24]

Kahkonen et al. (2003) isolated and identified the anthocyanins present in bilberry, blackcurrant and cowberry and evaluated their antioxidant activities through in vitro DPPH (2,2-diphenyl-1-picrylhydrazyl) assays with emulsified methyl linoleate and LDL (human low density lipoprotein). They found that the amounts of anthocyanins for bilberry, blackcurrant and cowberry were 6000, 2360 and 680 mgkg-1 of the fresh weight, respectively; all samples exhibited high activity in the DPPH tests and were effective antioxidants for the emulsion of methyl linoleate and human LDL. Rufino et al. (2010) studied the antioxidant strength of açaí (Euterpe oleraceae) with the aim of using it in functional foods and dietary supplements and found an antioxidant capacity for acai oil in the DPPH assay that was higher (EC50=646.3 g/g DPPH) than the value for virgin olive oil (EC50=2057.27 g/g DPPH), indicating its considerable potential for nutritional and health applications.

3. NON-DAIRY MATRICES AS VEHICLES FOR PROBIOTICS AND VIABILITY

Currently, there is increasing consumer interest in probiotic foods as an alternative to improve health. The majority of probiotic products found on the market are milk based, including milk drinks, yogurts, cheese and ice cream. Despite being an ideal substrate for the growth of these microorganisms, dairy products have several disadvantages, such as the need for refrigerated transportation, their cholesterol content and the restriction of their consumption to individuals who are not intolerant of or allergic to the products [27].

Thus, the development of new alternatives for consumption has increasingly earned the attention of the scientific and industrial communities, and new products, such as those based on soybeans, cereals, fruits, vegetables and meats, are being developed as potential carriers. In addition, these non-dairy matrices contain reasonable amounts of carbohydrates, fibres, proteins and vitamins, which may beneficially favour the growth and maintenance of the probiotics [27].

The viability and stability of probiotics have been a formidable market and technological challenge for food producers, given that probiotic foods should contain specific lineages and maintain an appropriate level of viable cells during the product's shelf life. Before they reach consumers, probiotics need to be produced under industrial conditions and maintain their functionality during storage in the form of a starter culture. Then, they need to be able to survive the processing of the food to which they are added. Finally, when ingested, the probiotics need to survive under the harsh conditions of the gastrointestinal tract and perform their beneficial effects in the host. In addition, they must be incorporated into the foods without producing unpleasant flavours or textures [28].

The application of probiotics in non-dairy matrices must be evaluated, given that several factors may influence the survival of these organisms and their activity when they pass through the gastrointestinal tracts of consumers. The following are among the factors that should be considered: the physiological state of the added probiotic organism as a function of the logarithmic or stationary growth phase; the appropriate concentration of viable cells in the product at the time of consumption; the physical conditions, such as low temperatures, during product storage; and the chemical composition of the product to which the probiotic is added, such as the pH, water content and amounts of carbon, nitrogen, minerals and oxygen [29].

Alternative vehicles for the incorporation of probiotic microorganisms may be fruits and fruit juices, but maintaining their viability is challenging because the pH of fruits and fruit juices is frequently low (< 4.0). They also contain antimicrobial substances. To minimise these factors, fruit juice may be formulated to have a higher pH value and smaller amounts of antimicrobial substances [30]. Sheehan, Ross and Fitzgerald (2007) evaluated the survival of several probiotic lineages in orange, pineapple and cranberry juices and observed that in addition to the juice's pH, the probiotic lineage and type of fruit also influenced the counts of the final product.

Pereira, Maciel and Rodrigues (2011) obtained a survival value of 8 log UFC/mL of L. casei in fermented cashew juice when the initial pH was 6.4 and the fermentation temperature was 30ºC. Yoon, Woodams and Hang (2004) also reported viable cell counts of more than 8.0 log CFU/mL in tomato juice. In addition, L. acidophilus, L. plantarum, L. casei and L. delbrueckii were capable of rapidly using this juice for cellular synthesis without nutrient supplementation. In another study by the same researchers, the fermentation of beets by probiotic bacteria was also investigated, and the authors observed a cellular survival of 109 UFC/mL of juice after 48 hours of fermentation in this substrate [34].

Lima, (2010) studied the behaviour of probiotic microorganisms in different tubers. Beetroot (Beta vulgaris) displayed the best survival results compared with sweet potato and arracacha. Upon adding pure betaine to the samples before fermentation and dehydration to supplement the amount of betaine already present in beetroot, the counts of L. plantarum and L. rhamnosus (LPRA Clerici-Sacco culture) were maintained at 8 log CFU/g of dehydrated sample (Figure 5).

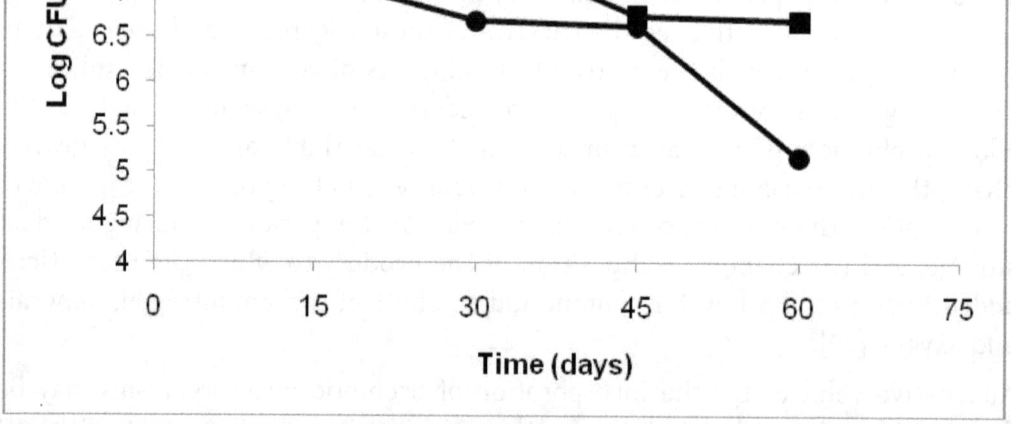

Figure 5. Behaviour of LPRA culture (composed of Lact. plantarum and Lact. rhamnosus) in assays 1, 2 and 3 during storage for 60 days at 25 °C. Assay 1(●): beet; assay 2 (▪): Beet + 0.5 mM betaine; assay 3 (▲): Beet + 2 mM betaine

To increase the robustness of the probiotic lineage of Lactobacillus salivarius UCC118, Sheehan et al. (2007) in a previous study cloned the betL gene of Listeria monocytogenes enables the system to capture or accumulate compatible solutes, such as betaine. BetL increases the tolerance to salt, low temperature and pressure stress as well as increases the viability of the probiotic in foods.

In addition to studies focusing only on the survival of probiotics in alternative matrices, for these products to be fit for human consumption and compatible with industrial production, evaluating the sensory quality of the formulated products is important. In this context, Ellendersen et al. (2012) developed and optimised a probiotic drink composed of apple juice and established the sensory profile using quantitative descriptive analysis (QDA). Sensorially, apple juice recently fermented with L. casei was characterised as having a thick texture and sweet flavour, but at 28 days of storage, a sour taste was observed by the tasters. When the fermented drink was tested by potential consumers, a rate of 96% acceptance was obtained, indicating that apple juice may be a medium for the inclusion of probiotics. Baptista, (2010) used orange peels with a pectin content of 19.3% (p/p) and a subsequent fermentation by a starter culture (Lyofast M36 LV) of kefir in milk serum and dehydrated the product, which was used to produce a cereal bar. An average acceptance rate of 6.97 (in a structured Hedonic scale of nine points) for samples without the peel and 6.90 for samples containing the dehydrated probiotic was obtained (the samples did not differ between one another at a level of $p<0.05$). The counts of Lactococcus lactis found in the product were 5.4x107 CFU/g.

Cereals are considered one of the most important sources of proteins, carbohydrates, vitamins, minerals and fibres. The traditionally fermented products of cereals exhibit modified textures, tastes, aromas and nutritional qualities and are widely consumed in Asia, Africa, South America and India. The fermentative process of these foods, in addition to improving the nutritional value, contribute to increasing its preservation via the production of alcohols and acids and reduction in the amount of toxic substances and cooking time for the cereal [29].

According to Charalampopoulos et al. (2002), the possible applications of cereals or cereal constituents in the formulation of functional foods may include the following: (a) as a fermentable substrate for the growth of probiotic microorganisms, especially lactobacilli and bifidobacteria; (b) as a dietary fibre promoting various beneficial physiological effects; (c) as a prebiotic due to its specific non-digestible carbohydrate content; and (d) as encapsulating materials for probiotics to increase their stability.

Thus, several studies have been performed to bind probiotic microorganisms to cereal matrices. Charalampopoulos, Pandiella and Webb (2003) verified the viability of Lactobacillus plantarum, L. acidophilus and L. reuteri in extracts of malt, barley and wheat for 4 hours in a phosphate buffer with an acidity of pH 2.5. They observed that these cereals displayed a significant protective effect toward the viability of these microorganisms, which may mainly be attributed to the amount of sugar present in these extracts. In 2010, Charalampopoulos and Pandiella evaluated the survival of Lactobacillus plantarum in extracts of barley, wheat and malt that were produced in suspensions of flour/water at concentrations of 5%, 20% and 30%, fermented for 24 hours at 37ºC and stored at 4ºC for 70 days. The authors observed that the cells displayed greater survival when they were stored in a medium containing malt extract, and this result was attributed to the higher concentration of sugar and the presence of other unidentified compounds.

Rathore, Salmerón and Pandiella (2012) used malt, barley and a mixture of malt and barley as substrates in the fermentation of Lactobacillus plantarum and Lactobacillus acidophilus with the objective of evaluating the influence of the lineages and the matrices used for the production of a probiotic beverage. The authors observed that a higher level of cellular growth was obtained in the medium that contained malt. In addition, these results suggested that the functional and sensory properties of probiotic beverages based on cereals may be considerably modified by changes in the composition of the substrate or the inoculum.

Oats, one of the major sources of beta-glucan, are commonly used in studies with probiotics. Guergoletto et al. (2010a) achieved a high level of survival for L. casei attached to oat bran when undergoing the vacuum drying process. Angelov et al. (2006), after optimising several factors such as the concentrations of the starter culture, oat flour and sucrose, developed an oat beverage fermented with L. plantarum, obtaining approximately 7.5x1010 CFU/mL of probiotics at the end of the process.

Another interesting application of probiotic microorganisms would be the enrichment of chocolate. With the development of technologies modified and adapted to maintain cells, this process may contribute toward increasing the benefit of this product for human health and increasing the consumption of probiotics by children, given that chocolate is one of their favourite products. For this combination to be successful, the sensory attributes of chocolate must remain unaltered, and the probiotic population must remain viable during commercialisation [44].

Finally, the application and development of new probiotic products of a non-dairy origin continue to grow. In light of the studies that have been presented by the scientific community, minimising the difficulties found in the application of these microorganisms in other food segments is possible.

4. NON-TRADITIONAL FERMENTED SOYBEAN-BASED PRODUCTS

Soybean is a plant that has been consumed since ancient times and is known worldwide for its nutritional benefits; it has a composition of approximately 40% proteins, 35% carbohydrates, 20% lipids and 5% ash [10]. In addition, soybean contains a considerable amount of components that are beneficial to health, such as fibres, isoflavones, essential fatty acids and oligosaccharides.

Traditionally, soybean-based products may be fermented by bacteria and/or fungi, with the most well-known being koji, shoyu, miso, tempeh, natto and sufu. These products are traditionally consumed by the East Asian population and represent an important source of dietary protein.

The search for foods that offer health benefits in addition to basic nutrition has promoted the development of new products that are based mainly on soymilk, which is obtained from the aqueous extraction of the bean's components. Soymilk possesses a chemical composition and appearance similar to those of animal milk and constitutes an appropriate substrate for fermentation; it contains, on average, 3.6% protein, 2.0% lipids, 2.9% carbohydrates and 0.5% ash [10]. Although studies have shown an increase in the consumption of soymilk, there are still technological limitations with regard to its sensory characteristics due to the perception of undesirable flavours that were inherent in the extract or that were formed during the processing [45,46]. Fermentation, especially by lactic bacteria, has been used to improve the flavour and increase the acceptability of soymilk and sometimes, to decrease the levels of saponin, phytate and oligosaccharides [47, 48].

Soybean-based products that are analogous to the products derived from milk have been developed and are widely consumed. In general, these products are not cheaper than dairy products, yet they meet the growing demand for lactose- and cholesterol-free products. The main products developed in this segment are beverages, yogurt and cheese made from soybean, which are sought by consumers looking for healthier foods. The fermentation of soymilk by lactic bacteria, in addition to increasing shelf-life, is aimed at obtaining products with flavours and textures that are more acceptable to consumers [47]. In general, the microorganisms that are utilised are capable of using soybean sugars, or sucrose may be added as a substrate for fermentation. Table 1 shows several non-traditional fermented soybean products and the respective microorganisms used in their production.

Table 1. Non-traditional fermented soybean products

Product	Microorganisms used	Reference
Soy yogurt	Streptococcus thermophilus, L. delbrueckii subsp. bulgaricus and L. johnsonii, L. rhamnosus and bifidobacteria	Farnworth et al. (2007)

	L. delbrueckii subsp. bulgaricus and Streptococcus thermophilus	Rinaldoni et al. (2012)
Fermented soy beverage	Bifidobacteria	Chou and Hou (2000)
	Streptococcus thermophilus and L. helveticus	Champagne et al. (2010)
Kefir	L. delbrueckii subsp. lactis, L. helveticus, L. rhamnosus and Bifidobacterium longum and Streptococcus thermophilus	Champagne et al. (2009)
	Lactococcus lactis ssp. lactis, Lactococcus lactis spp lactis biovar Baú (2012) diacetylactis, L. brevis, Leuconostoc and Saccharomyces cerevisiae.	Baú (2012)
Custard	Commercial kefir culture - Streptococcus lactis, Streptococcus cremoris, Streptococcus diacetylactis, L. plantarum, L. casei, Saccharomyces fragilis and Leuconostoc cremoris	McCue and Shetty (2005)
	Kefir grains	Sánchez-Pardo et al. (2010)
Soy cheese	L. rhamnosus	Liu et al. (2006)

Several studies on the survival of probiotic microorganisms indicate that soybean is an appropriate substrate for the growth of several probiotic species, such as bifidobacteria and several lactobacilli, such as L. casei, L. helveticus, L. fermenti, L. reuteri and L. acidophilus. Therefore, new probiotic products based on soybean are being continuously developed, exploring the potential that soybean has as a vehicle for functional ingredients.

4.1 Fibre in non-traditional fermented soybean-based products

The development of ingredients and products rich in fibre has significantly increased and involves the incorporation of fibre into a wide variety of products, including those made from soybean, with the aim of improving the dietary habits of the population.

In addition to performing physiological functions that are beneficial to the human body, when added to food products, fibre may change the sensory characteristics and consumer acceptance as well as the product's cost and stability. The addition of fibre may affect the processing and handling of the products, with changes in the viscosity, texture, creaminess, syneresis, acidity, colour and other characteristics [49]. In fermented foods, fibre may change the fermentative ability of the products and, in some cases, may protect probiotic microorganisms under stress conditions. Soluble fibre can also be fermented by bacteria in the colon, giving rise to short-chain fatty acids, mainly acetate, propionate and butyrate. In contrast, insoluble fibres are not very fermentable. Furthermore, some types of fibre may act as prebiotics, selectively stimulating the growth of some probiotic microorganisms.

The main types of soluble fibre added to products include pectin, inulin, oligofructose, gums, β-glucan and some non-digestible oligosaccharides. Insoluble fibre mainly comprises cellulose and hemicellulose, with the most common sources being legumes and cereals, such as soybean, rice, corn, oats and wheat. In general, some sub-products have been used as an alternative for the incorporation of fibre into products as in the case of okara, which is the residue from producing soymilk and has a significant amount of fibre and other important compounds, such as proteins and isoflavones.

In fermented soybean products, the addition of inulin and oligofructose in soybean yogurt has been reported [50], and soybean, oat and wheat fibre have been added to soybean kefir [49]. In the soybean product fermented with kefir, the soybean fibre stimulated the growth of a probiotic microorganism and promoted an increase in firmness and viscosity and a decrease in the synerisis of the product. Yeo and Liong (2010) supplemented WSSE with the prebiotics maltodextrin, pectin, inulin and fructooligosaccharides and observed an alteration in the lactic bacteria count and other characteristics.

Therefore, it is possible different uses of soybean in human food, including being a source of fibre and providing foods with high nutritional value to meet the population's demand for healthy foods.

5. PRODUCTS DEVELOPED FOR INDIVIDUALS WITH CELIAC DISEASE

Increasing our knowledge on the relationship between diet and health has caused consumers to look for high nutritional value, additional health benefits, convenience and pleasant sensory characteristics in processed products. In addition to this demand, a portion of the population is allergic to gluten. For this group, the treatment is essentially based on diet modification, which consists of eliminating gluten. The appropriate foods for individuals who are allergic to gluten are restricted and normally expensive, given that during processing, naturally gluten-free products may experience contamination that is unacceptable for those with celiac disease.

Celiac disease (CD) is an immune-mediated enteropathy triggered by the ingestion of wheat gluten (Triticum aestivum and T. durum) and similar proteins from rye (Secale cereale) in genetically susceptible individuals. During proteolytic digestion, prolamins (secalins) from rye and those in a subgroup of wheat (a-, b-, g- and w-gliadin) release a family of polypeptides rich in Pro and Gln that is responsible for the auto-immune response in celiac enteropathy [51]. The disease corresponds to hypersensitivity to gliadin (protein portion of gluten), which may be found in wheat, rye, barley and oat, and this hypersensitivity is marked by intense inflammatory processes. The consumption of cereals that contain gluten by individuals with celiac disease harms the small intestine [52], causing atrophy and a flattening of the intestinal villi, thereby leading to a limitation of the area available to absorb nutrients, among other manifestations. Situations such as travelling, eating outside the home and even enjoying relationships with friends and families may represent difficulties for celiac sufferers, thus interfering in their social lives [53]. With this disease, the processes of digestion and absorption may be compromised due to the increase in the immune activation of the intestinal tract. Celiac disease is one of the main causes of malabsorption in developed countries [54].

Therefore, there is a search for healthy foods that contain a variety of sensory attributes to allow for the possibility of providing a diverse selection of these foods. However, even with these possibilities, the celiac population is deprived of the consumption of many foods given that the formulations contain cereal-derived ingredients that contain gluten, such as oat flakes, wheat flour and malt.

Therefore, the development of new products for this population is essential, which may be performed by incorporating ingredients that contribute to an increase in mineral absorption, such as the fructans of inulin and oligofructose and other gluten-free bases. Fructans are soluble dietary fibres that may contribute to an increase in the absorption of minerals through colonic absorption [55, 56]; this effect may be especially important for those with celiac disease, given that the absorption of calcium in the small intestine is impaired in these individuals [57]. Capriles and Gomes Arêas (2010) developed amaranth bars with different flavours through the addition of inulin and oligofructose and observed that the amaranth bars enriched with these fructans may contribute to greater compliance by those with celiac disease to a gluten-free diet and help increase the absorption of calcium. These bars also have a reduced energy content and a high fibre content.

Other alternatives available for the celiac population include the substitution of the wheat flour that is present in several foods, such as breads, cakes, biscuits and pasta, with a mixture of flours that contain rice cream, tapioca flour, potato starch or corn starch, among other products.

Also notable is the use of soluble fibre such as Psyllium – Plantago ovate [59]. The main component of psyllium is mucilage (made up of slightly branched polysaccharides, found in algae and seeds), which represents 10 to 30% of its structure. These types of fibre also contain lipids, proteins, oxalic acid and the enzymes invertase and emulsin. Psyllium is considered to be a prebiotic food and is used either pure or in preparations to improve intestinal constipation [60]. With the double function of substituting for wheat in the development of special foods, psyllium has been added to bread dough, which is traditionally made with wheat flour, to improve the characteristics obtained via water retention and gelatinisation [61].

In a study performed by Zandonadi, 2006, psyllium was added to breads, biscuits, pasta, cake and pizza dough, and these products could be classified as foods for special purposes because they reduce the gluten fraction and exhibit good acceptability both by those with and without celiac disease. In addition, they reduce the lipid fraction and thus the product´s energy values.

Given the importance of seeking alternatives that promote sensory and functional characteristics that are similar to those of products prepared with gluten, Stork et al. (2009) studied two protein and transglutaminase sources in bread from rice flour to produce a better-quality bread. They observed that rice flour may be enriched with albumin and casein modified by transglutaminase to improve the bread's nutritional quality.

Figueira et al. (2011) evaluated the characteristics of gluten-free breads produced with rice flour and enriched with the microalga Spirulina platensis, which is a microalga that has an appropriate composition for use as a food complement, that have a possible use in combating malnutrition [65]. The dry composition of Spirulina platensis contains high amounts of proteins (64-74%), polyunsaturated fatty acids and vitamins [66] and contains antioxidant compounds [67]. This microalga is classified as GRAS ("generally recognised as safe") by the FDA, which ensures it can be used as a food without health risks [68]. The authors recommend the use of S. platensis for the enrichment of gluten-free breads made from rice flour using a suggested microalga concentration of 3%, and these bread are appropriate for celiac patients.

In a study on quinoa flour, Berti et al. (2004) evaluated the triglyceride and free fatty acid levels and glycaemic and insulinaemic responses in individuals with celiac disease and showed that the foods prepared with quinoa flour resulted in improved measures for all of these factors compared with the foods prepared with common flours. They also found that satiety was higher in the ingestion of products prepared from quinoa flour.

The use of kefir, which may act as an anti-inflammatory agent, may provide satisfactory results in patients with celiac disease. For these individuals, kefir may help to combat the nutritional deficiencies resulting from the reduction in intestinal villi because kefir is rich in vitamin B12, thiamine and potassium, which increase the absorption of the vitamin B complex [70,71].

Mixtures of several LAB were capable of hydrolysing 109 out of 129 ethanol-soluble polypeptides of rye, and De Angelis et al. (2006) concluded that long-term fermentation by selected LAB may be a potential tool to decrease the risk of contamination with rye in gluten- free products for patients with celiac disease.

Green banana flour may also be an alternative for the celiac population because the cost is not high, it is easy to prepare, and it exhibits a high amount of resistant starch, approximately 74% of its composition. This high level of starch is related to its glycaemic index and ability to reduce cholesterol levels and promote gastric fullness and intestinal regulation, and its fermentation by intestinal bacteria produces short-chain fatty acids that may prevent the emergence of cancer in intestinal cells [72]. Given these observations, Zandonadi (2012) evaluated the development of a gluten-free pasta alternative for those with celiac disease using green banana flour and demonstrated good acceptance without compromising the product while imparting important nutritional characteristics.

However, in foods that are thermally processed, especially breads, the lack of gluten represents a challenge in maintaining good sensory qualities, especially in the structure or softness during storage. The use of fermented dough (sourdough) by baker's yeast resulted in an improvement of the texture and effectively delayed the hardening of the gluten-free breads [74]. Fermented doughs also provided the breads with characteristics such as starch digestibility and low glycaemic responses, thereby proving to be a promising procedure in the improvement of the texture of gluten-free breads for those with celiac disease [75]. Galle et al. (2012) studied the influence of the in situ formation of EPS from LAB on the rheology of the dough for gluten-free sorghum bread. Among the EPSs, dextran improved the texture quality of the bread in addition to contributing to the nutritional benefits.

Therefore, the search for and development, market availability, diversification, cost compatibility and even improvement of already existing products for the celiac population all need to increase not only to improve the selection or consumption of these foods but also to ensure a better quality of life for the individuals who require a gluten-free diet.

6. PROBIOTICS, METABOLIC ACTION AND VEHICLES OF BIOACTIVE COMPOUNDS

Through fermentation, toxic compounds may be hydrolysed and transformed into derivatives that are more or less absorbable or less toxic. Several studies describe the reduction of toxic or mutagenic compounds following fermentation or in the presence of microorganisms. In most cases, the microbial cells adsorb these compounds, and this process is normally increased with thermal treatment of the cells; the result is the possibility of reducing or degrading the compounds, but this latter mechanism is still not yet completely understood. Franco et al. (2010) observed a gradual increase in the reduction of the percentage of deoxynivalenol in solution depending on whether the LAB cells were viable or thermally inactivated (Table 2).

Other toxic compounds that could be degraded with this approach are toxins produced by algae. Considering the increase in the occurrence of cyanobacterial blooms and the possibility of metabolites being released into water supply sources used for human consumption, Guergoletto et al. (2010b) studied the microcystin (MC) biodegradation activities of microorganisms in water (Figure 6). Their work evaluated the use of the probiotic bacteria Lactobacillus acidophilus (La-5) and Lactobacillus casei (LC-1) and kefir grains for MC degradation over time. The mixtures were maintained at 27°C and 100 rpm, and samples were collected at 0, 12, 24, 48, 72 and 96 h to determine the level of MCs by immunoassay ELISA. The results indicated that the highest degradation percentage was obtained for kefir grains, reaching 60% and 62% of the total MC degradation for Microcystis sp. and NPLJ4 extracts, respectively, followed by the La-5 strain with levels of 43% and 51%. For LC-1, the degradation activities were 20% and 34% for Microcystis sp. and NPLJ4 extracts, respectively, but significant cellular growth was not verified when compared with the La-5 strain (Figures 7A and 7B).

Table 2. Reduction of deoxynivalenol level by LAB viable cells and cells that were heat inactivated (unviable) by pasteurisation or sterilisation

Microorganism	Reduction Percentage (%)*		
	Viable cells**	Unviable cells following pasteurisation***	Unviable cells following sterilisation****
Lyofast LPRA	52.07 ± 0.1aB	53.18 ± 2.06cB	70.32 ± 1.65aA
Lyofast BG 112	52.62 ± 4.95aB	67.45 ± 2.95aA	71.19 ± 2.77aA
Lyofast LA3	39.23 ± 2.22bC	60.67 ± 1.34bB	66.71 ± 1.82aA
LC 01	40.61 ± 1.19bB	64.03 ± 0.07abA	66.56 ± 2.43aA
Yo flex YC 180	31.25 ± 0.89cC	57.43 ± 0.95bB	65.64 ± 0.77aA
Florafit LP 115	32.61 ± 1.38cC	40.81 ± 0.95deB	58.51 ± 1.29cA
Yo mix	40.67 ± 0.76 bB	41.98 ± 0.45deB	48.75 ± 1.81cA
Choozit Helv A	55.30 ± 1.35aB	59.05 ± 0.45bAB	63.84±0.16abA
L. plantarum TG VIII	29.86 ± 1.18cC	50.38 ± 0.46cdB	56.05 ± 1.86bA
L. plantarum FT VI	34.88 ± 0.94bcB	38.58 ± 1.66eB	55.74 ± 1.25bA
L. plantarum GT III	56.12 ± 1.02aB	62.67 ± 1.09abA	66.79 ± 0.43aA

L. plantarum FTQ VII	39.70 ± 1.93bC	51.37 ± 1.36cB	65.26 ± 1.27aA
L. plantarum FB VII	16.41 ± 5.35dC	48.34 ± 1.46cdB	59.62 ± 1.02bA
L. plantarum FI IX	39.71 ± 0.30bB	44.73 ± 0.29bB	57.68 ± 0.41cA
L. pentosus S I	19.51 ± 4.63dC	35.95 ± 1.57eB	47.48 ± 1.59dA
L. paracasei K VI	29.51 ± 1.16cC	44.98 ± 1.77dB	57.19 ± 1.04cA
The results correspond to the average of duplicates ± standard deviation. Averages ± standard deviation in the same column followed by the same lowercase letter do not differ at p≤0.05. Averages ± standard deviation in the same line accompanied by the same uppercase letter do not differ at p≤0.05.**Viable cells were separated by centrifugation (5 ºC/ 3000 g/ 10 minutes), washed in PBS pH 7.2 and ultrapure water, resuspended in DON solution with ultrapure water at a concentration of 1500 ng ml-1 and incubated at 37±1 ºC for 4 hours.***Nonviable cells following pasteurisation (100 ºC/ 30 minutes) were separated by centrifugation, washed in PBS pH 7.2 and ultrapure water, resuspended in DON solution with ultrapure water at a concentration of 1500 ng ml-1 and incubated at 37±1 ºC for 4 hours.****Nonviable cells following sterilisation (121 ºC/ 15 minutes) were separated by centrifugation, washed in PBS pH 7.2 and ultrapure water, resuspended in DON solution with ultrapure water at a concentration of 1500 ng ml-1 and incubated at 37±1 ºC for 4 hours.			

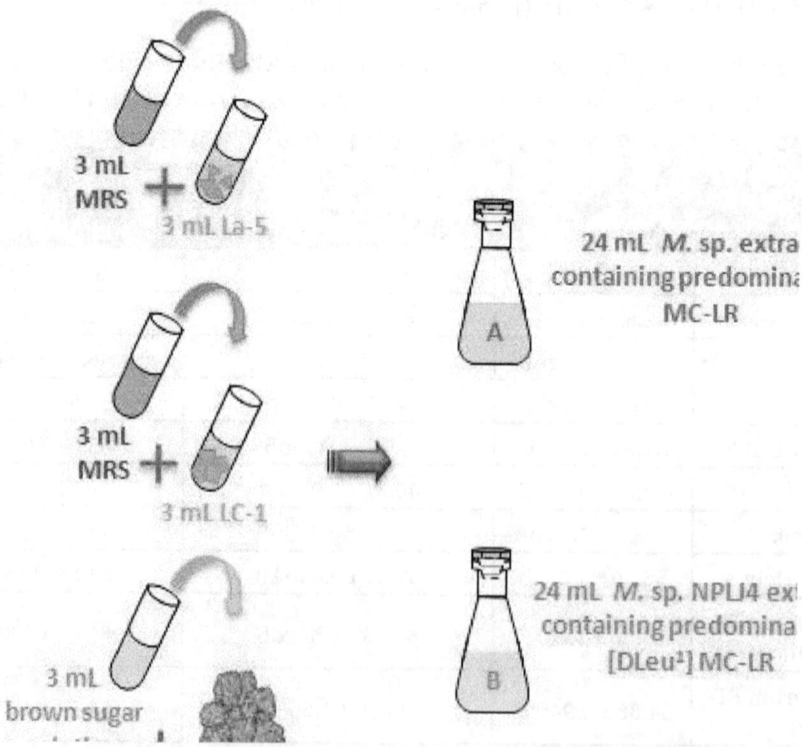

Figure 6. Scheme for the microcystin biodegrading activity experiment Lactobacillus acidophilus (La-5), Lactobacillus casei (LC-1) and kefir grains

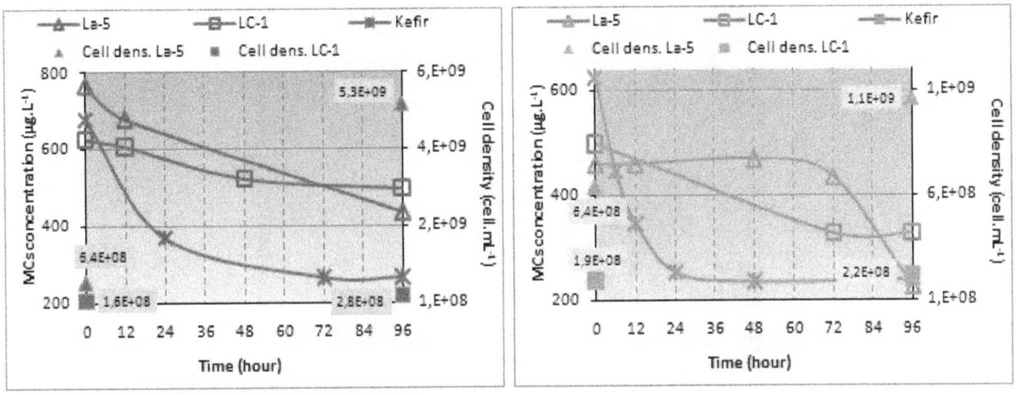

Figure 7. A Degradation kinetics of total MCs by La-5 and LC-1 bacteria and kefir grains during 96 hours of incubation with Microcystis sp. (A) extract B Degradation kinetics of total MCs by La-5 and LC-1 bacteria and kefir grains during 96 hours of incubation with NPLJ4 (B) extract

Mutagenic or carcinogenic activity in the caecal or urinary structures may be reduced by the consumption of L. casei shirota (LcS). A mechanism to explain the production of mutagenic substances was described by an in vitro study [79] in which the LcS was capable of strongly adsorbing and inactivating mutagenic pathogens and carcinogens, such as 3–amino-1,4 dimethyl-5H-pyrido (4,3-b) indole-trp-P-1 and 3-amino-1-methyl 5H pyrido (4,3-b) indole-trp-P-2. LcS also has the ability of binding aflatoxin, a known carcinogen produced by fungi [80].

Multi-functional polysaccharide molecules of plant, algal, bacterial or fungal origins have been extensively studied in recent decades for applications as thickeners, stabilisers, gelling agents, prebiotics and bioremediators or anti-pollutants [81-83]. Until now, plant macromolecules have dominated the market due to their ease and availability and because their purification is cost efficient, as plants are superior primary sources of polysaccharides, including starch, cellulose, pectin and gums. However, because polysaccharides of microbial origin are renewable, have little cost variation and have reproducible physical-chemical properties, they may be of value in certain situations, although they are still not widely marketed and represent an unexplored market [82]. Prasanna et al. (2012) studied the growth, acidification, EPS production and viscosity potential of 22 lineages of Bifidobacterium spp of intestinal origin, and EPSs were produced by Bifidobacterium bifidum ALM 35, B. breve NCIMB 8807 (UCC 2003), B. longum subsp. infantis CCUG 52486 and Bifidobacterium infantis NCIMB 702205 in concentrations varying from 25 to 140 mg L-1, producing an increase in the viscosity of dairy products with a low fat content.

Figure 8. Scanning electron micrographs at magnifications of 20,000x Agar-agar and yam (Dioscorea sp.) microspheres containing Saccharomyces cerevisiae Laboratory of Electron Microscopy and Microanalyses - State University of Londrina.

Laurenti, E. (2010) studied the controlled release of probiotic S. cerevisiae (Biosaf SC-47) from microspheres of agar-agar added to mucilage (Figure 8) and gums and evaluated the potential application of these new natural materials for the protection of probiotics using gastrointestinal simulation tests. Okra and flaxseed showed the greatest retention of yeast cells in the microspheres and, consequently, a lower percentage of release at 66.97% and 72.96%, respectively (Figures 9 and 10). abc Means between columns followed by different lowercase letters are significantly different ($p < 0.05$). ARA acacia gum; YAM Dioscorea sp; TARO Colocasia esculenta; I-CAR iota carrageenan; A-A agar-agar; ALG alginate; LIN Linun usitatissimum; OKR Hibiscus esculentus.

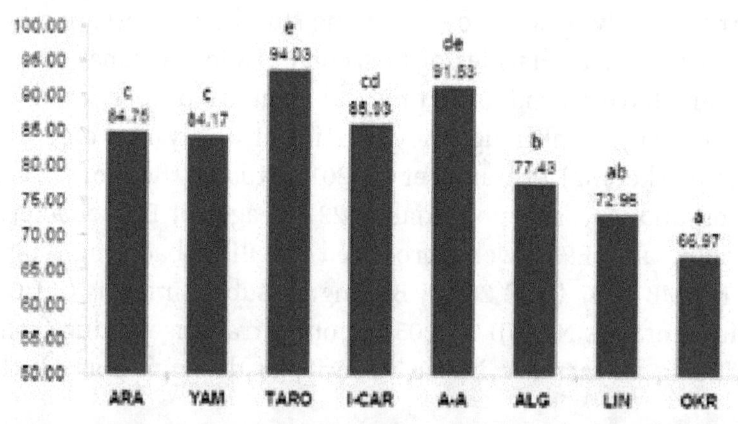

Figure 9. Mean release (% log CFU/g) of the probiotic S. cerevisiae encapsulated in agar-agar and different gums and mucilages after digestion in vitro

Mean release (% log CFU/g) of the probiotic S. cerevisiae encapsulated in agar-agar and different gums and mucilages after digestion in vitro Free radicals, especially those belonging to the family of ROS, are increasingly implicated or recognised as the cause of aging and in the pathogenesis of different diseases, such as cancer.

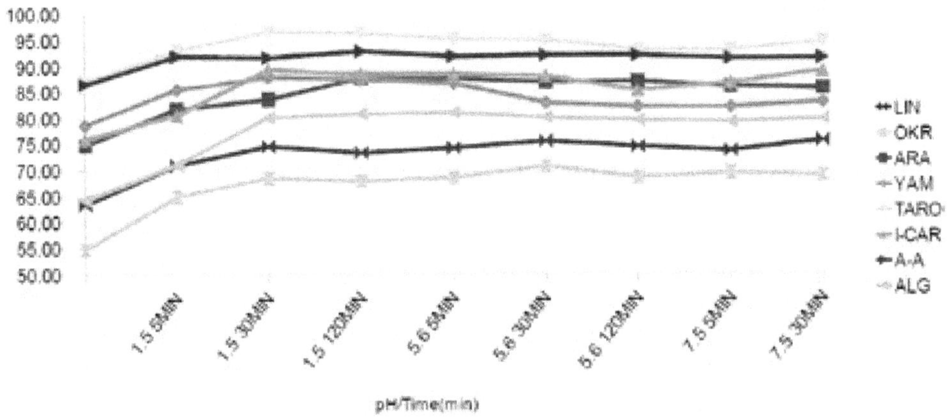

Figure 10. Digestion in vitro of the probiotic S. cerevisiae encapsulated in mucilages and gums. ARA acacia gum; YAM Dioscorea sp; TARO Colocasia esculenta; I-CAR iota carrageenan; A-A agar agar; ALG alginate; LIN Linun usitatissimum; OKR Hibiscus esculentus

Oxidative damage to the cellular molecules caused by chain reactions of free radicals may be combatted by antioxidants or by free-radical-sequestering agents. The use of natural antioxidants with less harmful effects and better bio-acceptability is gradually be coming important. Many plant or microbial polysaccharides have been demonstrated to exhibit sequestering or antioxidant ability due to the abundance of functional groups in the molecule [86]. Pan and Mei (2010) described the antioxidant action of EPS from Lactococcus lactis subsp lactis 12.

New evidence increasingly suggests the correlation of human IADS with the microbial organisms in the gastrointestinal tract. Specific lineages with physiological and antioxidant activities have a major impact on the management of the levels of oxidative stress in the lumen, among the mucosa cells and even in blood to support the functionality of the IADS in the human body. A lineage of Lactobacillus fermentum ME-3 (LfME-3) with antioxidant, antimicrobial and antiatherogenic properties was patented by the University of Tartu and proven to be 80 to 100 times more potent in vitro in sequestering the superoxide anion radical than Trolox or ascorbic acid. This lineage expresses Mnb superoxide dismutase (Mn SOD) activity, which effectively eliminates hydroxyl and peroxyl radicals and has the complete glutathione system (GSH, GPx, glutathione reductase – Gred) necessary for the recycling, transportation and synthesis of glutathione [5].

According to estimates by the World Health Organization [88], 3.2 million deaths per year are associated with physical inactivity. A sedentary lifestyle, a term derived from the Latin root "sedere", meaning to be seated, includes physical activities with low energy expenditure that are correlated with obesity, metabolic syndrome, type 2 diabetes and cardiovascular diseases (CVD) [89]. Therefore, new approaches are necessary to reduce the risk of CVD, for which prevention via anti-inflammatory agents and antioxidants is considered to be the "third great wave" [90].

In contrast to the traditional action of probiotics involving a direct interaction with the host, the action of LAB in the cardiovascular system occurs via the release of bioactive peptides from proteins by L. helveticus during the fermentation process. A functional dairy product, Cardi-04TM, was developed to reduce blood pressure [91]. A functional cheese with L. plantarum lineage TENSIA (DSM 21380, property of the Bio-competence Centre of Healthy Dairy Products LLC) may reduce blood pressure, both diastolic and systolic (a dose of 1010UFC of viable probiotic cells per daily portion), in adults with high blood pressure or healthy adults and elderly individuals [5].

Recently, novel bioactive chemicals have been incorporated into several medical and therapeutic uses. These compounds have been utilized for their antioxidant, anti-tumor, anti-inflammatory, and antiviral properties. The EPSs stimulate cytosine and interferon activity, impede platelet aggregation, and regulate the immunological system [81]. Polysaccharides derived from Lactobacillus species confer health benefits. Kefir can be categorized as a functional food because of its effects on several biological levels in animals. Kefiran, administered at doses ranging from 100 to 300 mg/kg in rats, decreased blood pressure, blood sugar, and cholesterol levels, while also exhibiting a beneficial effect on constipation [92]. Additional features were observed after the oral administration of this polysaccharide, including anti-inflammatory and anti-tumor actions, as well as the stimulation of immunoglobulin secretion. Diosgenin, a steroid saponin found in yams (Dioscorea sp.) and fenugreek, exhibits qualities that may be beneficial for future medical uses in lowering blood sugar and cholesterol levels, as well as in the treatment of cholestasis [93].

Hobbs et al. (2012) investigated the impact of beet juice and beet-enriched breads on blood pressure alterations, providing substantial evidence for a cardioprotective benefit and a reduction in blood pressure attributable to nitrate-rich plants. Recent hypotheses suggest that the cardioprotective properties of green-leafed plants like beets may be attributed to their elevated nitrate levels. Considering that hypertension correlates with a reduction in endogenous nitric oxide (NO) synthesis and that dietary nitrates can facilitate NO generation, innovative and economical approaches to dietary nitrate incorporation are of significant interest.

REFERENCES

1. Simões CMO, Schenkel EP, Gosmann G, Mentz LA, de Mello Palazzo JC, Schenkel EP. Farmacognosia: da planta ao medicamento. Porto Alegre/Florianópolis: Editora da Universidade UFRGS / Editora da UFSC. 1999 819.
2. Zheng W, Wang SY. Antioxidant activity and phenolic compounds in selected herbs. Journal Agricultural and Food Chemistry. Chicago: 2001; 49 5165-5170.
3. Wang SY, Zheng W. Effect of plant growth temperature on antioxidant capacity in strawberry. Journal Agricultural and Food Chemistry. Chicago: 2001; 49 4977-4982.
4. Yildirim A, Mavi A., Kara AA. Determination of antioxidant and antimicrobial activities of Rumex crispus L. extracts. Journal Agricultural and Food Chemistry. Chicago: 2001; 49 4083-4089.
5. Kullisaar T, Songisepp E, Zilmer M. Probiotics and Oxidative Stress. In: Oxidative Stress –Environmental Induction and Dietary Antioxidants. Lushchak, V.L. Croatia,Intech, 2012 p.203
6. Halliwell B, Aeschbach R, Loliger J, Aruoma OI. The characterization of antioxidants. Food Chemistry and Toxicology. 1995; 33(7) 601-617.
7. Brenna OV, Pagliarini, E. Multivariate analyses of antioxidant power and polyphenolic composition in red wines. Journal Agricultural and Food Chemistry. Chicago: 2001; 49 4841-4844.
8. Sousa CMD M, Silva HRE, Vieira Jr GM. et al. Total phenolics and antioxidant activity of five medicinal plants. Quimica Nova. 2007; 30(2) 351–355.
9. BRASIL, Decreto nº 50.040, de 24 de janeiro de 1961. Dispõe sobre as Normas Técnicas Especiais Reguladoras do emprego de aditivos químicos a alimentos. Agência Nacional de Vigilância Sanitária (ANVISA), Brasília, 1961.
10. Liu K. Soybeans: chemistry, technology and utilization. ITP., 1997.
11. Dornas WC, Oliveira TT, Rodrigues-das-Dores RG, Santos AF, Nagem TJ Flavonóides: potencial terapêutico no estresse oxidativo. Revista de Ciências Farmacêuticas Básica e Aplicada. 2007; 28(3) 241- 249.
12. Silva JB, Carrão-Panizzi MC, Prudêncio SH. Chemical and physical composition of grain type and food type soybean for food processing. Pesquisa Agropecuária Brasileira. 2009; 44(7) 777-784.
13. Kudou S, Shimoyamada M, Imura T, Uchida T, Okubo K. A new isoflavone glycoside in soybean seeds (Glycine Max Merrill), glycitein 7-O-beta-D-(6”-O-acetyl)-glucopiranoside. Agricultural and Biological Chemistry, 1991; 55(3) 859-860.

14. Setchell KD. Phytoestrogens: The biochemistry, physiology, and implications for human health of soy isoflavones. American Journal Clinical of Nutrition. 1998; 134(6) 1333-1343.
15. Schwartz SJ, Elbe JHV, Giusti MM. Corantes. In: DAMODARAN, Srinivasan; PARKIN, Kirk L.; FENNEMA, Owen R. Química de Alimentos de Fennema. 4 ed. Porto Alegre: Artmed, 2010. p. 445-498.
16. Arora A., Nair MG, Strasburg GM. Antioxidant activities of isoflavones and their biological metabolites in a liposomal system. Archives of Bichemistry and Biophysics, 1998; 356(2) 133-141.
17. Chaiyasut C, Kumar T, Tipduangta P, Rungseevijitprapa W. Isoflavone content and antioxidant activity of Thai fermented soybean and its capsule formulation. African Journal of Biotechnology. 2010; 9 4120-4126.
18. Barbosa ACL, Hassimotto NMA, Lajolo FM, Genovese MI. Teores de isoflavonase capacidade antioxidante da soja e produtos derivados. Science and Food Technology. 2006; 26(4) 921-926.
19. Jeng TL, Shih YJ, Wu MT, Sung JM. Comparisons of flavonoids and anti-oxidative activities in seed coat, embryonic axis and cotyledon of black soybeans. Food Chemistry. 2010 1112 –1116.
20. Qin Y, Jin Xiao-nan, Dong PH. Comparison of Antioxidant Activities in Black Soybean Preparations Fermented with Various Microorganisms. Agricultural Sciences in China. 2010 1065-1071.
21. Alkema S, Seager SL; The chemical pigments of plants. Journal of Chemical Education. 1982; 59(3) 183.
22. Francis FJ. Food Colorants: Anhtocyanins. Critical Reviews in Food Science and Nu trition. 1989 28(4) 273.
23. Lee J, Durst RW, Wrolstad RE. Determination of total monomeric anthocyanin pigment content of fruit juices, beverages, natural colorants, and wines by the pH differential method: Collaborative study. Journal AOAC International. 2005; 88(5) 1269-1278.
24. Levi MAB et al. Three-way chemometric method study and UV-vis absorbance for the study of simultaneous degradation of anthocyanins in flowers of the Hibiscus rosa-sinensys species. Talanta. 2004 62(2) 299-305.
25. Kahkonen MP, Heinamaki J, Ollilainen V, Heinonen, M. Berry anthocyanins: isolation, identification and antioxidant activities. Journal of the Science of Food and Agriculture. 2003; 83 1403–1411.
26. Rufino MSM et al. Açaí (Euterpe oleraceae) 'BRS Pará': A tropical fruit source of antioxidant dietary fiber and high antioxidant capacity oil, Food Research International. 2010; doi:10.1016/j.foodres.2010.09.011.

27. Yeo, S. K.; Ewe, J. A.; Tham, C. S. C.; Liong, M. T. Carriers of Probiotic Microorganisms. In: Probiotics: Biology, Genetics and Health Aspects. LIONG, M.T. Malasia: Springer, 2011; p. 191-215.

28. Mattila-Sandholm T, Myllarinen P, Crittenden R, Mogensen G, Fonden R, Saarela M. Technological challenges for future probiotic foods. International Dairy Journal. 2002 173-183.

29. Rivera-Espinoza Y; Gallardo-Navarro Y. Non-dairy probiotic products. Food Microbiology, 2010; 27 1-11.

30. Ouwehand AC, Svendsen LS, Leyer G. Probiotics: from Strain to Products. In KNEIFEL, Wolfgang, SALMINEN, Seppo. Probiotics and Health Claims. Wiley-Blackell, 2011, p. 37-48.

31. Sheehan V, Ross P, Fitzgerald GF. Assessing the acid tolerance and the technological robustness of probiotic cultures for fortification in fruit juices. Innovative Food Science and Emerging Technologies, 2007; 8 279-284.

32. Pereira ALF, Maciel TC, Rodrigues S. Probiotic beverage from cashew apple juice fermented with Lactobacillus casei. Food Research International. 2011; 44 1276-1283.

33. Yoon KYE, Woodmans EE, Hang YD. Fermentation of tomato juice by lactic acid bacteria. Journal of Microbiology. 2004; 42(4) 315-318.

34. Yoon KYE, Woodmans EE.; Hang YD. Fermentation of beet juice by beneficial lactic acid bacteria. Lebensmittel-Wissenschaft und-Technologie. 2005; 38 73-75.

35. Lima IR de. Desenvolvimento de produto vegetal desidratado a base de raízes tuberosas fermentadas com bactérias probióticas protegidas pela presença de betaína. Dissertação de Mestrado em Ciência de Alimentos, State University of Londrina. Brazil; 2010.

36. Ellendersen L de SN, Granato D, Guergoletto KB, Wosiacki G. Development and sensory profile of a probiotic beverage from apple fermented with Lactobacillus casei. Engineering in Life Sciences. 2012; 12(4) 1-11.

37. Baptista EV. Desenvolvimento de ingrediente simbiótico por fermentação de soro de leite e do subproduto da agroindústria de suco de laranja por grãos de Kefire cultura probiótica. Dissertação de Mestrado em Ciência de Alimentos. State University of Londrina. Brazil; 2010.

38. Charalampopoulos, D, Wang R, Pandiella SS, Webb C. Application of cereal and cereal components in functional foods: a review. International Journal of Food Microbiology. 2002; 79 131-141.

39. Charalampopoulos D, Pandiella SS, Webb C. Evaluation of the effect of malt, wheat and barley extracts on the viability of potential probiotic lactic acid bacteria under acidic conditions. International Journal of Food Microbiology. 2003; 82 133-141.

40. Charalampopoulos D, Pandiella SS. Survival of human derived Lactobacillus plantarum in fermented cereal extracts during refrigerated storage. LWT - Food Science and Technology 2010; 53(3) 431-435.

41. Rathore S, Salmerón I, Pandiella SS. Production of potentially probiotic beverages using single and mixed cereal substrate fermented with lactic acid bacteria cultures. Food Microbiology. 2012; 30 239-244.

42. Guergoletto KB, Magnani M, Martin JS; Andrade CG T de J, Garcia S. Survival of Lactobacillus casei (LC-1) adhered to prebiotic vegetal fibers. Innovative Food Science and Emerging Technologies. 2010a; 11 415-4212.

43. Angelov A, Gotcheva V, Kuncheva R, Hristozova T. Development of a new oat-based probiotic drink. International Journal of Food Microbiology. 2006; 112 75-80.

44. Bispo E da S, Guimarães AG, Miranda M S. Cacau e café e a aplicação de probióticos e prebióticos. In: Saad SMI, Cruz Adriano, Gomes da; FARIA, José de Assis Fonseca. Probióticos e Prebióticos em Alimentos: Fundamentos e Aplicações Tecnológicas. São Paulo: Editora Varela, 2011. p. 583-598.

45. Cuenca MM, Quincazán MC. Comparación de la fermentación de bebida de soya e leche de vaca utilizando un cultivo láctico comercial. Ingeniería y competitividad. 2004; 5(2)16-22.

46. Cruz NS, Capellas M, Jaramillo DP, Trujillo AJ, Guamis B, Ferragut V. Soymilk treated by ultra high-pressure homogenization: Acid coagulation properties and characteristics of a soy-yogurt product. Food Hydrocolloids. 2009; 23(2) 490-496.

47. Beasley S, Tuorila H, Saris PEJ. Fermented soymilk with a monoculture of Lactococcus lactis. International Journal of Food Microbiology. 2003; 81(2) 159–162.

48. Favaro Trindade CS, Terzi SC, Trugo LC, Dello Modesta RC, Couri S. Development and sensory evaluation of soy milk based yoghurt. Archivos Latinoamericanos de Nutrición. 2001; 51(1) 100-104.

49. Baú TR. Desenvolvimento, caracterização e estabilidade de produto de soja fermentado com cultura de kefir e adição de fibras. Dissertação (Mestrado em Ciência de Alimentos). State University of Londrina. Brazil; 2012.

50. Fucks RHB, Borsato B, Bona E, Hauly MC. O. Iogurte de soja suplementado com oligofrutose e inulina. Ciência e Tecnologia de Alimentos. 2005; 25(1) 175-181.

51. De Angelis M, Coda R, Silano M, Minervini F, Rizzello CG, Di Cagno R, Vicentini O, De Vincenzi M, Gobbetti M. Fermentation by selected sourdough lactic acid bacteria to decrease coeliac intolerance to rye flour. Journal of Cereal Science 2006; 43 301– 314.

52. Casellas F, et al. Factors impact health-related quality of life in adults with celiac disease: A multicenter study. World Journal of Gastroenterology. 2008; 14(1) 46-52.

53. Araújo HMC, Araújo WMC, Botelho RBA, et al. Doença celíaca, hábitos e práticas alimentares e qualidade de vida. Revista de Nutrição. 2010; 23(3) 467-474.

54. Sundar, N.; Crimmins,R.; Swift,G.L. Clinical presentation and incidence of complications in patients with coeliac disease diagnosed by relative screening. Postgraduate Medical Journal. 2007; 83, 273–276.

55. Abrams S, Griffin I, Hawthorne K, et al. A combination of prebiotic short- and long- chain inulin- type fructans enhances calcium absorption and bone mineralization in young adolescents. The American Journal of Clinical Nutrition. 2005; 82(2) 471-476.

56. Abrams S, Hawthorne K, Aliu O, et al. An inulin-type fructan enhances calcium absorption primarily via an effect on colonic absorption in humans. Journal of Nutrition. 2007; 137(10) 2208-2212.

57. Capriles Martini LA, Arêas JAG. Metabolic osteopathy in celiac disease: importance of a gluten-free diet. Nutrition Reviews. 2009; 67(10) 599-606.

58. Dias Capriles V, Gomes Arêas J A. Archivos Latinoamericanos de Nutricion. 2010; 60(3).

59. Packer SC, Dornhorst A, Frost GS. The glycaemic index of range of gluten-free foods. Diabetic Medicine. 2000; 17(9) 657.

60. Dukas L, et al. Bowel movement, use of laxatives and risk of colorectal adenomatous polyps among women. United States. Cancer Causes Control. 2000; 11 907-914.

61. Haque A, Morris ER, Richardson RK. Polysaccharide substitutes for gluten in non- wheat bread. Carbohydrate Polymers. 1994; 25(4) 337-344.

62. Zandonadi RP. Psyllium como substituto de Glúten. Dissertação (Mestrado em Nutrição Humana) – University of Brasilia. Brazil; 2006.

63. Storck CR, Pereira JM, Pereira GW, et al. Características tecnológicas de pães elaborados com farinha de arroz e transglutaminase. Brazilian of Food Journal Technology. II SSA; 2009.

64. Figueira F da S. Pão sem glúten enriquecido com a microalga Spirulina platensis. Brazilian Journal of Food Technology. 2011; 14(4) 308-316.

65. Fox RD. Spirulina Production & Potencial. Aix-en-Provence: Edisud. 1996; 89.

66. Cohen Z. The chemicals of Spirulina. In: VONSHAK, A. Spirulina Platensis (Arthrospira) Physiology, Cell-Biology and Biotechnology. London: Taylor & Francis, 1997. 233 p.

67. Colla LM, Reinehr CO, Reichert C, Costa JAV. Production of biomass and nutraceutical compounds by Spirulina platensis under different temperature and nitrogen regimes. Bioresource Technology. Oxford. 2007; 98(7) 1489- 1493.

68. Morais MG, Miranda MZ, Costa JAV. Biscoitos de chocolate enriquecidos com Spirulina platensis: características físico-químicas, sensoriais e digestibilidade. Food and Nutrition. 2006; 17(3) 323-328.

69. Berti C, Riso P, Montini LD, Porrini M. In vitro starch digestibility and in vivo glucose response of gluten-free foods and their gluten counterparts. European Journal of Nutrition. 2004; 43 198-204.

70. Vecchi D, Hamsananda SS. Kefir: "uma benção milagrosa para nosso século". 1999. http://www.angelfire. com/ab6/om/kefir.html (Accessed 13 july 2012).

71. Silva SMCS, Mura JDPM. Tratado de alimentação, nutrição e dietoterapia. São Paulo: Rocca; 2007. p515-533.

72. Fasolin LH, Almeida GC, Castanho PS, Netto-Oliveira ER. Chemical, physical and sensorial evaluation of banana meal cookies. Science and Food Technology. 2007; 27(3) 787-792.

73. Zandonadi RP, Botelho RBA, Gandolfi L, et al. Journal of the Academy of Nutrition and Dietetics. 2012; 112(7) 1068-1072.

74. Moore MM, Juga B, Schober TJ, et al. Effect of lactic acid bacteria on properties of gluten-free sourdoughs, batters, and quality and ultrastructure of gluten-free bread. Cereal Chemistry. 2007; 84 357-364.

75. Poutanen K, Flander L, Katina K. Sourdough and cereal fermentation in a nutritional perspective. Food Microbiology. 2009; 26 693-699.

76. Galle S, Schwab C, Dal Bello F, Coffey A, Ganzle MG, Arendt EK. Influence of in-situ synthesized exopolysaccharides on the quality of gluten-free sorghum sourdough bread. International Journal Food Microbiology. 2012; doi:10.1016/j.ijfoodmicro. 2012.01.009.

77. Franco TS, Garcia S, Hirooka EY, Ono ES, dos Santos JS. Lactic acid bacteria in the inhibition of Fusarium graminearum and deoxynivalenol detoxification. Journal of Applied Microbiology. 2011; 111(3) 739-748.

78. Guergoletto KB, Kuriama F, Kuroda EK, Garcia S, Hirooka EY. Potencial of degradation of microcystins by probiotic bacteria and kefir grains. In: 4th International Congress on Bioprocesses in Food Industries, 2010, Curitiba, Brazil 4th International Congress on Bioprocesses in Food Industries, 2010b.

79. Morotomi M, Mutai M. In vitro binding of potent mutagenic pyrolyzats to intestinal bacteria. Journal of the National Cancer Institute. 1986; 77 195-201.

80. Sako T. The world's oldest probiotic: perspectives for health claims. In: Probiotic and health claims. Ed. Kneifel, W & Salminen, S. Blackwell Publishing Ltd, p.17, 2011.

81. Chowdhury SR, Basak RK, Sen R, Adhikari B. Optimization, dynamics, and enhanced production of a free radical scavenging extracellular polysaccharide (EPS) from hydrodynamic sediment attached Bacillus megaterium RB-05. Carbohydrate Polymers. 2011; 86, 1327– 1335.

82. Donot F, Fontana A, Baccou JC, Schorr-Galindo S. Microbial exopolysaccharides: Main examples of synthesis, excretion, genetics and extraction. Carbohydrate Polymers. 2012; 87 951– 962.

83. Piermaria J, de la CANAL M, Abraham A. Gelling properties of Kefiran, a food grade polysaccharide obtained from kefir grain. Food Hydrocolloids. 2008; 22 1520-1527.

84. Prasanna PHP, Grandison AS, Charalampopoulos D. Screening human intestinal Bifidobacterium strains for growth, acidification, EPS production and viscosity potential in low-fat milk. International Dairy Journal. 2012 23 36-44.

85. Laurenti E. Materiais encapsulantes naturais na obtenção de esferas de S. cerevisiae para incorporação em ração extrusada de frangos de corte. Dissertação de Mestrado em Ciência de Alimentos, Estadual University of Londrina. Brazil, 2011.

86. Liu J, Luo J, Ye H, Sun Y, Lu Z, Zeng X. Production, characterization and antioxidant activities in vitro of exopolysaccharides from endophytic bacterium Paenibacillus polymyxa EJS-3. Carbohydrate Polymers. 2009 78 275–281.

87. Pan D, Mei X. Antioxidant activity of an exopolysaccharide purified from Lactococcus lactis subsp. lactis 12. Carbohydrate Polymers. 2010; 80 908–914.

88. World Health Organization (2011). New physical activity guidance can help reduce risk of breast, colon cancers: http://www.who.int/mediacentre/news/notes/2011/ world_cancer_day_20110204/en/index.html

89. Hamilton MT, Hamilton DG, Zderic TW. Role of low energy expenditure and sitting in obesity, metabolic syndrome, type 2 diabetes, and cardiovascular disease. Diabetes. 2007; 56(11) 2655–2667.

90. Bhatt DL. Anti-inflammatory agents and antioxidants as a possible "third great wave" in cardiovascular secondary prevention. America Journal of Cardiology. 2008; 101 (10A) 4D-13D.

91. Flambard B, Johansen E. Developing a functional dairy product: from research on Lactobacillus helveticus to industrial application of Cardi-04™ in novel antihypertensive drinking yoghurts. In: Functional Dairy products Vol 2 Ed. Saarela, M. CRC Press, Cambridge, England, 2007, p.506. ISBN 978-1-84569-310-7 (e-book).

92. Badel S, Bernardi T, Michaud P. New perspectives for Lactobacilli exopolysaccharides. Biotechnology Advances. 2011; 29 54-66.

93. Rajul J, Rao CV. Diosgenin, a Steroid Saponin Constituent of Yams and Fenugreek: Emerging Evidence for Applications in Medicine. In: BIOACTIVE COMPOUNDS IN PHYTOMEDICINE Ed. Rasooli, I.; Croatia, Intech 2011 p. 127

94. Hobbs DA, Kaffa N, George TW, Methven L, Lovegrove JA. Blood pressure-lowering effects of beetroot juice and novel beetroot enriched bread products in normotensive male subjects.British Journal of Nutrition, page 1 of 9 doi:10.1017/S0007114512000190.

95. Webb AJ, Patel N, Loukogeorgakis S, et al. Acute blood pressure lowering, vasoprotective, and antiplatelet properties of dietary nitrate via bioconversion to nitrite. Hypertension. 2008; 51 784–790.

Chapter 9

QUALITY MANAGEMENT: CRITICAL CONSIDERATIONS FOR THE FOOD INDUSTRY

1. INTRODUCTION: AN OVERVIEW

Indeed, the emergence of globalization has intensified market competition by facilitating the entry of new competitors. This does not inherently signify a threat to the survival of local enterprises, but rather a problem that they must contemplate. This challenge pertains to the necessity of fostering enhanced consumer loyalty to products and services, improving product alignment with consumer wants, and increasing awareness of the company's social impact. This global context gives chances for corporations to engage in new markets. This action will primarily depend on the quality of their products and services.

Nevertheless, the notion of product quality is not immediately apparent or self-evident. While not widely acknowledged, the definition of quality that enjoys broader support is "suitability for consumer use." This definition is thorough as it encompasses two elements: attributes that contribute to product satisfaction and the lack of deficiencies. The primary element comprises the qualitative attributes of the product features that fulfill consumer needs, hence ensuring satisfaction. These needs pertain not only to the inherent attributes of the product, such as its sensory qualities, but also to its market availability at a competitive price and in appropriate packaging. The other aspect is the lack of defects, which pertains to the product's attributes in accordance with its specifications, instilling confidence in the consumer regarding the product's reliability; that is, the consumer is assured of obtaining a safe product, devoid of health hazards, and possessing the properties asserted on the label.

Achieving these objectives necessitates effective quality management, which entails ongoing improvement initiatives at every operational level and throughout all functional areas of the company. Quality management integrates commitment, discipline, and a collective effort from all participants in the manufacturing process, alongside essential management and administrative approaches, aimed at the ongoing enhancement of all processes. To achieve this, industries must be organizationally structured, implement policies and quality programs, assess customer satisfaction, and employ additional quality tools and methodologies. Particularly within the food business, it encompasses the understanding and implementation of techniques and processes for product safety.

The objective of this chapter is to delineate the prospective application of quality instruments within food enterprises. The study aims to contextualize quality management within the food business and its associated quality function activities. Furthermore, quality control support tools will be proposed, accompanied by actual application examples.

2. EVOLUTION OF QUALITY MANAGEMENT: A CONCISE HISTORY

It can be said that each company has a particular stage of maturity on the issue of quality management. In general they tend to evolve in four stages, the similarity of ages or how the quality management in the world has evolved over the years. Thus, it is important to highlight these stages of evolution of quality that began with the inspection of products, have passed the statistical quality control, the stage of quality systemic management until the strategic quality management.

Garvin, a scholar of quality management, highlights four ages or stages through which the way to manage the quality has evolving over time in the U.S [1]. The first stage of development was called "era of inspection." In this stage the quality control of products was limited to a focus on corrective inspection, i.e., was a way to check the uniformity of the final product by separating the non-conforming products. According to Garvin in the U.S. only in 1922 the inspection activities were related more formally with quality management, after the publication of the book "The Control of Quality in Manufacturing". For the first time, the quality was seen as managerial responsibility having distinct and independent function in the companies.

Later, the year of 1931 was a milestone in the quality movement and the beginning of the second phase, the Statistical Quality Control. This phase had a preventive approach, centered on the monitoring and control of process variables that could influence in the final product quality through the development of statistical tools for sampling and process control.

The next phase was called Quality Assurance, that was associated with broader control and prevention, which sought through systematic management, ensure quality at all stages of obtaining the product. The quality management became a practice restricted to industrial production management applied to all production support functions. In the U.S., this time started in the late 50's when the quality of the instruments have expanded far beyond the statistics, now covering the quantification of quality costs, total quality control, reliability engineering and zero defect.

Finally, quality management has been incorporated within the strategic scope of organizations, this phase called Strategic Management of Quality. It represented a vision of market- oriented management, i.e., with a view of opportunities before the competition and customer satisfaction, where market research has become more important for evaluating the market needs and how the competition stands. The strategic approach is an extension of its predecessors, but with a more proactive approach.

Several scholars of quality management are unanimous in emphasizing that the companies in general, and also the food industry, through its organizational structure, the policies adopted, the focus given to the business and the practice of quality control, demonstrate a certain degree of maturity in how to manage quality. Some companies may present practices related to more advanced stages, mature, such as quality assurance and strategic quality management, others may prove more practices related to inspection and process control. Through observation of tools and methods currently adopted in the food industry, it can be inferred that this quality management company is based on the characteristics of a particular stage of the quality evolution.

For example, the control of the raw material and products for inspection, with special attention to satisfy the governmental health rules, is a characteristic of the inspection stage. Likewise, the product control only by laboratory analysis is a feature of this stage. Moreover, quality control practices in process, application of statistical methods for quality control and the adoption of Good Manufacture Practices (GMP) and Hazard Analysis and Critical Control Points (HACCP) denote that the company has a slightly broader approach that inspection, i.e., a more preventive approach control in the production process. But when practice inspection and process control are well established in the company and efforts are directed towards continuous improvement, it can be inferred that the company is evolved into a system of quality assurance. Practices consistent with this era are shown by performing quality audits in different sectors of the company, adoption of quality systems across the supply chain and also implementation of programs for the development of quality suppliers of products and services. Companies that take a strategic quality management are those that use market research and specific indicators to measure customer satisfaction, such as consumer complaints, returns by wholesalers for the time of the product in the inventory and sales below target. Further, evaluate their products compared to competitors' products and apply techniques of sensory analysis to compare products and find sensory qualities required by the market. Concerned to improve their production processes, automate production lines and constantly launch new products into the market.

3. TASKS QUALITY OF THE SECTOR IN THE FOOD INDUSTRY

In general, the operating system of quality control in the food industry must meet some specific tasks. One of the tasks is to ensure compliance with sanitary standards and compliance requirements of the legislation, including with regard to food safety standards, the Good Manufacturing Practices (GMP) and the system Hazard Analysis and Critical Control Points (HACCP). For this, there is need for procedures to control insects, rodents, birds and other pests, and procedures for cleaning and sanitizing equipment, industrial plant and storage areas. Still, personal hygiene of staff working on process lines and proper habits on food handling should be implemented and monitored to ensure that food safety standards are met. In cooperation with the departments of production, research and development, engineering or operations, the department of quality control analyzes manufacturing processes to "Hazard Analysis and Critical Control Points." The integrity and safety of food products should be ensured through the identification and assessment of all unit operations of the process in order to prevent potential contamination and adulteration that could expose consumers to health risks.

In cooperation with the department of research and development (R&D), production, purchasing and sales, should be prepared written specifications for raw materials, ingredients, packaging materials, other supplies and finished products. Furthermore, should be established in writing form and in cooperation with the departments of production and R&D the procedures for each unit operation of all manufacturing processes of the fashion industry that can be implemented in processing lines. The participation of staff from other departments of the company occurs by the virtue of their expertise in relation to consumer demands or knowledge of product technology and process, and the participation of the operators of the process, because of its experience in the production.

The quality control personnel works in different laboratories performing physical, chemical, microbiological and sensory properties of raw materials, ingredients, packaging materials and finished products. They also work in the factory or processing areas, collecting samples for performance evaluation processes, unit operations, sanitary conditions or levels, verifying compliance with the requirements of food safety and all other operating specifications. It is the responsibility of the department of quality control implementation of Statistical Quality Control (SQC), in which statistical techniques are applied to assessments of control for scientific analysis and interpretation of data. The SQC's functions include the selection of sampling techniques, control charts for attributes and variables, the use of analysis of variance and correlation, among other statistical tools. The methods, procedures and selection of instruments used to measure quality attributes of products and processes are the responsibility of the department of quality control. These techniques can be developed for specific purposes within the production process, to product development or troubleshooting and optimization standards.

The quality control personnel must interact cooperatively with the personnel of the standards and inspection agencies to ensure that the official food law is understood and met. It should also watch the production department in its efforts to increase revenues, reduce losses and improve efficiency of operations. It should also develop, conduct and assist in an organized program, training of supervisors, operators and workers in general, into specific concepts of quality.

The development of an appropriate plan of "recollect" adulterated or defective product in marketing channels and the planning of internal traceability of products is also a function of the quality control department. Another assignment of quality control includes reviewing and responding to consumer complaints.

Thus, faced with so many responsibilities, it remains to note that the dynamics of intervention and performance of those who are responsible for the quality department is paramount to the success of the food industry and customer satisfaction.

4. METHODOLOGIES IN SUPPORT OF THE QUALITY MANAGEMENT IN THE FOOD INDUSTRY

The quality management applies systems and tools that are intended to assist the implementation of quality-oriented way to improve the product and the process, increasing the levels of quality business and ensuring customer's satisfaction.

The purpose of this topic is to describe some tools, techniques and systems that have been more widely used in quality management in the food industry. Besides the methods mentioned, there are others that could be employed by companies. The choice of which implement depends on the company's strategies and know-how of its employees.

4.1 Food security programs

The issue of food safety has been in the public eye as never before. Foodborne disease has an enormous public health impact, as well as significant social and economic consequences. It is estimated that each year foodborne disease causes approximately 76 million illnesses, 325,000 hospitalizations and 5,000 deaths in the U.S., and 2,366,000 cases, 21,138 hospitalizations and 718 deaths in England and Wales [2]. Thus, many food safety programs have been published in order to ensure safe food production and consumer protection.

Safety food programs can be set as the measures to be taken to ensure that food can be eaten without adversely affect to the consumer's health. These measures aim to prevent food contamination, such contamination are chemical, physical or microbiological. The programs commonly used in this area are Good Manufacturing Practices (GMP), Hazard Analysis and Critical Control Points (HACCP), British Retail Consortium (BRC) and Global Food Safety Initiative (GFSI), frequently found in the food industry, are obligatory by law, and others are implemented voluntarily by the food chain members [3].

4.1.1 Good Manufacturing Practices (GMP)

The Good Manufacturing Practices program is composed of a set of principles and rules to be adopted by the food industry in order to ensure the sanitary quality of their products. The GMP program came at the end of the last century when the U.S. pharmaceutical industry began to define optimal manufacturing practices based on technological knowledge available. In the late 60's, organizations such as the WHO (World Health Organization) and the Food and Drug Administration of the United States, the FDA (Food and Drug Administration) adopted the program as a minimum criterion recommended to the manufacture of food products under adequate sanitation conditions and routine inspection. Later in 2002, FDA forms Food GMP Modernization Working Group and announces effort to modernize food GMP´s [4].

The rules establishing the so-called Good Manufacturing Practices involves requirements for industry's installations, through strict rules of personal hygiene and cleanliness of the workplace to the description in writing form of all procedures involved in the product. These standards are characterized by a set of items summarized below.

The projects and industry facilities, in addition to requirements engineering/ architecture, must meet requirements to ensure food safety, such as the installation of devices to prevent the entry of pests, contaminated water, dirt in the air, and still be designed to avoid the accumulation of dirt or physical contamination of food that is being manufactured. The equipment and the entire apparatus of materials used in industrial processing should be designed from materials that prevent the accumulation of dirt and must be innocuous to avoid the migration of undesirable particles to foods. On the production line, the procedures and steps for handling the product have to be documented, in order to ensure the standardization of safety practices. Also running records should be implemented as evidence that the job was well done.

Otherwise, the cleaning and sanitizing phases are inherent to the processing and handling of foods, and thus programs for execution on a routine and efficiently must be implemented. Similarly, is required a plan for integrated pest control in order to minimize access vector and reduce the number of possible focus of insects, rodents and birds.

Regarding food handlers, the GMP recommend that training should be given and recycled so the concepts of hygiene and proper handling are assimilated as a working philosophy and fulfilled to the letter.

A control of raw materials should be developed with suppliers, not only in the laboratory, but in a gradual and continuous improvement work, where food security is split with suppliers. Guidelines for the safe packaging of raw materials, inputs and finished products should be followed and extended to the storage and loading area, and to the transportation that reach the consumer.

The Good Manufacturing Practices have wide and effective application when all the elements cited are effectively deployed.

4.1.2 Hazard Analysis and Critical Control Points (HACCP)

HACCP is a system based on prevention of hazards to the industry to produce safe food to consumers. The HACCP involves a complete analysis of the dangers in the systems of production, handling, processing and consumption of a food product. HACCP is widely acknowledged as the best method of assuring product safety and is becoming internationally recognized as a tool for controlling food-borne safety hazards [3].

In short, this system has a systematic and scientific approach to process control, designed to prevent the occurrence of failures, ensuring that the controls are applied in processing steps where hazards might occur or critical situations. For this, the HACCP system combines technical information updated with detailed procedures to evaluate and monitor the flow of food into an industry.

The new sanitary requirements and quality requirements dictated by the main international markets, led since 1991, to the deployment experimental stage of the HACCP. There are new rules governing the international market, established during the Uruguay Round of Trade Negotiations and applicable to all member countries of the World Trade Organization (WTO). The Codex Alimentarius has become the regulatory body for matters of hygiene and food safety in the WTO. The Codex Alimentarius reflects an international consensus regarding the requirements for protection of human health in relation to the risks of foodborne illness. This measure is accelerating the process of harmonization of food laws of the countries, process that is oriented concerning food security, with the recommendation of the use of the system Hazard Analysis and Critical Control Point, to ensure food safety.

Generally the HACCP system initially involves the creation of a multifunctional team, supported by senior management of the company, and the characterization of all food products that will be included in the system. Also a set of programs, such as Good Manufacturing Practices (GMP) and Sanitation Standard Operating Procedures (SSOP) are universally accepted as prerequisites for the implementation of the HACCP system and therefore should be consolidated. Only then each step of the production process of a product will be analyzed for the possibility of a chemical, physical and microbiological contamination. Thereafter preventive measures are described and identified the Critical Control Points (CCPs). For each critical point is necessary to establish critical control limits, which allow the monitoring of hazards. As there is always a possibility of failure, it is essential to provide corrective measures in order to ensure the process return into a controlled situation. It should also establish procedures for verification of CCP´s and their respective records. After the HACCP plan drawn up, it is validation occur through discussions among team members [5].

Finally, the HACCP plan is disseminated to the production employees and for those responsible for assessing the products quality on the factory floor. Internal and external audits are recommended for periodic maintenance and continuous improvement of the system [5].

4.2 Standardization of processes

Standardization is a management tool involved in the preparation, training and control standards within the company. Such standards are documents containing technical specifications or specific criteria that will be used as a guide in order to ensure that products, processes and services are designed with quality [6]. The main objective of a program of standardization for the food industry is to minimize the variations in quality of production. For this, it is necessary to provide means to standardize both the operational and analytical procedures, as raw materials, machinery and equipment used in the manufacturing process.

The patterns are instruments that indicate the goal and procedures for accomplishment of the work and can be classified as follows:

- Standards of Quality (SQ): refer to the parameters related to quality of products, raw materials and inputs.
- Operation Standards: describe the manufacturing process of a product, the technical parameters of control by the operators and operating procedures. These are divided into Standard Process Technician (SPT) and Operational Procedure (OP). The first document describes the process of manufacture of a product, the quality characteristics and the control parameters. Operating procedures standards are prepared by managers and operators to achieve the objectives proposed in the SPT and SQs.
- Standards Inspection: describe methods and criteria for assessing the degree of success achieved in carrying out an activity, compared to planned levels of quality for the product. The inspection may occur in the process, the finished product and in the raw material.

Through standardization it is achieved greater standardization of products, improved productivity and product quality, cost reduction, simplification and optimization of production processes, increase the technical capacity of operators of process, greater job security, reduction of inventory levels of raw materials and inputs, reducing the preparation time of the machines and self-management by the workers.

Also noteworthy is that the patterns facilitate the transfer of knowledge since all the people and functional units involved in a particular pattern should collaborate, as far as possible, be trained in their preparation and for their use.

4.3 PDCA cycle

The PDCA originated in the 30´s in the laboratories of the United States, becoming known in the fifty decade due to the expert quality, Deming, who was responsible for implementing and disseminating tools of control and quality management in several countries. The PDCA cycle is a method of managerial decision-making to ensure the achievement of goals related to a process, product or service [7].

The letters that form the acronym PDCA mean Plan, Do, Check, Action. The Plan (P) consists in establishing goals, and procedures to achieve them. The stage Do (D) consists in performing the tasks as planned and collect data that will be used in the step control. Thus, in the stage of "implementation" are essential trainings at work. Check (C) consists of comparing the results achieved with the planned goals through quality control tools. Finally, Action (A) is to act correctively in the process in order to correct an unexpected result.

As can be seen in Figure 1, a schematic representation of PDCA cycle translates the dynamism steps purposes. The conclusion of a turn in the cycle continues back to the beginning of the next cycle, and so on. Following in the spirit of continuous quality improvement, the process can always be renewed and a new change process can be started. Continuous improvement occurs the more times the PDCA cycle is run, and optimizes the execution of processes, enables cost reduction and increases productivity. Moreover, the gradual and continuous improvements add value to the project and ensure customer satisfaction.

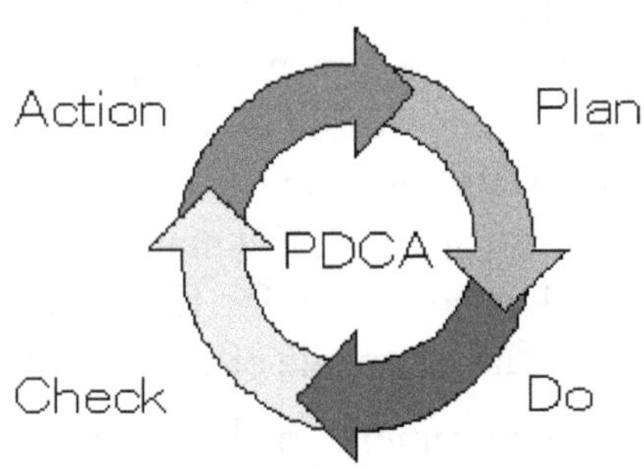

Figure 1. PDCA Cycle

In using the PDCA method may be necessary to use various tools, such as the basic tools for process control as stratification, check sheet, Pareto chart, cause and effect diagram, and scatter plots, histograms, control charts. Other techniques could include analysis of variance, regression analysis, design of experiments, process optimization, multivariate analysis and reliability [8].

Within the food industry, the PDCA cycle can be applied to the standardization or improvement of any product, process or activity the support the production, such as the standardization of procedures for cleaning and sanitizing, pest control, production processes, or improvement in the set-ups of equipment, reduction in losses in production, among others.

4.4 Traceability

The concept of traceability of products originated in the aeronautical and nuclear industries and it is widely practiced in industries. The tool aims to locate the source and the root causes of a particular problem of quality or safety, by the information recorded from a particular product, regardless of the stage of production where it is - whether raw material, in-process product or finished product. Through the traceability of products is possible to develop prevention and improvement actions, so that a specific problem does not occur again.

Traceability can cover only internal actions of the company, or otherwise, may be complete, when it involves the entire chain of production, allowing identifying even basic raw material that led to the final product and locations outside the company where finished products are stored. Consideration as the consumer safety, as the demands of the institutional environment and the costs of implementation of the traceability system will define the scope more suited to be deployed by the company.

4.5 Statistical quality control

The Statistical Quality Control uses statistical tools to control a product or process. To do this, it works with data collection and the interpretation thereof, acting as a fundamental tool to solve problems in critical product and process. Thus, ensures the quality sector the product conformity with the specifications defined as ensures the production sector the information needed for effective control of manufacturing processes providing subsidies to decision making in purchasing processes, receiving raw materials and shipment of products and also in reducing cost and waste. From the identification of the market requirements it is collected sufficient statistical information necessary for the development of new products and assists in monitoring the quality profile of competing products.

Although not a mandatory requirement in the food industry, statistical quality control can prove beneficial to organizations in the sector regardless of their particular specialism and size [9]. According Grigg, the initiatives of training of new graduates entering the industry in the principles of quality assurance and statistical methods and training the existing workforce and management in applying statistical control procedures to processes will make this methods more use of it than they are [9, 10].

The industrial statistic includes descriptive statistics, process capability analysis, measurement system analysis, basic graphics as histogram, scatter, box-plot, Pareto diagram, cause and effect, design of experiments, linear regression and correlation, multiple regression, hypothesis testing, confidence intervals, analysis of variance, analysis of process capability, among other tools [8]. It also covers the sampling techniques and control charts that will be described below, to be very useful to inspection and process control.

4.5.1 Inspection by sampling

The inspection process is to analyze or examine units of a product in order to verify with its quality characteristics are in accordance with technical or contractual specifications. Upon inspection of the product by sampling units are randomly selected to compose the sample batch. Depending on the number of defectives in the sample or the level of quality, that lot is accepted or rejected. Thus, sampling allows, by analysis of a small part of the whole or lot it is possible to draw conclusions about the rest not inspected. Therefore, in the sampling inspection an absolute conclusion about the quality of the lot will never be achieved, there is always a risk rate inherent in the sampling plan and dependent on its discriminatory power.

The current continuous improvement programs that evolve throughout the production chain, call for reducing the use of inspection techniques for the evaluation of the product or process, based on the idea that efforts should focus on "getting it right" in the first time and not in check it, then add value to the product, if it was done properly. However, these inspection techniques for acceptance have restored the importance of quality of audits.

There are two types of sampling plans, sampling plans by attributes and sampling plans by variables. The sampling rate by attributes consists in classifying units of a product just as acceptable or unacceptable based on the presence or absence of a particular feature in each unit qualitative inspected. The results of the inspection by attributes are expressed in terms of defective/not defective, conforming/ nonconforming. In the inspection by variable the characteristics or indicators of quality of the product unit are analyzed and the results are expressed by some continuous numeric scale. While inspection by attributes takes values from the set of integers, inspection by variable takes values in the set of real numbers [11, 12]. Upon inspection by attributes the probability of acceptance of the lot is based on Poisson Probability Distribution. The Poisson Probability Distribution is sometimes used to approximate the binomial distribution when the sample size (n) is too large and the proportion of defectives (p) is small. Otherwise, the use of sampling plans by variable assumes that the Normal Probability Distribution fits well with the distribution of the values of the quality characteristic under study.

Inspections by sampling can be used in finished products, raw materials, manufacturing operations, products in intermediate stages of processing, stored materials, among others. There are situations when only one plan by variable applies, for example, when the buyer will accept the product, but will pay different prices depending on the level of product quality. Also when the analysis result of the product will be expressed as quantitative values. For example, in the determination of chemical composition, weight, volume, and physical and rheological measurements. Therefore, measures such as pH, acidity by titration, soluble solids, fat, objective measurements of color and texture, among others, are typical of the sampling variable. The sampling by attributes can be implemented when it wanted to analyze a quality parameter in qualitative terms. Thus they are quite applied, for example, in visual analysis of packaging, the presence of dirt and physical damage in fruit and vegetables.

The following hypothesis test is linked to inspection for acceptance:

$$\begin{aligned} &H0: p = p0 \\ &H1: p > p0 \end{aligned} \tag{1}$$

Being "p" the proportion of defectives that the process produces. If the process is in control properly, this ratio is around p0 (hypothesis H0 true). The risk α, also known as producer's risk is likely rejection of a batch of a process whose average is equal to p0 defective, that is, the risk that the producer suffers as a result of inspection or analysis of sample can lead to a rejection of a good plot (which meets the specifications). The risk β, also known as consumer's risk is the probability of acceptance of a batch of a process in which the proportion of defectives is greater than p0, i.e., the result of inspection or analysis of the sample can lead to the acceptance of a batch inadequate; i.e., which does not meet the specifications [13].

A single sampling plan by attributes is defined by two parameters: sample size and acceptance number. The likelihood of acceptance of batches relates to the sample size, the severity in the acceptance criterion and the quality level of the products being analyzed in relation to the predetermined quality parameter [11]. In the sampling plans by variables, the probability of acceptance is related to the quality level of the product under examination and depends on the average of the quality parameter in question and its variability. It also depends on the severity criterion for acceptance of the lot [12].

Finally, it is worth noting that the Codex Alimentarius recommends the use of the ISO 2859 series relating to the procedures for sampling by attributes and the ISO 3951 series for the procedures for sampling by variables [14].

4.5.2 Control chart

The formal start of statistical process control occurred around 1924, when Shewhart developed and applied control charts at Bell Telephone Laboratories, a telephone company in the United States [1, 7, 13]. As in the entire production process variability occurs, Chart Control or Control Chart, or Map Control, aims to monitor these changes in processes, as well as to evaluate the stability of this process and eliminate or control the causes of variations. A Control Chart (Figure 2) consists of a Central Line (CL), is a pair of control limits: one above Upper Control Limit (UCL) and one below, Lower Control Limit (LCL), and characteristic values marked on the graph. If these values are within limits, without any particular trend, the process is considered under control. But if the points relate outside the control limits or submit an atypical arrangement, the process is judged out of control.

Variability in process may be classified into two types: the variability caused by random or common cause, which are inherent in the process and will be present even considered that this process is fully standardized. If only this kind of cause is acting in the process, it is said that the manufacturing process remains in statistical control. The other type of variability is caused by remarkable and special causes that arise sporadically due to a particular situation which causes the process to behave in a completely different way than usual, which can result in a displacement of the quality level. Thus, it is said that the process is out of statistical control.

The manufacturing control is exercised by the manufacturer during the industrialization process. The goal is to maintain the quality of the product satisfactorily uniform, preventing the production of items outside specification. The proofing that the process is in control or not is, made by examining unit samples taken periodically out of the production line. If the process is under control, samples that present variability corresponding to samples taken from a normal population, i.e., the variability is attributable only to product that is the sample. The "under control process" supposes, therefore, that the quality characteristic of all units produced has Normal Probability Distribution (Figure 3). Moreover, it also implies that this distribution remains stable, i.e., that its two parameters, medium (μ) and standard deviation (Σ), remain constant, which is verified by extracting a sequence of samples. So it is said that in a process under statistical control, the variability is attributed solely to random causes. These causes of variation do not cause appreciable variation in product quality; its elimination is impossible or anti-economical, and therefore, random causes are considered a natural part of the manufacturing process [8].

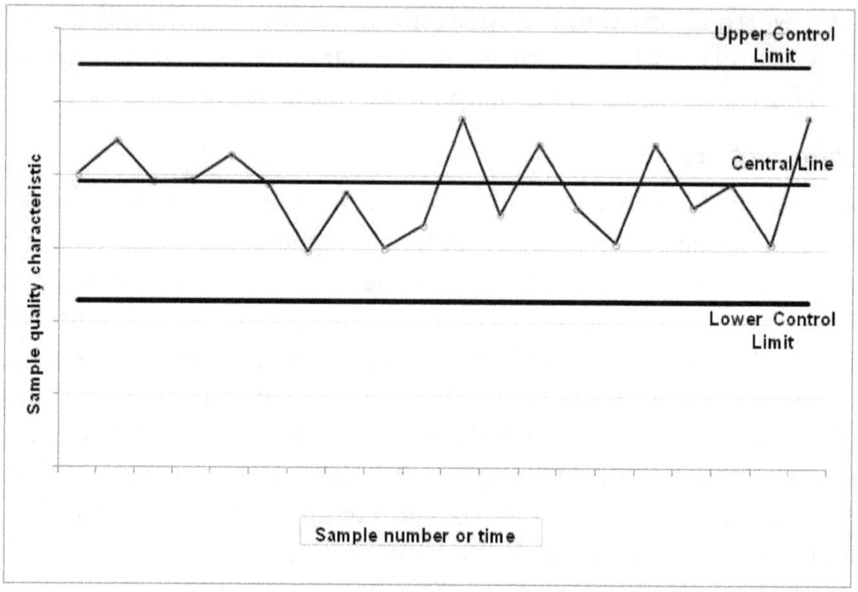

Figure 2. A typical control chart [8]

The Normal Distribution consists of an essential notion in statistical quality control rational. It is known that the items of a Normal Distribution (average μ and standard deviation Σ) are distributed around the average, approximately by the following proportions: 68% of the values in the range μ ± Σ, 95% in interval μ ± 2Σ and 99.7% in the range μ ± 3Σ. Consequently, differences between an observed value X and the average μ, greater than ± 3Σ are separated, three times to every 1000 observations, and therefore, the range of variability "normal" in the process under control is μ-3Σ and μ+3 Σ (Figure 3).

When the variability becomes "abnormal" changes in the quality characteristics of the product are sensitive. The causes of modification can be discovered and are therefore called "identifiable causes". These causes require prompt corrective action, in order to eliminate them. In these situations the samples indicate that the manufacturing process has changed and that the units were produced out of control. Some typical situations in process out of control occur when can be seen points outside the control limits. This is the clearest indication of lack of control of a process, which requires an immediate investigation of the cause of variation. Also can happened of points of the chart represent a trend, which consists of a continuous motion of the points of the control chart in one direction (ascending or descending). Also there is a configuration in sequence in several successive points of the control chart shown in only one side of the center line (eight or more consecutive points on one side of the center line). Another approach is the normality of the control limits, where 2 out of 3 consecutive points are outside the limits of 2Σ [8].

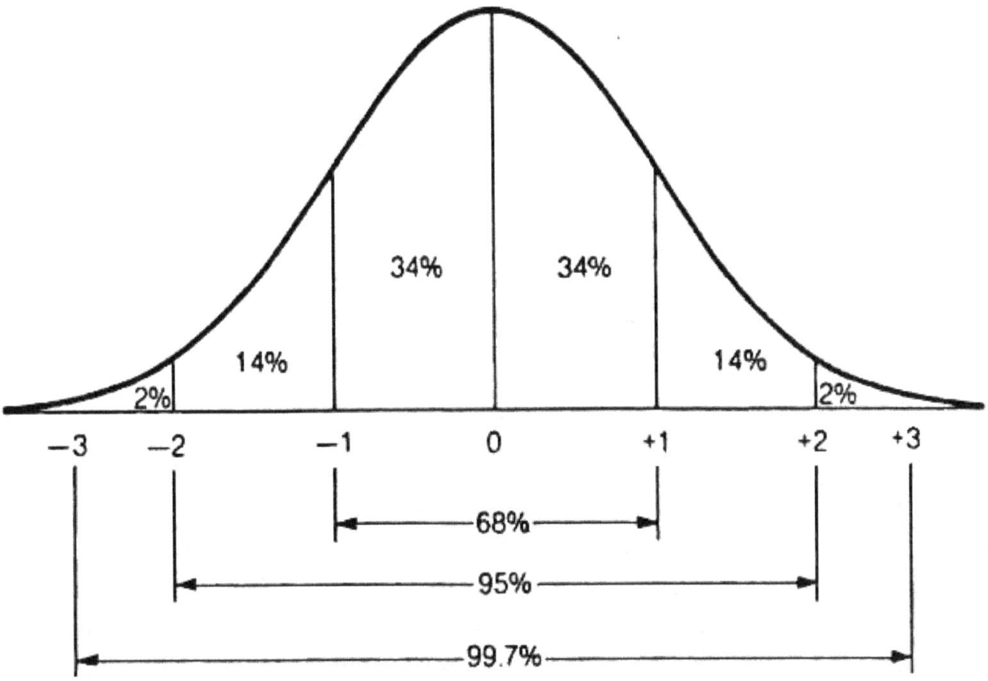

Figure 3. Scheme of Normal Probability Distribution

The food industry use control charts in different ways depending upon their level of maturity in statistical thinking [15]. In a survey conducted in UK food industry, revealed that while there are large differences in process types, quality priorities and key measures among different sub-sectors of the industry, the use of control charts was broadly similar. This generally extended to the use of control charts for recording or monitoring product net weight and volume data [15].

There are two types of quality control charts: control charts for variables and control charts for attributes, which will be described below.

4.5.2.1 Control charts for variables

Control charts for variables are named due to the fact that the quality characteristic being analyzed is expressed by a number on a continuous scale measures. Some examples of control charts are to yield a formulation, to verify the volume of a drink during their bottling, the soluble solids of a sweet after its cooking and the time to deliver a product to the customer.

Some control charts for variables most commonly used are: chart of the average (x), chart of amplitude (R), chart of standard deviation (s). When a quality characteristic of interest is expressed by a number on a continuous scale of measurement, the two control charts most used are the chart of the average (x) and a chart of variability (R or s). The two charts should be employed simultaneously.

Although the benefits of the application of control charts can be obtained in various situations of the food industry, the construction of the charts by variables will be exemplified by a typical situation of the food industry, in a packing operation. Imagine that a poultry slaughterhouse want to control the process of packaging of poultry cuts. In practice, the parameters average μ and standard deviation Σ are unknown and must be estimated from sample data. The procedure to estimate μ and Σ is to take m preliminary samples, each containing n observations of quality characteristic considered. These samples, known as rational subgroup should be taken when one believes that the process is under control and the operating conditions kept as uniform as possible. It is usual to consider m = 20 or 25 at least and n = 4, 5 or 6 [7,8].

The procedure for construction of the chart is:

1. Collect the data

Table 1 shows the values xij, weight of "j" cutting belonging to "i" sample, for 25 rational subgroup size of 4 (m = 25 and n = 4). Therefore, "i" varies from 1 to 25 and "j" from 1 to 4. The sections were collected when the machine was operating within normal procedure, i.e. no stops or apparent defects.

Table 1. Values of xij and Ri.

Samples	xi1	xi2	xi3	xi4	Ri
1	250,11	250,30	249,50	248,60	1,70
2	248,00	248,60	249,78	250,15	2,15
3	249,19	250,02	250,84	250,84	1,65
4	251,29	248,86	251,00	249,39	2,43
6	249,33	251,80	249,65	248,31	3,49
7	250,26	248,56	250,43	251,21	2,65
8	250,31	249,11	249,54	249,95	1,20
9	250,72	250,80	249,35	249,35	1,45
10	250,21	248,78	248,99	250,20	1,43
11	251,21	251,45	249,34	250,55	2,11
12	249,22	250,43	250,45	250,78	1,56
13	251,89	250,87	249,65	249,00	2,89
14	250,98	249,01	249,51	249,51	1,97
15	249,00	249,00	251,45	250,00	2,45
16	249,98	249,55	249,67	249,23	0,75

17	248,88	250,43	249,76	249,11	1,55
18	251,65	249,76	249,12	250,32	2,53
19	248,65	248,32	249,00	250,12	1,80
20	248,12	248.15	249,45	249,67	1,55
21	251,13	250,21	249,11	247,88	3,25
22	250,44	251,17	250,01	250,01	1,16
23	250,12	251,98	251,13	251,93	1,86
24	248.56	248.90	248,20	248,98	0,78
25	248,12	248,45	248,90	250,16	2,04

2. Calculate the amplitude of each sample Ri

$$Ri = \text{highest sample value} - \text{lowest value of the sample} \quad (2)$$

See the values of Ri in Table 1.

3. Calculate the average amplitude of the sample R

$$R = \frac{R_1 + R_2 + \ldots + R_m}{m} \quad (3)$$

Thus the value of R (average amplitude) is R = 1,93.

4. Establish the boundaries of the amplitude chart (Chart of R):

$$UCL = D4 \times R$$
$$CL = R \quad (4)$$
$$LCL = D3 \times R$$

The values of D4 and D3 are tabulated [7, 8]. Thus, D4 = 2,282 and D3 = 0.

Therefore:

$$UCL = 2,282 \times 1,93 = 4,41$$
$$CL = 1,93 \quad (5)$$
$$LCL = 0 \times 1,93 = 0$$

5. Build the chart of amplitude (Figure 4).
6. Analyze the chart.

Analyze the behavior of the points on the chart of amplitude and verify if the process is in statistical control. If necessary, recalculate the chart boundaries after the abandonment of the points there are out of control. Repeat this procedure until the control state is reached.

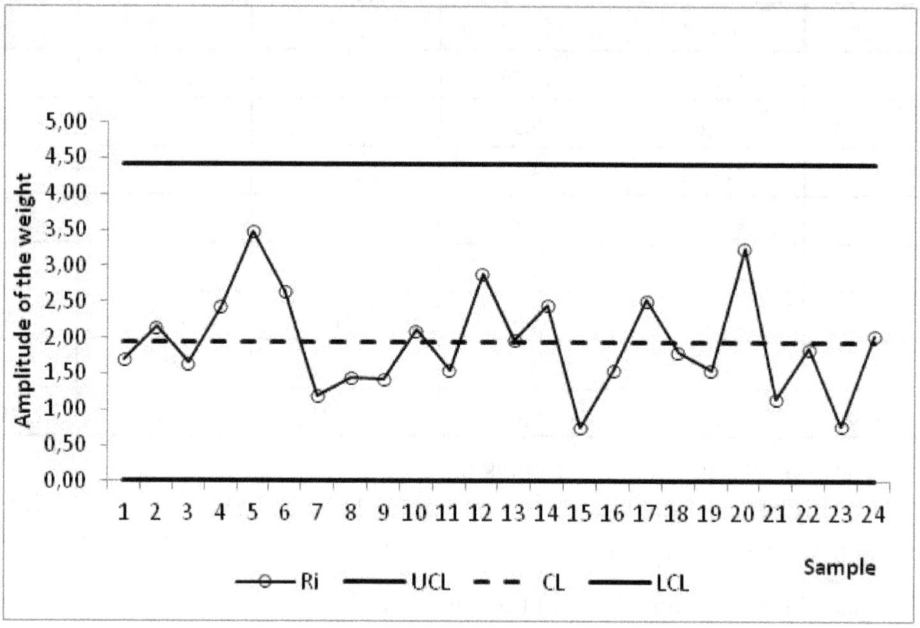

Figure 4. Chart of Amplitude (R) (25 points)

Analyzing the Figure 4, it can be seen that all points present within normal behavior. Now it is necessary to build the chart of average (x). To do this:

7. Calculate the average xi of each sample (Table 2).

$$x_i = \frac{x_{i1} + x_{i2} + \ldots + x_{in}}{n} \qquad (6)$$

8. Calculate the global averageX¯.

$$\overline{X} = \frac{x_1 + x_2 + \ldots + x_m}{m} = 249{,}83 \qquad (7)$$

9. Calculate the control limits of the chart average.

$$ULC = \overline{X} + A_2R$$
$$CL = \overline{X} \qquad (8)$$
$$LCL = \overline{X} - A_2R$$

The value of A2 is a constant tabulated [7, 8]. Thus, A2 = 0,729.X¯ is the average of averages and R is the average amplitude found in the last chart of amplitude. Thus:

$$ULC = 249{,}83 + 0{,}729 * 1{,}93 = 251{,}24$$
$$CL = 249{,}83 \qquad (9)$$
$$LCL = 249{,}83 - 0{,}729 * 1{,}93 = 248{,}42$$

10. Construct of the average chart (Figure 5).

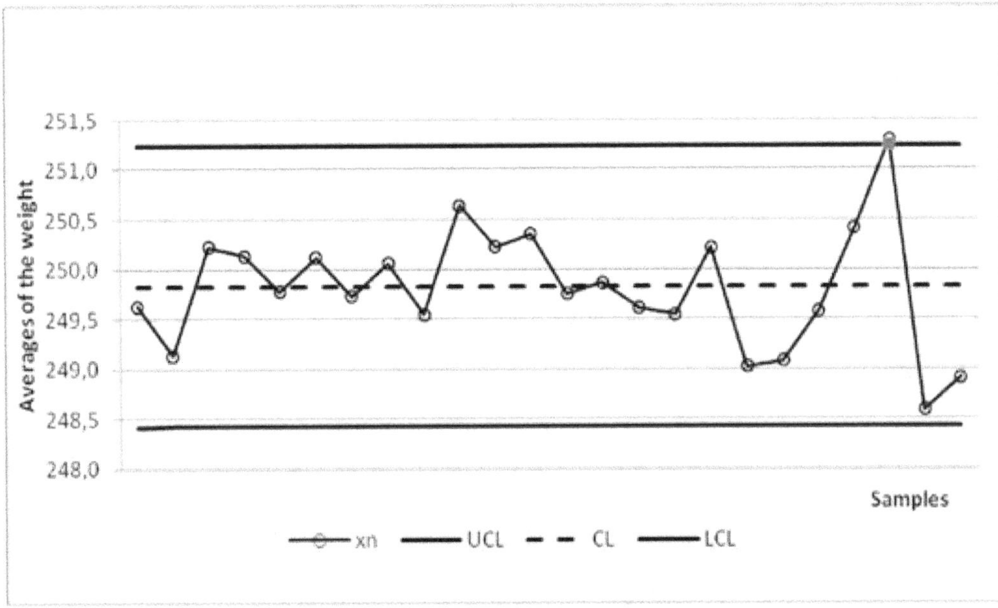

Figure 5. Chart of Average (x)

11. Interpret the chart of average built.

Analyze the behavior of the points on the average chart and whether the process is in statistical control. If necessary, recalculate the chart boundaries after the abandonment of the points there are out of control. Repeat this procedure until the control state is reached.

Analyzing the Figure 5, it can be seen that point 23 is above the UCL and therefore should be eliminated. The boundaries must be recalculated and a new chart of amplitude must be drawn (Figure 6).

New limits of the graph of the average (x) after removal of the subgroup 23.

$$ULC = 249{,}76 + 0{,}729 * 1{,}93 = 251{,}17$$

$$CL = 249{,}76 \qquad (10)$$

$$LCL = 249{,}76 - 0{,}729 * 1{,}93 = 248{,}36$$

Table 2. Values of xij and xn.

Samples	Xi1	Xi2	Xi3	Xi4	x_n
1	250,11	250,30	249,50	248,60	249,63
2	248,00	248,60	249,78	250,15	249,13
3	249,19	250,02	250,84	250,84	250,22
4	251,29	248,86	251,00	249,39	250,14

6	249,33	251,80	249,65	248,31	249,77
7	250,26	248,56	250,43	251,21	250,12
8	250,31	249,11	249,54	249,95	249,73
9	250,72	250,80	249,35	249,35	250,06
10	250,21	248,78	248,99	250,20	249,55
11	251,21	251,45	249,34	250,55	250,64
12	249,22	250,43	250,45	250,78	250,22
13	251,89	250,87	249,65	249,00	250,35
14	250,98	249,01	249,51	249,51	249,75
15	249,00	249,00	251,45	250,00	249,86
16	249,98	249,55	249,67	249,23	249,61
17	248,88	250,43	249,76	249,11	249,55
18	251,65	249,76	249,12	250,32	250,21
19	248,65	248,32	249,00	250,12	249,02
20	248,12	248.15	249,45	249,67	249,08
21	251,13	250,21	249,11	247,88	249,58
22	250,44	251,17	250,01	250,01	250,41
23	250,12	251,98	251,13	251,93	251,29
24	248.56	248.90	248,20	248,98	248,59
25	248,12	248,45	248,90	250,16	248,91

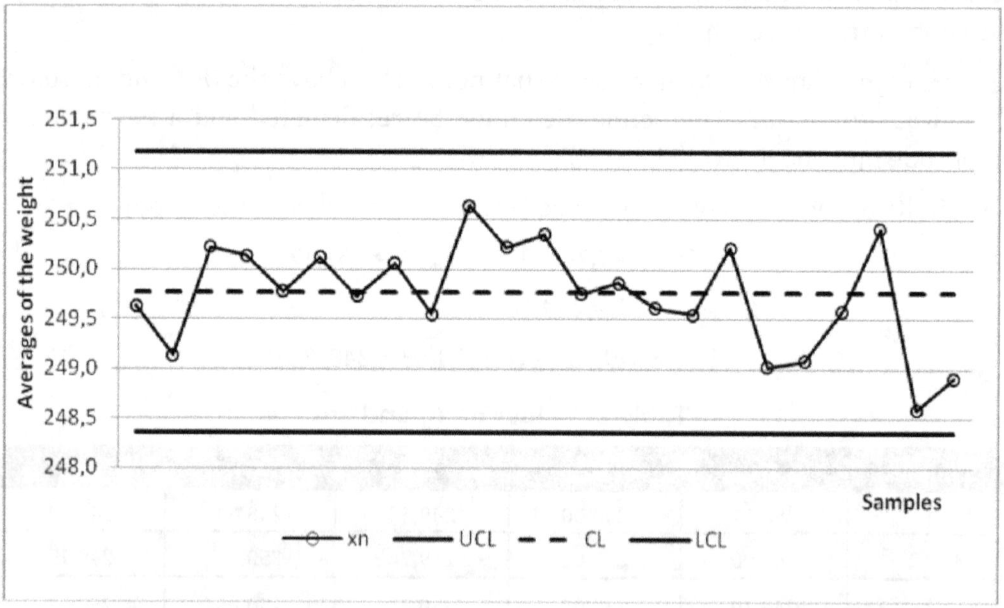

Figure 6. Chart of Average (x) (without the 23th subgroup)

12. Place the final charts of amplitude and average in the production line.

Note that for control of the packaging process of cuts of poultry, it chart has to be placed without padding, only with the UCL, CL and LCL, so that operators or responsible for quality control of packaging can monitor the process.

13. Periodically review the values of the control limits.

4.5.2.2 Control charts for attributes

It is not always by means of measurements that assess the quality of a product. For example, the color of a biscuit or of a sweet can be evaluated sensorially and the result is expressed as conforming or not conforming to a specified standard. Or, a PET bottle can be classified as not defective if it is whole in its structure or defective if it is crushed or broken.

Control charts for attributes can be: chart of the proportion of defective items (Chart p), chart of the total number of defects (Chart np), chart of number of nonconformities in the sample (Chart C) and the chart of number of nonconformities by inspection unit (Chart u) [8, 13].

Also here the construction of a chart for attributes will be exemplified. Suppose a manufacturer industry of biscuits decides to build a control chart p to visually check whether the product color after baking, was established as a standard for quality control. The number of defective products is presented in Table 3 and is important to note that the samples were numbered according to the date of production.

Table 3. Number of defective biscuits in samples of 100 units

Date	Nº. Biscuit Lot	inspectionated	Defective items (xi)	Proportion of defective items (p)
01/mai	1	200	7	0,035
02/mai	2	200	9	0,045
03/mai	3	200	4	0,02
04/mai	4	200	5	0,025
05/mai	5	200	6	0,03
06/mai	6	200	9	0,045
07/mai	7	200	5	0,025
08/mai	8	200	6	0,03
09/mai	9	200	6	0,03

10/mai	10	200	4	0,02
11/mai	11	200	6	0,03
12/mai	12	200	7	0,035
13/mai	13	200	4	0,02
14/mai	14	200	6	0,03
15/mai	15	200	7	0,035
16/mai	16	200	8	0,04
17/mai	17	200	8	0,04
18/mai	18	200	4	0,02
19/mai	19	200	7	0,035
20/mai	20	200	6	0,03

1. Collect the data

Collect m samples of size n. In general m = 20 or 25 at least. Collect the samples at successive intervals and record observations in the order they were obtained (Table 3).

2. Calculate the average proportion of defective items p (average).

$$\overline{p} = \frac{1}{mn}\sum_{i=1}^{n} X_i \qquad (11)$$

Xi is the number of defective items in the "i" sample.

3. Calculate the control limits.

$$\begin{aligned} UCL &= \overline{p} + 3\sqrt{\overline{p}(1-\overline{p})/n} \\ CL &= \overline{p} \\ LCL &= \overline{p} - 3\sqrt{\overline{p}(1-\overline{p})/n} \end{aligned} \qquad (12)$$

The LCL is not considered when the value is negative.

4. Draw the control limits. Mark left-hand vertical axis in the scale for horizontal axis p and the number of samples. Draw lines to represent full UCL, CL and LCL (Figure 7).

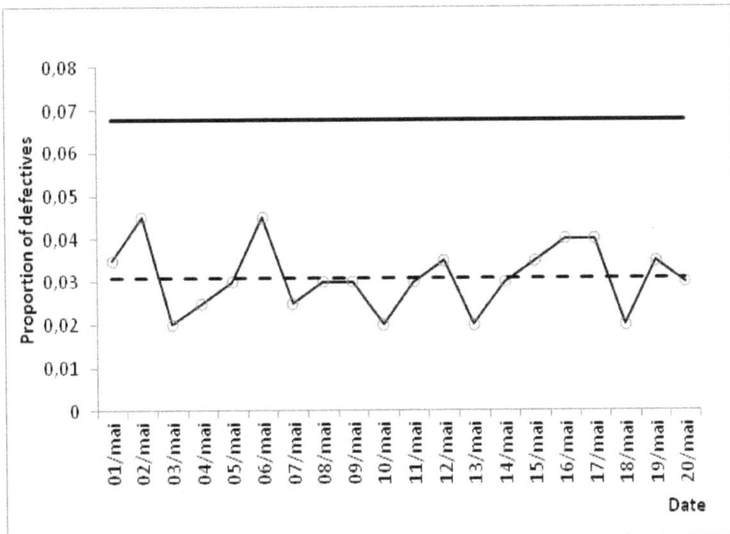

Figure 7. Chart p (proportion of defective products in the sample)

5. Mark the points on the chart.

Represent on the chart the m values of p (Figure 7).

6. Interpret the graph constructed.

To analyze the behavior points on the graph, and verify that the process is in statistical control. If necessary, recalculate the chart boundaries after the abandonment of the points there are out of control. Repeat this procedure until the control state is reached.

7. Check if the control state reached is appropriate to the process. If so, adopt the current control chart. Note that for control of the biscuit color, it chart has to be placed without padding, i.e., only with the UCL, CL and LCL.
8. Periodically review the values of the control limits.

5. QUALITY MANAGEMENT SYSTEMS

5.1 Total Quality Control (TQC)

In the '50s, the quality control was employed in Japan, by an intensive use of statistical techniques. However, the excessive emphasis on statistical techniques led to some problems, such as low interest shown by senior management of companies, by the quality control, which remained a movement of ground and plant, i.e., to engineers and workers [16].

In 1954 JUSE invited the engineer Juran, one of the masters of quality management, to deliver seminars to senior management. From the visit of Juran, the Quality Control came to be understood and used as an administrative tool, which represented the beginning of the transition of Statistical Quality Control for Total Quality Control as is currently practiced, involving the participation of all sectors and employees[16].

The quality management system proposed by the Japanese model shows how basic features to the participation of all sectors and all company employees in the practice of quality control, constant education and training for all levels of the organization, circles activity of quality control, audits, use of basic and advanced statistical techniques and national campaigns to promote quality control.

The TQC ideas developed by the Japanese were broadcast around the world, being this model capable of being deployed in companies of various sectors, with appropriate adjustments to the corporate culture.

5.2 ISO 9000 series

While the movement occurred in Japan by TQC, in Europe there was a movement around an organizational structure whose purpose was to develop standards for manufacturing, trade and communication in European countries for the increased levels of quality of activities. Thus, in 1947 the International Organization for Standardization was founded, based in Geneva, Switzerland. And in terms of quality control there was difficulty to unify standards that ensure that a product had been manufactured under quality criteria, after several trials, in 1989, was published the standard ISO 9000. The goal was to establish requirements for a quality management system, the implementation of which would extend to all types and business segments. The requirements of the series represented the consensus of different countries of the world.

More specifically, ISO 9001 deals with the requirements of the quality management system for an organization to produce compliant products and get customer satisfaction. Within the rules of the certification ISO 9001, there are specific requirements regarding the responsibility and involvement of management with the quality system, requirements for preparing and controlling of the documentation, for the critical analysis of contracts and selection of suppliers, to traceability and processes control, for measurement, for inspection and testing, analysis of nonconformities and for continuous improvement, for audits and training.

As ISO 9001 is a rule of general character it contains requirements to serve the most various sectors, it is necessary, once adopted by the food industry, some aspects can be considered in some cases insufficient. There's not in the standard, explicit references to the risks to consumer health, the safe products, the nutritional values, the critical control points, the good manufacturing practices. Food security can be seen as failures risk of deterioration and damage as a result of careless handling and storage inconvenient and not because of contamination and loss of sensory and nutritional values. Thus management systems for food safety have also been employed to address this need [17].

5.3 ISO 22000 series

Aiming to harmonize the international level, the various guidelines related to food safety systems, it was developed the ISO 22000:2005 - Food Safety Management systems - Requirements for any organization in the food chain. This applies the principles of a plan Hazard Analysis and Critical Control Points (HACCP) programs along with prerequisites, such as Good Manufacturing Practices (GMP) and Good Hygiene Practices (GHP). The standard has a similar format to the standard of ISO 9001 Quality Management. This similarity allows organizations to implement the specifics of food management system integrated to the quality management system. In this context the ISO 22000 presents as fact the benefits of being recognized internationally, to apply to all elements of the food chain and fill for the food sector, the gap between ISO 9001 and HACCP.

The ISO 22000 standards specifies the requirements to a safety management system that combines elements of food management system to ISO 9001 templates, as already said, and interactive communication, since communication along the supply chain is essential to ensure that all relevant safety hazards of food are identified and controlled. Finally, through concrete measures, tangible and that can be checked in audits, ISO 22000 combines the HACCP plan with prerequisite programs (PRP), since they are keys to an effective management system of food safety.

The ISO 22000 considers that the safety of food is related to the presence of hazards in food at the time of consumption. And because of the dangers that can occur at any stage of the supply chain, the security must be ensured at all levels of the supply chain. So it should be applied to producers of animal feeds and other agricultural products, food manufacturers, packaging, transportation and food warehouses to suppliers of retail and food services. So for its strong integrator character, the success of the implementation depends largely on the acceptance of the various links in the supply chain. Other barriers may arise in terms of local practices and investment cost.

5.4 Six sigma

The concept of 6-Sigma system was developed by Motorola in the mid 80′s. The 6-Sigma program involves the application of statistical methods to business processes, guided by the goal of eliminating defects. The 6-Sigma focuses on quality improvement (eg, waste reduction) to help organizations produce better, faster and more economical. More generally, the program focuses on defect prevention, reduction of cycle times and cost savings. Unlike careless cost cutting, which reduce the value and quality, Six Sigma identifies and eliminates costly waste, i.e., that do not add value to the customers. With this, the company increases operational efficiency reduces costs, improves quality, increases customer satisfaction and increases profitability [18, 19].

Sigma (Σ) is a letter of the Greek alphabet used by statisticians to measure the variance in any process. The performance of a company is measured by the sigma level of their business processes. Organizations that employ the Six Sigma method aim to achieve 3.4 defects per million on manufactured products. This methodology is based on the implementation of a system based on the measurement and monitoring of processes so that deviations from 'normality' are avoided as much as possible.

The Six Sigma methodology is composed by a broad set of tools and techniques for quality improvement, among which there is a strong application of statistical tools and techniques. The cycle of phases, called DMAIC (Define, Measure, Analyze, Improve, Control) is used as a guide for professionals (mainly black belts and green belts) to implement projects that meet the goals most daring and radical pre-set by the company. The DMAIC can be resumed as follows:

- Define: define problems and situations to be improved, including the goals of the activities, as they will be the company's strategic objectives.
- Measure: to establish valid and reliable measurements for information and data.
- Analyze: analyze the information captured in order to identify ways to eliminate the gap between the current performance of the system or process and the desired goal. It should apply statistical tools to aid analysis.
- Increment: deploy processes, it can use management tools of projects or planning and managing to deploy a new approach,
- Control: control the improved processes in order to generate a continuous improvement cycle.

The statistical aspects of six sigma must complement business perspectives and challenges to the organization to implement six sigma projects successfully.In the list of tools and statistical techniques of DMAIC, are included: descriptive statistics, principles of sampling, control charts, process capability analysis, measurement system analysis, basic charts (histogram, scatter, box-plot, Pareto, etc..), cause and effect diagram, statistical process control (SPC), design of experiments, linear regression and correlation, multiple regression, hypothesis testing, confidence intervals, analysis of variance, capability process analysis, among others [18-20].

Factors influencing successful six sigma projects include management involvement and organizational commitment, project management and control skills, cultural change, and continuous training. It is a methodology that crosses the entire company, i.e., it is not the isolated involvement of a team, but the involvement of all in the pursuit of the implementation of continuous improvement and customer satisfaction [19, 20].

The adoption of the Six Sigma methodology as a quality program in all agribusiness chain in general is still new, but it is important to highlight the potential of this method for improving the quality of food products and reduce production costs.

6. CONCISE SUMMARY

A company's competitiveness reflects the techniques employed to adapt to the existing competitive standards in its operating marketplaces. Quality is a crucial aspect for the food business operating in an increasingly worldwide market. Companies must formulate competitive strategies and develop a suitable organizational structure.

This chapter discusses the critical elements pertinent to quality management within the food business based on these assumptions. The financial, cultural, organizational, and motivational realities of each company will dictate the level of maturity and efficiency in quality management. It may be argued that competitive advantage is achieved by the continual pursuit of innovative tools and learning management systems that enhance the quality of processes and services, hence improving the products provided by the food business..

REFERENCES

1. Garvin, D. A. Managing Quality: The Strategic and Competitive Edge. Boston: Harvard Business School Press; 1988.
2. Gurudasani, R., Sheth, M. Food safety knowledge and attitude of consumers of various food service establishments. Journal of Food Safety 2009;29: 364–380.
3. Fotopoulos C, Kafetzopoulos D, Gotzamani K. Critical factors for effective implementation of the HACCP system: a Pareto analysis. British Food Journal 2011; 113(5): 578-97.
4. Food and Drug Administration. FDA. Return to Good Manufacturing Practices (GMPs) for the 21st Century-Food Processing. www.fda.gov/Food/ Guidance Compliance Regulatory Information / Current Good Manufacturing PracticesCGMPs (accessed 10 August 2012).
5. Codex Alimentarius Commission. Recommended international code of practice general principles of food hygiene. CAC/RCP-1 (1969); Rev.4; 2003.
6. Campos VF. TQC: Controle da Qualidade Total (no estilo japonês). Nova Lima: INDG; 2004.
7. Werkema MCC. Ferramentas estatísticas básicas para o gerenciamento de processos. Belo Horizonte: Fundação Christiano Ottoni; 1995.
8. Montgomery, D.C. Introduction to statistical quality control. 5th edition, New York: Wiley; 2005.

9. Grigg N. Statistical process control in UK food production: an overview. International Journal of Quality & Reliability Management 1998; 15(2): 223-38.
10. Grigg N, Walls L. Developing statistical thinking for performance improvement in the food industry. International Journal of Quality & Reliability Management 2007; 24(4): 347-69.
11. International Standard Organization. Sampling procedures for inspection by attributes - Part 10: Introduction to the ISO 2859 series of standards for sampling for inspection by attributes. ISO 2859-10:2006(E).
12. International Standard Organization. Sampling procedures for inspection by variables - Part 2: General specification for single sampling plans indexed by acceptance quality limit (AQL) for lot-by-lot inspection of independent quality characteristics. ISO 3952-2:2006(E).
13. Costa, AFB, Epprechi EK, Carpinetti LC. Controle estatístico de qualidade. São Paulo: Atlas; 2004.
14. Codex Alimentarius Commission. General guidelines on sampling. CAC/GL 50; 2004.
15. Grigg N, Walls L. The role of control charts in promoting organizational learning: new perspectives from a food industry study. The TQM Magazine 2007; 19(1): 37-49.
16. Mizuno S. Company-wide quality control activities in Japan. Reports of Statistical Application Research 1969, 16(3): 68-77.
17. Grigg N, McAlinden C. A new role for ISO 9000 in the food industry? Indicative data from the UK and mainland Europe. British Food Journal 2001; 103(9): 644-56.
18. Tjahjono B, Ball P. Six Sigma: a literature review. International Journal of Lean Six Sigma 2010; 1(3): 216-33.
19. Werkema, C. Criando a cultura Seis Sigma. Nova Lima: Werkema Editora; 2004.
20. Kwak YH, Anbari FT. Benefits, obstacles, and future of six sigma approach. Technovation 2006; 26(5-6): 708-15.

SECTION 3
FOOD SAFETY

10 Chapter

SOCIOECONOMIC CONCERNS – GENETICALLY MODIFIED ORGANISMS

1. INTRODUCTION: AN OVERVIEW

Food is a fundamental necessity for humans; we consume it to sustain life, and while most individuals are fortunate enough to have at least one meal per day, others may afford three or more. Regardless of our culture and customs, dining remains an essential component of various celebrations globally, among family and friends. Moreover, we desire a wholesome and nutritious supper; nonetheless, the inquiry remains: "How safe is the food we are consuming?"

The enhancement of flora and fauna for food production, alongside the implementation of various conservation measures, has been practiced since humanity ceased migration and adopted agriculture for sustenance. In the pursuit of producing greater quantities and superior quality food to satisfy the demands of our rapidly expanding global population, genetic engineering of crops has emerged as a novel approach alongside traditional plant breeding.

Molecular genetics has been and continues to be an invaluable tool for elucidating the genes responsible for quantitative features linked to enhanced crop yields and improved food quality. The desire to enhance food yields has led to the genetic modification of plants, provoking significant controversy including political, ethical, and social issues. Genetically modified food refers to the alteration of the original DNA (deoxyribonucleic acid) structure of plants. Given that DNA serves as the fingerprint of every organism, modifications to the genetic code may potentially result in alterations to the quality or characteristics of the plant in question.

Despite a consistent increase in the overall area dedicated to genetically modified (GM) crop farming, there has been a significant deceleration in recent years.. The most extensively cultivated GM crops include soybean, corn and cotton. Europe is known to grow less than 0.5% of the world's GM crops, primarily because of the very rigorous EU regulations imposed on GMO crops in Europe until 2003 and the refusal of European consumers to buy GM products.

Nonetheless, the fundamental comprehension of cellular function and heritability, coupled with genetic engineering that facilitates the transfer and modification of DNA between organisms, has empowered governments in numerous countries to adequately supply food to their expanding populations for the first time. These breakthroughs have led to the creation of effective vaccinations and pharmaceuticals, innovative food technology, and numerous other goods that enhance the overall quality of life. Genetic engineering of crops can enhance traditional plant breeding to meet contemporary agricultural demands. Many of these enhancements can be categorized under the term "biotechnology," which seeks to utilize organisms, cells, or cellular components in technical or industrial applications.

2. REGULATIONS AND WHY?

Because genetically modified foods have been one of the most controversial topics that have made news in the last years. Many European environmental organizations, NGOs and public interest groups have been actively protesting against GM foods for months. Beside, recent controversial studies about the effects of genetically-modified food have brought the issue of genetic engineering to the forefront of the public consciousness (Fonseca, Planchon, Renaut, Oliveira, & Batista, 2012; Losey, Rayor, & Carter, 1999; Nykiforuk, Shewmaker, Harry, Yurchenko, Zhang, Reed, et al., 2012). Generally in Europe, the idea of introducing GM food products in the market for human consumption and or as animal feed has not been welcome for health reasons (Maga & Murray, 2010). Although there are no clear research results suggesting the negative effects of GM food to human health, the distancing from GM foods is more or less preventive. Nevertheless, with the growing interest in the use of biofuels as one of the sources of alternative sources energy, genetic engineering then comes in to play for economic reasons.

As a reaction to the growing public concern on GM food and products, many governments across the world have taken different approaches to tackle this hot topic on GM foods. This has resulted in the creation of GMO regulations which are most often country or region specific. The European parliament and council for example have set up regulations regarding GM foods to protect human health and well-being of citizens, and European social and economic interests (McCabe & Butler, 1999). The EU regulations segregates between GM food and feed, it further gives specific instructions on how GM products should be labelled in terms of the amount of modifications involved.

EU GMO regulations suggest for example that it is appropriate to provide the combined level of adventitious or technically unavoidable presence of genetically modified materials in a food or feed or in one of its components is higher than the set threshold, such presence should be indicated in accordance with this regulation and that detailed provisions should be adopted for its implementation (Ramon, MacCabe, & Gil, 2004). The possibility of establishing lower thresholds, in particular for foods and feed containing or consisting of GMOs or in order to take into account advances in science and technology, should be provided for. In my opinion, the European GM food regulations are the most stringent in the world and it is not quite clear whether or not there is any room for GM products due to the complexity in understanding and implementation of the said regulations. Nonetheless, the EU GMO regulations could be summarized as it is meant to provide the basis for ensuring a high level of protection of human life and health, animal health and welfare, environment and consumer interests in relation to genetically modified food and feed, whilst ensuring the effective functioning of the internal market; lay down community procedures for the authorisation and supervision of genetically modified food and feed; and to lay down provisions for the labelling of genetically modified food and feed.

Similarly, the United States regulation process is confusing because there are three different government agencies that have jurisdiction over GM foods. The Food and Drug Administration (FDA) evaluate whether the plant is safe to eat; the U.S. Environmental Protection Agency (EPA) evaluates GM plants for environmental safety, and the United States Department of Agriculture (USDA) which evaluates whether the plant to be grown is safe (Pelletier, 2005; Strauss, 2006). The USDA has many internal divisions that share responsibility for assessing GM foods. Among these divisions are, the Animal Health and Plant Inspection Service (APHIS), which conducts field tests and issues permits to grow GM crops, the Agricultural Research Service which performs in-house GM food research, and the Cooperative State Research, Education and Extension Service which oversees the USDA risk assessment program (Whitman, 2000). This implies there is a combination of regulations from these three agencies to be followed in order to carry on with GM food. Nevertheless, it is estimated that up to 70% of processed food on US supermarkets shelves ranging from soda to soup, crackers to condiments contain genetically engineered ingredients. Currently, up to 85% of U.S. corn is genetically modified as are 91% of soybeans and 88% of cotton (cottonseed oil is often used in food products) (Whitman, 2000).

In many developing countries whereby due to seasonal changes, there are usually a season of plenty and that of starvation, GM food is less a problem because the goal is to feed the starving population. Although, some of them might have GMO regulations, when food aid is coming into their countries in the moment of disaster, their rules and regulations are not important at that moment. This is understandable because the ultimate goal is saving lives before thinking of any qualms.

Plants have always been able to developed mechanisms over the years to endured environmental stress (drought, predation and pollutions just to name a few) and consequently adapted to the changing environment by developing genes resistant to the different factors. This is supported by the fact that, historically it was assumed that changes in plants as a result of genetic modification in breeding are generally safe and not harmful. Nevertheless, this was eventually challenged with the arrival of rDNA (ribosomal deoxyribonucleic acid) technology in the early 1970s when Cohen and Boyer successfully linked two different pieces of DNA (McHughen & Smyth, 2008).

The scientific world did not acknowledged the positive potentials of genetic engineering to crop breeding but the risks associated with these techniques (Berg & et al., 1974; McHughen & Smyth, 2008).

Over the last century, agriculture in general and plant breeding in particular have enjoyed fast dynamic research, which have been speedy and valuable developments. Traditional forms of crop genetic improvements, such as selection and cross-pollination, remain the standard tools in the breeder's toolbox, but have been supplemented with a range of new and specialized innovations, such as mutation breeding using ionizing radiation or mutagenic chemicals, wide crosses across species requiring human interventions such as embryo rescue and transgenic, commonly called genetic modification.

3. GM FOOD AND HUMAN HEALTH

Food choice is influenced by a large number of factors, including social and cultural factors. One method for trying to understand the impact of these factors is through the study of attitudes. Research is described which utilizes social psychological attitude models of attitude- behaviour relationships, in particular the Theory of Planned Behaviour. This approach has shown good prediction of behaviour, but there are a number of possible extensions to this basic model which might improve its utility. One such extension is the inclusion of measures of moral concern, which have been found to be important both for the choice of genetically-modified foods and also for foods to be eaten by others.

It has been found to be difficult to effect dietary change, and there are a number of insights from social psychology which might address this difficulty. One is the phenomenon of optimistic bias, where individuals believe themselves to be at less risk from various hazards than the average person (Paparini & Romano-Spica, 2004).

This effect has been demonstrated for nutritional risks, and this might lead individuals to take less note of health education messages. Many children in the US and Europe have developed life-threatening allergies to peanuts and other foods. There is a possibility that introducing a gene into a plant may create a new allergen or cause an allergic reaction in susceptible individuals. There is a growing concern that introducing foreign genes into food plants may have an unexpected and negative impact on human health. A recent article published in Lancet examined the effects of GM potatoes on the digestive tract in rats (Brunner & Millstone, 1999).

Another concern is that individuals do not always have clear-cut attitudes, but rather can be ambivalent about food and about healthy eating. It is important, therefore, to have measures for this ambivalence, and an understanding of how it might impact on behaviour (Shepherd, 1999).

One measure of how far we have travelled down that road is that it hardly matters any more whether objections to GMO are based on alleged environmental risks of cultivating GM crops or alleged toxicological hazards of eating them. GMO like 'radioactivity' has become an odious, generic shibboleth. Given that millions of people throughout the world are al ready benefiting from pharmaceuticals made by GM organisms, this is bizarre (Dixon, 2003).

Among the next generation of genetically modified (GM) plants are those that are engineered to produce elevated levels of nutritional molecules such as vitamins, omega-3 fatty acids, and amino acids. Based upon the U.S. current regulatory scheme, the plants and their products may enter our food supply without any required safety testing. The potential risks of this type of GM plants are discussed in the context of human health, and it is argued that there should be very careful safety testing of plants designed to produce biologically active molecules before they are commercially grown and consumed. This will require a mandatory, scientifically rigorous review process (Schubert, 2008).

Nevertheless, advances in our understanding of molecular biology, biochemistry, and nutrition may in future allow further improvement of test methods that will over time render the safety assessment of foods even more effective and informative (Konig, Cockburn, Crevel, Debruyne, Grafstroem, Hammerling, et al., 2004).

4. GM FOOD AND ENVIRONMENT

Genetic modification and "biosafety" are concepts that have not been well understood by, or accessible to, the non-geneticists working in the fields of conservation science, law, administration and management, and in the scientific, legal, administrative and management aspects of sustainable use.

Genetically modified (GM) plants represent a potential benefit for environmentally friendly agriculture and human health. Although, poor knowledge is available on the potential hazards posed by unintended modifications occurring during genetic manipulation processes, the increasing amount of reports on ecological risks and benefits of GM plants stresses the need for experimental works aimed at evaluating the impact of GM crops on the natural and agro-ecosystems. One of the major environmental risks associated with GM crops include their potential impact on non-target soil microorganisms which plays a fundamental role in crop residues degradation and in biogeochemical cycles (Giovannetti, Sbrana, & Turrini, 2005).

Transformed corn plants with genetic material from the bacterium Bacillus thuringiensis (Bt) have been reported to represent a risk because most hybrids express the Bt toxin in pollen which could be further deposited on other plants near such corn fields causing non-target organisms that consume these plants (Yu & Shepard, 1998). It is thought that genetically modified plants could be harmful to the environment by depleting soil microorganism which are very important for soil fertility and or influence the micro-environments of other organisms (Giovannetti, Sbrana, & Turrini, 2005). The cultivation of GM seeds and plants could be detrimental to the environment (Losey, Rayor, & Carter, 1999).

The biodiversity debate is at the forefront of the larger question of how humanity can, in an integrated, congruent way, address human livelihoods, while at the same time fulfilling its international mandates to conserve and sustainably use the environment. In a world focused on issues such as poverty and food security, as well as species loss and ecosystem destruction, these questions are among the most important and the most difficult on the planet.

5. GM FOOD AND ECONOMIC ISSUES

Bringing a GM food to market is a lengthy and costly process, and of course agro-biotechnological companies wish to ensure a profitable return on their investment. Thus many new plant genetic engineering technologies and GM plants have been patented, and patent infringement is a big concern of agribusiness.

Although, genetically modified (GM) plants represent a potential benefit for environmentally friendly agriculture and human health, poor knowledge is available on the potential hazards posed by unintended modifications occurring during genetic manipulation. The major economic fears are the risk of patent enforcement which may oblige farmers to depend on giant engineering companies such as Monsanto for strains when their crops are cross pollinated. Consumer advocates are equally worried that patenting these new plant varieties will raise the price of seeds so high that small farmers and third world countries will not be able to afford seeds for GM crops, thus widening the gap between the wealthy and the poor. It is hoped that in a humanitarian gesture, more companies and non-profits will follow the lead of the Rockefeller Foundation and offer their products at reduced costs to impoverished nations.

These plants would be viable for only one growing season and would produce sterile seeds that do not germinate. Farmers would need to buy a fresh supply of seeds each year, consequently will have to be dependent on the few agric-biotech companies with patent rights. However, this would be financially disastrous for farmers in third world countries who cannot afford to buy seed each year and traditionally set aside a portion of their harvest to plant in the next growing season.

6. SOCIAL AND CULTURAL ASPECTS ON GM FOODS

With the emergence of transgenic technologies, new ways to improve the agronomic performance of crops for food, feed, and processing applications have been devised. In addition, ability to express foreign genes using transgenic technologies has opened up options for producing large quantities of commercially important industrial or pharmaceutical products in plants. Despite this high adoption rates and future promises, there is a multitude of concerns about the impact of genetically modified (GM) crops on the environment (Paparini & Romano-Spica, 2004). Potential contamination of the environment and food chains has prompted detailed consideration of how such crops and the molecules that they produce can be effectively isolated and contained. One of the reasonable steps after creating a transgenic plant is to evaluate its potential benefits and risks to the environment and these should be compared to those generated by traditional agricultural practices (Poppy, 2004). The precautionary approach in risk management of GM plants may make it necessary to monitor significant wild and weed populations that might be affected by transgene escape. Effective risk assessment and monitoring mechanisms are the basic prerequisites of any legal framework to adequately address the risks and watch out for new risks. Several agencies in different countries monitor the release of GM organisms or frame guidelines for the appropriate application of recombinant organisms in agro-industries so as to assure the safe use of recombinant organisms and to achieve sound overall development. We feel that it is important to establish an internationally harmonized framework for the safe handling of recombinant DNA organisms within a few years (Singh, Ghai, Paul, & Jain, 2006).

7. CONCISE SUMMARY

Genetically modified foods possess the capacity to address numerous global hunger and malnutrition issues, while also aiding in environmental protection and preservation by enhancing yield and diminishing dependence on chemical pesticides and herbicides. Numerous issues persist for governments, particularly with safety testing, regulation, international policy, and food labeling. A multitude of individuals believe that genetic engineering is an unavoidable future trend, and that we cannot disregard a technology with such significant potential advantages. Nevertheless, we must exercise caution to prevent inadvertent harm to human health and the environment stemming from our enthusiasm for this potent technology.

In this context, numerous assertions regarding genetically modified organisms (GMOs) emerge, suggesting that they may serve as a foundation for enhancing food production without necessitating the conversion of more area for cultivation. These assertions, nonetheless, are challenged by the contention that GMOs may have diverse effects on humans and animals, particularly on ecosystems and uncultivated regions, alongside apprehensions about the actualization of GMO benefits in developing nations.

Additionally, some inquiries must be addressed to enhance our comprehension of GMOs, including:

- Are the existing scope and aims of the GMO legislation aligned with societal demands, particularly those of biotechnology businesses and consumers?
- Are the procedures related to the legislative framework adequate in both definition and execution?
- Are the methods for the risk evaluation of GMOs and their application current, efficient, time-bound, and transparent?
- Are the laws controlling risk management of GMO marketing in design and execution current, efficient, transparent, and aligned with the overarching objectives of our legislation?
- Is the communication of danger about the release of GMOs into the environment and its implementation understood?

REFERENCES

1. Berg, P., & et al. (1974). Potential biohazards of recombinant DNA molecules. Science, 185(4148), 303.
2. Brunner, E., & Millstone, E. (1999). Health risks of genetically modified foods. Lancet, 354(9172), 71.
3. Dixon, B. (2003). Genes in food--why the furore? Biochemical Society transactions, 31(2), 299-306.
4. Fonseca, C., Planchon, S., Renaut, J., Oliveira, M. M., & Batista, R. (2012). Characterization of maize allergens - MON810 vs. its non-transgenic counterpart. Journal of proteomics, 75(7), 2027-2037.
5. Giovannetti, M., Sbrana, C., & Turrini, A. (2005). The impact of genetically modified crops on soil microbial communities. Rivista di biologia, 98(3), 393-417.
6. Konig, A., Cockburn, A., Crevel, R. W., Debruyne, E., Grafstroem, R., Hammerling, U., Kimber, I., Knudsen, I., Kuiper, H. A., Peijnenburg, A. A., Penninks, A. H., Poulsen, M., Schauzu, M., & Wal, J. M. (2004). Assessment of the safety of foods

derived from genetically modified (GM) crops. Food and chemical toxicology: an international journal published for the British Industrial Biological Research Association, 42(7), 1047-1088.

7. Losey, J. E., Rayor, L. S., & Carter, M. E. (1999). Transgenic pollen harms monarch larvae. Nature, 399(6733), 214.

8. Maga, E. A., & Murray, J. D. (2010). Welfare applications of genetically engineered animals for use in agriculture. Journal of animal science, 88(4), 1588-1591.

9. McCabe, H., & Butler, D. (1999). European Union tightens GMO regulations. Genetically modified organisms. Nature, 400(6739), 7.

10. McHughen, A., & Smyth, S. (2008). US regulatory system for genetically modified [genetically modified organism (GMO), rDNA or transgenic] crop cultivars. Plant biotechnology journal, 6(1), 2-12.

11. Nykiforuk, C. L., Shewmaker, C., Harry, I., Yurchenko, O. P., Zhang, M., Reed, C., Oinam, G. S., Zaplachinski, S., Fidantsef, A., Boothe, J. G., & Moloney, M. M. (2012). High level accumulation of gamma linolenic acid (C18:3Delta6.9,12 cis) in transgenic safflower (Carthamus tinctorius) seeds. Transgenic research, 21(2), 367-381.

12. Paparini, A., & Romano-Spica, V. (2004). Public health issues related with the consumption of food obtained from genetically modified organisms. Biotechnology annual review, 10, 85-122.

13. Pelletier, D. L. (2005). Science, law, and politics in the Food and Drug Administration's genetically engineered foods policy: FDA's 1992 policy statement. Nutrition reviews, 63(5), 171-181.

14. Poppy, G. M. (2004). Geneflow from GM plants--towards a more quantitative risk assessment. Trends in biotechnology, 22(9), 436-438.

15. Ramon, D., MacCabe, A., & Gil, J. V. (2004). Questions linger over European GM food regulations. Nature biotechnology, 22(2), 149.

16. Schubert, D. R. (2008). The problem with nutritionally enhanced plants. Journal of medicinal food, 11(4), 601-605.

17. Shepherd, R. (1999). Social determinants of food choice. The Proceedings of the Nutrition Society, 58(4), 807-812.

18. Singh, O. V., Ghai, S., Paul, D., & Jain, R. K. (2006). Genetically modified crops: success, safety assessment, and public concern. Applied microbiology and biotechnology, 71(5), 598-607.

19. Strauss, D. M. (2006). The international regulation of genetically modified organisms: importing caution into the U.S. food supply. Food and drug law journal, 61(2), 167-196.

20. Whitman, D. B. (2000). Genetically Modified Foods: Harmful or Helpful? CSA Discovery Guides.

21. Yu, D. W., & Shepard, G. H., Jr. (1998). Is beauty in the eye of the beholder? Nature, 396(6709), 321-322.

11 Chapter

ACIDIFIED FOODS: FOOD SAFETY CONSIDERATIONS FOR FOOD PROCESSORS

1. INTRODUCTION: AN OVERVIEW

The food processing industry is one of the United States' largest manufacturing sectors, accounting for more than 10 percent of all manufacturing shipments. Concerns over food safety have increased as the industry has been hit by several high profile and large-scale food recalls. Thus, commercial food processors must be vigilant about ensuring the safety of their products. If inadequate or improper manufacturing, processing or packaging procedures are used in the production of low-acid or acidified canned foods serious health hazards, especially Clostridium botulinum, could result. To prevent this, processors must be in compliance with regulations established by the U.S. Food and Drug Administration (F.D.A., U.S. Department of Agriculture) and state agriculture and health departments across the United States (Barron, 2000).

2. ACIDIFIED FOODS

The term "acidified foods" means low-acid foods to which acid(s) or acid food(s) are added. These products include, but are not limited to:

- Pickled beets, cocktail onions, and cherry peppers (normally pickled by the addition of acid);
- Red bell peppers treated in an acid brine;
- Some pears and tropical fruits that have a natural pH greater than 4.6 and are acidified to a pH of 4.6 or below;
- Fermented green olives subjected to processes (such as lye treatment or washing with low- acid foods) that raise the pH above 4.6, with subsequent addition of acid or acid foods to reduce the pH to 4.6 or below;
- Tomato salsa made from tomatoes with a pH of 4.6 or below and low-acid ingredients, when the amount of low-acid ingredients is not a small amount and/or the resultant finished equilibrium pH differs significantly from that of the predominant acid or acid food; and

- Cold-pack pickles that are subjected to the action of acid-producing microorganisms but require the addition of acid or an acid food to achieve a pH of 4.6 or below.

All acidified foods must have a water activity (aw) greater than 0.85 and a finished equilibrium pH of 4.6 or below within the time designated in the scheduled process. These parameters must be maintained in all finished foods as outlined in 21 CFR 114.80(a). These foods may be called, or may purport to be, "pickles" or "pickled." However, some barriers exist in the preparation of acidified foods, including inadequate acid in the cover brine to overcome buffering capacity of the food, the presence of alkaline compounds from peeling or other processing aids, and the peels, waxing, piece size or oil in the product which can cause a barrier to penetration of the acid. These barriers may cause the failure to achieve the final equilibrium of a pH value of 4.6 and raise concerns about the growth of pathogens and production of toxins in the finished product.

After proper acidification, all acidified foods must then be heat processed to destroy the vegetative cells of pathogenic microorganisms or other microorganisms that cause spoilage and to inactivate enzymes that might affect color, flavor, or texture of the product. Acidified foods can be heat processed in a boiling water canner or by low-temperature pasteurization. The processing time, temperature, and procedure necessary to safely preserve acidified foods are determined by factors such as level of acidity (pH), size of food pieces (density) and percentage salt. An FDA recognized process authority must review the product and process and make the appropriate recommendations about time and temperature requirements. Processing temperatures higher than 185°F (85°C) could break down pectin and cause unnecessary softening of acidified foods (FDA 2010a).

All commercial establishments engaged in the manufacture of Acidified Foods and Low-Acid Canned Foods (LACF) offered for interstate commerce in the United States are required by 21CFR Parts 108, 113 and 114 to register their facility with form FDA 2541, "Food Canning Establishment Registration," and file scheduled processes for their products with forms FDA 2541a, "Food Process Filing for all Methods Except Low-Acid Aseptic," and FDA 2541c, "Process Filing for Low-Acid Aseptic Systems." The following items are not considered to be acidified foods or low-acid foods.

- Acid foods (naturally acid foods have a pH of 4.6 or less)
- Acid foods (including such foods as standardized and non-standardized food dressings and condiment sauces) that contain small amounts of low-acid food(s) and have a resultant finished equilibrium pH that does not significantly differ from that of the predominant acid or acid food
- Alcoholic beverages

- Carbonated beverages
- Standardized jams, jellies and preserves (21 CFR 150)
- Tomatoes and tomato products having a finished equilibrium pH less than 4.7
- Foods that are NOT packaged in hermetically sealed containers
- Any food prepared under the continuous inspection of the meat and poultry inspection program of the Animal and Plant Health Inspection Service of the Department of Agriculture under the Federal Meat Inspection Act and the Poultry Products Inspection Act
- Foods that are stored distributed and retailed under refrigeration
- Foods with water activity of 0.85 or below
- Food that are not thermally processed

Because these foods are not recognized as acidified foods, commercial processors do NOT have to file and register their processing information for these products with the Food and Drug Administration (FDA 2010b).

3. PATHOGENS OF CONCERN

In 1979, the Code of Federal Regulations (CFR) published the acidified regulations identified today as 21 CFR Part 114. Since then, new food processing technologies and methodologies have been developed and are frequently used in the industry. Furthermore, pathogens, such as E. coli 0157:H7 and Salmonella spp. have been shown to survive and grow in acidic environments. As a result of changing technologies and emerging pathogens, actions by federal agencies have motivated researchers to investigate new ways to eliminate pathogens, such as E. coli and Salmonella spp. The following are several research citations that provide a brief history of developments related to pathogens in acidified foods.

In 1996, an outbreak of E. coli 0157:H7 was identified when an individual contracted hemolytic uremic syndrome after drinking apple juice packaged in sealed containers. The outbreak affected 45 individuals across the USA and Canada. The product was voluntarily recalled by the manufacturing company (Centers for Disease Control and Prevention, 1996)

In 1999, an outbreak of Salmonella Muenchen serotype in the United States and Canada caused 298 cases of illness, which were attributed to unpasteurized orange juice. The outbreak affected 17 states, primarily in the Midwest, as well as regions of Canada. The product was voluntarily recalled after unopened product tested positive for the causative serotype (Centers for Disease Control and Prevention, 1999).

A study performed on the relative safety of pickled cucumbers from Clostridium botulinum infection, as a response to a 1976 study in which the organism was found in sealed containers previously believed to be safe. The study involved introducing C. botulinum spores into experimentally packed pickles artificially adjusted to a target pH and checking for growth of the organism. It was reported that any pH less acidic than 4.8 was insufficient to effectively kill C. botulinum spores, thus establishing a minimum safe pH for pickled cucumbers. (Ito et al, 1996).

A study investigating the effects of acetic acid on E. coli O157:H7 in apple juice and pickle brine found that increasing the pH of the food product yielded an increased inhibitory effect on pathogen growth. The study also demonstrated that acetic acid, a key component in vinegar, had a significant effect on the aforementioned inhibition over other methods of manipulating pH. (Breidt et al, 2004).

A study investigating the thermal resistance of E. coli O157:H7 found evidence for the phenomenon known as cross-protection, or the ability of a bacterium to apply resistance to one negative condition against another. These authors reported that microorganisms grown in an acidic environment display increased resistance to killing via thermal methods, indicating an increased threat by these types of organisms against current food safety methods involving both heat and acid (Buchanan and Edelson, 1999).

Recently, a study found that the Breidt model could be used to measure five-log reduction times in a less conservative manner, allowing for a more encompassing approach to determining safe preparation times for various foods. Acidified vegetable products with a pH above 3.3 must be pasteurized to assure the destruction of acid resistant pathogenic bacteria. The times and temperatures needed to assure a five log reduction by pasteurization have previously been determined using a non-linear (Weibull) model. Recently, the Food and Drug Administration has required that linear models be used with online electronic process filing forms for acidified foods. A linear model was developed that is based on the existing safe processing data. The processing times and temperatures meet or exceed the established heat processing conditions needed to assure safety (Breidt et al, 2010).

4. CONTROL MEASURES FOR ENSURING FOOD SAFETY OF ACIDIFIED FOODS

Control measures for ensuring the safety of acidified foods are well documented in the scientific literature. A simple overview of appropriate measures includes:

- Acidified foods must be properly acidified to a pH below 4.6, but most foods are acidified to a pH of 4.2 or below.
- To assure quick and proper acidification, the food is normally cooked or heated with the acid before being filled into the final container.

- A thermal process or heating step is required to kill all pathogens and any other non- pathogenic microorganisms that could grow during storage of the product. Thermal processing must be completed by hot-filling the product or by the boiling water bath process. The heating temperature and time must be validated by an FDA recognized process control authority and be monitored, controlled and documented.
- The final equilibrium pH must be checked, controlled and documented after the product has completed the thermal processing step. A pH meter with two decimal places accuracy must be used to measure the pH if the final pH is 4.0 or above; other methods can be used such as pH paper or a pH meter with one decimal place, if the final pH is below 4.0.
- Containers for acidified foods should be such that a hermetic seal is obtained. Vacuum is a good indicator of a hermetic seal and helps to keep the quality of the product.

5. ACIFIED FOOD GUIDANCE

Probably the most comprehensive guide to assist food processors in determining what constitutes an acidified food is a document prepared by the FDA in 2010 titled “Guidance for the Food Industry: Acidified Foods.” This guidance document provides nonbinding recommendations but nevertheless presents step by step guidelines to determine if a food can be classified as an acidified food. In this document standardized and non-standardized food dressings, such as mayonnaise, and condiment sauces, such as ketchup, are considered acid foods, which have a natural pH of 4.6 or below.

Processors who are not sure if a particular food is classified as an acidified or not, can voluntarily submit the respective FDA forms for a preliminary evaluation. The draft guidance reminds processors that jams, jellies and preserves are excluded from the 21CFR114 as long as these products meet the applicable standard of identity under 21CFR150; otherwise, the non- standardized products are covered by 21CFR114 based on the pH of the fruit, the pH of the final product and the water activity level of the finished product.

Another important aspect to be considered by a food processor is the use of acid foods and small amounts of low amounts of low acid foods as ingredients to produce an acidified food.

There are two basic criteria needed to exclude any food from being subject to 21 CFR Part 114. The first is that acid foods contain small amounts of low acid foods and the second is that acid foods have a resultant finished equilibrium pH that does not significantly differ from that of the predominant acid or acid food.

Fermented foods with a water activity level above 0.85, such as cucumber pickles and green olives, are considered low acid foods subject to the action of acid producing microorganisms to reduce the pH of the food to 4.6 or below. As such, these products are subject to the requirements of 21CFR114. Processors repacking and reprocessing previously acidified foods are also subject to 21CFR114.

Common questions of food processors new to the food processing industry are precisely related to this matter of reprocessing or repacking a previously acidified food and to procedures to determine a finished equilibrium pH. The draft guidance reminds processors about the meaning of equilibrium pH. It is recommended to use a reference temperature of 25oC, commonly used in laboratory measurements. Equilibrium means the acid is fully diffused throughout the food (especially solid particles) and any successive measurements produce the same results. Further recommendations about food preparation for pH measurements and indicated to follow 21CFR114.90 and to ensure the pH of an in process batch to be reduced and reach the 4.6 within 24 consecutive hours. The likelihood that spores of C. botulinum will germinate and grow increases with the length of time it takes to reduce the equilibrium pH of a food to 4.6

There are three very important terms embedded in the definition of acidified foods (21 CFR par 114): (1) small amount of low acid food(s), (2) predominant acid or acid food, and (3) pH that does not significantly differ. Regarding the small amount of low acid food(s), it has been recommended to be no more than 10% by weight in the finished product. This recommendation is based on FDA experience when evaluating filed processed. This recommendation has been identified by FDA as the 'small amount provision" which means that acid foods that contain small amounts of low acid food(s) AND have a resultant finished equilibrium pH that does not significantly differ from that of the predominant acid or acid food are excluded from complying with 21CFR114. Some examples under this provision may be products such as tomato puree with added spices, or a salad dressing where the predominant acid is the mixture of all acid ingredients, such as mayonnaise, lemon juice, vinegar and tomato paste, and the small amount of low acid foods are red peppers, onion and garlic.

The acid ingredient, such as vinegar has a pH of 4.6 or below; the acid food such as tomatoes has a natural pH of 4.6 or below. These acid ingredients need to be at least 90% of the total weight of the finished product to be considered predominant.

Regarding the term pH that does not significantly differ from that of the predominantly acid or acid foods, FDA recommends the following criteria:

If the equilibrium pH of the predominant acid or acid food is:	Then one should consider a shift in pH to be significant when:
>4.2	Any shift in pH is present
4.2	The shift in pH is >0.2
≥ 3.8 and < 4.2	The shift in pH is >0.3
<3.8	The shift in pH is >0.4

It is important to consider variability factors, such as the accuracy of the pH meter and variations in the finished equilibrium pH of the food itself. Also, as a reminder to processors, water, being an important ingredient in many acidified foods, it is a low acid food and if it is a predominant ingredient in the finished product, this product is considered a water-based acidified food. Apple juice, bended juices, reconstituted juices and vegetable juices are all considered to be water-based liquids. When the finished equilibrium pH of a water-based liquid that contains acid(s) or acid food(s) is 4.6 or below, the product is subject to 21CFR114, unless the liquid is a carbonated beverage.

The draft guidelines recommend the use of decision tables to determine if a given food, including fermented foods to which low acid foods are added fall under the coverage of 21CFR114. These tables are a step-by-step series of questions that lead to the most probable correct answer about a food being an acidified or not product; however, it is recommended to consider other factors related to the product and the manufacturing process to make the final decision. The guidelines indicate that most acidified foods would require a heat treatment step. This thermal process is to be developed based on the most resistant microorganism that must be controlled under the given pH conditions. For example for a pH range of 4.0 to 4.6 the spores of acid tolerant spoilage microorganisms such as B. licheniformis need to be destroyed, while at a pH range below 4.0, the vegetative cells of yeasts, molds and non spore forming bacteria such as lactobacillus need to be destroyed.

The thermal destruction of spores and microorganisms can be expressed in terms of heat resistance parameters. The adequate combination of time and temperature (extent of thermal processing) to safely manufacture a commercial food product and is also resistant to spoilage is called thermal process lethality. The draft guidance document provides a table demonstrating relationships between finished equilibrium pH of products and the thermal process lethality of acidified foods. For example, for a pH range between 3.3 and 3.5 the F value of 1 minute is recommended. F being the destruction time desired at reference temperature of 195 F and a Z value of 10 F. This is typically written as F 10/195 =1.0 minutes. The thermal process lethality is part of a scheduled process required by FDA to prevent the growth of microorganisms of public health significance in the thermally processed food. This process need to be established by a competent process authority as defined in 21CFR114.3(e).

The draft guidance also includes final recommendations to address spoilage problems through quality control procedures such as systematically implementing written plans to investigate signs of spoilage and their causes, as well as corrective actions to solve the problem.

6. RECALLS

A commercial processor engaged in the processing of acidified foods is also required by 21CFR108.25 to prepare and maintain a written recall plan. Guidelines for product recalls are contained in 21CFR7. This plan will provide a current procedure for implementation, including:

- notifying FDA of any recalls
- a procedure for distributors to follow to recall products which may be injurious to health
- a procedure for identifying, collecting, warehousing and controlling products and a method for determining the effectiveness of any recalls.

Recall is a voluntary action taken by manufacturers and distributors to remove food that is in violation of laws administered by the FDA and USDA. These agencies may request a recall, but cannot order one without a court order. Product recovery is only classified as a recall when the product is violative.

Product Identification. Each batch or production lot must be properly coded. This code will allow the product lot to be identified as to date, batch product personnel production records, and ingredient records.

Records. Records are key to the recall plan and must be maintained for three years. They include:

- Records of examination of raw materials, packaging materials, and finished product along with any supplier guarantees or certifications.
- Processing and production records showing adherence to scheduled processes, including records of pH measurement and other critical factors.
- A log of all departures from scheduled processes, actions taken to rectify them, and disposition records of the portion of product involved.
- Records of initial distribution of the finished product adequate to facilitate separation of food lots which may have become contaminated or otherwise unfit for use.

Notification. Persons to be notified in the event of a recall include FDA and USDA, key company personnel, and distributors. The notification should include the product, container size, and code of affected lots. The extent of the hazard and the level of the recall will be as determined by FDA and USDA. Based on this determination, FDA will approve the recall strategy. The notification will include instructions for consumers and distributors for product recovery and information feedback. The contact person should be listed on all notification forms.

Product Recovery. Plans for recovery include procedures for segregation of affected lots, storage, warehousing, and control. Procedures in place shall allow determination of the effectiveness of the recall. The recall is concluded when FDA and USDA determine that recovery is adequate and there is no longer any threat to the public.

REFERENCES

1. Barron, F. H. 2000. Acid, Acidified and Low-acid Foods Canning Guidelines for Food Processors. Bulletin EC 705, Food Nutrition and Packaging Science Department, Clemson University, Clemson, SC, USA.
2. Breidt F., Sandeep, K.P., and Arrit D.M., 2010. Use of Linear Models for Thermal Processing of Acidified Foods. Food Protections Trends, 30(5), 268-272.
3. Breidt, F., Jr., J. S. Hayes, and R. F. McFeeters. 2004. Independent Effects of Acetic Acid and PH on Survival of Escherichia Coli in Simulated Acidified Pickle Products." Journal of Food Protection 67(1):12-18.
4. Buchanan, R. L., and S. G. Edelson. 1999. pH-dependent Stationary-phase Acid Resistance Response of Enterohemorrhagic Escherichia Coli in the Presence of Various Acidulants." Journal of Food Protection. 62(3):211-18.
5. Centers for Disease Control and Prevention. Outbreak of Salmonella Serotype Muenchen Infections Associated with Unpasteurized Orange Juice -- United States and Canada, June 1999. http://www.cdc.gov/mmwr/preview/mmwrhtml/mm4827a2.htm.Accessed October 7, 2012.
6. Centers for Disease Control and Prevention. 8 Nov. 1996. Outbreak of Escherichia coli O157:H7 Infections Associated with Drinking Unpasteurized Commercial Apple Juice -- British Columbia, California, Colorado, and Washington, October 1996. http:// www.cdc.gov/mmwr/preview/mmwrhtml/00044358.htm. Accessed October 7, 2012
7. Food and Drug Administration (FDA), 2010a. Acidified and Low-Acid Canned Foods (LCAF). http://www.fda.gov/Food / Food Safety/Product-Specific Information / AcidifiedLow - Acid Canned Foods / default.htm. Accessed October 11, 2012.
8. Food and Drug Administration (FDA), 2010b. Guidance for the Food Industry: Acidified Foods. http://www.fda.gov/Food/Guidance compliance regulatory information/guidance documents/acidified and low acid foods/ucm222618.htm. Accessed September 10, 2012.
9. Ito, K. A., J. K. Chen, P. A. Lerke, M. L. Seeger, and J. A. Unverferth. 1976. Effect of Acid and Salt Concentration in Fresh-pack Pickles on Growth of Clostridium botulinum Spores. Applied Environmental Microbiology. 32(1):121-24.

12 Chapter MICROBIAL CONTAMINATION OF HOMEMADE CUISINE

1. INTRODUCTION: AN OVERVIEW

The consumption of healthy food is a consumer's right and the duty of the manufacturing industry. Health authorities are duty bound to prepare and enforce laws to protect the population's health. The supply of food free from health risks to the population is actually a challenge. In fact, contaminated food may cause serious infections and jeopardize the health of the population.

Owing to their frequency, food-caused infections are a very grave issue to public health. They may cause hazards ranging from a simple intestine discomfort to cases that are more serious, such as neurological disorders and death, because of the high number of microorganisms involved in a simple epidemic event.

Fresh or processed animal-derived food may harbor several pathogenic microorganisms that cause physiological disorders in people who consume them. When food eventually contaminated by disease-causing microorganisms is consumed, pathogens or their metabolites invade the host's fluids or tissues and trigger serious types of diseases, such as tuberculosis. They are conveyed by non pasteurized milk or by cheese contaminated by bacterial populations of Mycobacterium bovis and M. tubercolosis or by Brucella abortus, gram negative bacteria, intracellular pathogen that cause undulant fever and arthritis in human beings.Bacteria, fungi, protozoa and viruses are the main microorganism groups that cause food disorders. Due to their diversity and pathogenesis, bacteria are by far the most important microbial group commonly associated with food-transmitted diseases. High rated agents in food infections are Salmonella sp., Campylobacter sp and Listeria monocytogenes due to their importance in eventual sequelae. The microbiological health risks in fowl consumption and its raw products include contamination by the above food pathogens.

Besides being one of the principal causes of food-derived diseases since its attack generally involves a great number of people, the genus Salmonella is associated with economic liabilities, commercial damage and decrease in production due to its frequency and extension. These facts occur because of the great number of food products that may be contaminated by this bacterium, namely, food with high humidity, protein and carbohydrate rates, such as beef, pork, chicken, eggs, milk and their derived products, highly liable to deteriorate. The contamination process by pathogenic bacteria in humans may be caused by poor hygiene conditions during processing involving sick people and animals or involving feces from infected agents. Bacteria- contaminated food may also be hazardous to public health due to the excessive growth in bacteria populations at food surface or within the food. These bacteria may come from the environment and cause toxins that develop into serious health problems on intake.

Hand-manipulated meat, sausages, salamis and cheese are among the most consumed products worldwide. They are also liable to high microbiological contamination due to their manufacturing process.

The World Health Organization and the Food and Agriculture Organization of the United Nations have published reports and studies developed in several regions of the planet highlighting the pathogen risks to populations and suggested the protection of food consumers through special industrial, operational, commercial and residence care. The need for great attention in food safety is a self-evident topic. In fact, improvements in food processing methods and conscience-awareness with regard to food safety by all involved in the food production chain will surely reduce the incidence of food-originated diseases.

2. MICROBIOLOGICAL CONTAMINANTS OF MILK AND HOMEMADE FRESH CHEESE

Milk is one of the most complete food featuring high levels of protein and mineral salts. However, due to the availability of nutrients and almost neutral pH, milk is highly perishable. It is highly liable to microbial growth and requires thermal treatment for its conservation [1]. Pasteurization prolongs milk conservation time, conserves its natural characteristics and preserves it safe for human consumption. High temperatures are involved so that the product's pathogenic microbiota are eliminated with no changes in its physical and chemical constitution. However, people in rural regions still drink milk in natura and use it thus as prime matter for the manufacture of derived products.

The hygienic obtaining of milk is the first critical factor within the manufacturing process of cheese and other products. In fact, the animal, equipments and environment at milking may be an important contamination source by microorganisms [2]. Faults during milking and processing coupled to inadequate conservation temperatures at the selling outlets are factors that contribute towards the commercialization of milk products with microbiological characteristics that go against health norms and legislation [3]. The quality of milk and that of its products is a highly relevant factor for positive industrialization success since both the dairy and the consumer are interested in the outcome. In some case, however, a significant increase in the price of milk ensues. Milk is a product that should come from healthy herds, with good meals and managements, and from farms with proper technical installations that guarantee conservation during transport up to the dairy factory [4].

Since the number of milk contaminants increases at a slow rate from the moment of their introduction, the importance of adequate conservation of recently obtained milk should be underpinned as a basic practice for the maintenance of its quality. Milk should be submitted at low temperatures immediately after the milking process, with the consequent avoidance of the proliferation of unwanted microorganisms [5].

As a milk-derived product, cheese is frequently a food-originating pathogen vector. This is especially true for handmade fresh cheese manufactured from raw milk, lacking any maturation process. The product's microbial contamination is relevant for the industry because of financial liabilities, and for public health because of the risks in food-transmitted diseases.

Several studies [6] have shown that a product's quality and durability largely depend on the prime matter used in manufacturing. It is practically impossible to improve the qualities of a derived product, such as cheese, with a high number of microorganisms present in raw milk.

3. FRESH MINAS CHEESE

Fresh Minas cheese (traditionally manufactured in the state of Minas Gerais, Brazil, whence its name) is defined by the Brazilian Ministry of Health (Decree 146) as fresh cheese obtained by enzyme coagulation of milk with curds and other appropriate coagulant enzymes, supplemented or not by the activity of specific lactic bacteria. According to the Technical Rules for the Identification and Quality of Milk Products [7], fresh Minas cheese may be classified as cheese with low moisture or semi-hard cheese with moisture ranging between 36 and 45.9%; cheese with high moisture or moderate mass cheese with 46 to 54.9% moisture; and very high moisture cheese or soft mass cheese, with not less than 55% moisture.

The processing of fresh Minas cheese comprises the following stages: milk pasteurization, coagulation, cutting, draining, milling, salting, packing and cooling [8]. Since the manufacturing of this type of cheese is highly simple, many small, medium-sized and large dairies are interested in its fabrication. In fact, it is the most common type of cheese found in fairs, bars and grocers. The cheese is normally placed in a common non-vacuum plastic bag and closed by a metal seal [9].

According to the Brazilian Association of Cheese Industry (ABIQ), Brazil produces 400,000 tons of cheese per year, of which 240,000 tons are produced under federal, state and municipal inspection. Most production (95%) is consumed by common people [10].

The intake of fresh cheese may be risky for the consumer's health. However, Decree 861/1984 basically prohibits the sale of fresh cheese manufactured from the raw milk of cows, goats or sheep, pure or mixed. Milk should undergo pasteurization or other equivalent thermal treatment. Current legislation was published after several registers of human brucellosis caused by fresh cheese. In defiance of the law, the homemade manufacture of cheese in certain regions of Brazil is not done with pasteurized milk. Consequently, the consumption of homemade cheese brings to the fore old dangers such as brucellosis (Maltese fever) and other infectious diseases.

In spite of the legal prohibition against the commercialization of fresh and tender cheese manufactured from raw milk, the sale of homemade fresh Minas cheese occurs openly and everywhere in Brazil [11]. This is partially due to a greater yield, simpler processing and lack of product's maturation in the fabrication of this type of cheese, with low costs for the consumer and a fast return of expenditure to the manufacturer [12].

Food protection authorities classify microbial biological contamination as a main danger to public health. Who has constantly raised its voice on the need to restrict food contamination by health-impairing biological agents. Although microbial quality of food is of paramount importance, registration at the Federal Inspection Service does not guarantee lack of pathogens in food [13].

Food-derived diseases may be caused by several microorganism groups that include bacteria, fungi, yeasts, protozoa and viruses. Due to their diversity and pathogenesis, bacteria are by far the most important microbial group and commonly associated with food-transmitting diseases

Bacteria are microorganisms largely spread throughout the natural world and may be found in every type of environment [14]. They cause diseases in humans, animals and plants and deteriorate food and other materials. On the other hand, they may be useful too when they compose the human being's normal microbiota and are used in the production of food as symbiotic in agriculture and medicine.

In spite of certain unreliable Brazilian statistics, it is believed that food-derived diseases in Brazil are high [15]. In fact, several studies estimate that 12% of hospitalization cases in Brazil occur because of infectious intestinal diseases [16].

Occurrences of food-derived diseases are normally associated with certain risk factors, or rather, procedures that benefit toxin infections. The following may be highlighted: faults in food refrigeration; conservation of warm food at room temperature; food prepared many hours earlier for later consumption with inadequate conditioning during the interval; faults in the cooking process; handling of food by people with inadequate personal hygiene practices, or with lesions or with contaminating diseases; usage of contaminated prime matter; faults in the hygiene of utensils and other equipments in food preparation; favorable environmental conditions for the growth of etiological agents; food obtained from unreliable sources; inadequate storage; use of utensils which release toxic residues; intentional or accidental addition of toxic chemicals to the food; usage of water with uncontrolled drinkability features; water contamination from damages in the supply system [17].

Problems in the manufacture of cheese in Brazil are related to precarious conditions of milk, bad conditions during the manhandling of cheese and the lack or deficiency of refrigeration throughout the production chain. These factors worsen the situation and establish contamination conditions which favor the development of microorganisms at several places [18].

Whereas some microorganisms contribute beneficently towards the processing, safety and quality of certain food products, other organisms are involved in processes with unwanted effects in food and for the consumers' health. There are two categories of food-transmitted microbial diseases: food intoxication and infection by food. In food intoxication, the person ingests toxins that are pre-formed by microorganisms in the food. The toxin causes damage to the organism. Examples comprise botulinum toxin that binds itself to the nerve terminals at the muscle level and impedes the release of acetylcholine neurotransmitter, and staphylococcus toxin that acts on the brain's vomiting-center [19]. Infection by food occurs when the pathogen, such as by Salmonella typhy and other serotypes, is ingested and multiplies itself, causing diseases in the intestine tract and often in other organs [20].

The sale of animal-derived food in fair stalls without any refrigeration and without any protection against dust and insects may alter their quality. In the case of cheese, it is sold in portions or slices and thus the external incorporation of biological or non-biological foreign matter is dangerous due to faults in the handling of the product during commercialization, poor hygiene of the stalls and utensils used, and crossed contamination between exposed products [21].

Food microbial contamination is unwanted and dangerous within food microbiology. This aspect should be faced with great strictness. The acknowledgement of possible hygiene deficiency implying in food contamination brings to the fore microorganism groups, comprising indicators, and pathogenic microorganisms that find an excellent environment in food for their development and even for the release of toxic substances [22]. Total and thermotolerant coliforms, such as Staphylococcus aureus, fungi, yeasts and even Salmonella spp., should be highlighted among the microorganisms whose presence and numbers indicate the quality of the product.

The above mentioned microorganisms, causes of several types of pathogenesis, are transmitted to humans because of lack of hygiene, bad habits of handlers, inefficient production processes, maintenance or re-heating of food at inadequate temperatures and also by non-adequate conditions in industries where the food is produced [23].

Most microorganisms, whose pathogenicity in humans depends on their variegated presence in food, are relatively sensitive to high temperatures. In fact, they are destroyed by the adequate cooking of eventually contaminated food or by pasteurization processes.

The Brazilian Agency for Health Vigilance (ANVISA) established, by Decree RDC 12 of the 2nd January 2001[24], the microbiological Standards for several types of food, described in Table 1.

So that food-caused disease cases and events could be characterized, the populations should be informed on the symptoms of each, such as mild diarrheas and vomiting since these are considered as a "passing illness" and not necessarily associated with food consumption [25].

Table 1. Microbiological Standards for Food: cheese with high moisture (55%).

Microorganism	Quantity
Coliforms at 45ºC	5x102 MPN/g
Staphylococcus aureus	5x102 CFU/g
Salmonella sp.	Absence in 25g
* MPN (most probable number), CFU (colony forming unit). Source: ANVISA/2001[24]	

According with registers, more than a billion cases of acute diarrhea are detected in less- than-5-year-old children in developing countries yearly, with 5 million deaths. Between 1999 and 2001, in the state of Paraná, Brazil, 67.1% of food epidemics were caused by bacteria. Moreover, out of 1389 notified epidemics, 38.6 were confirmed in the laboratory; 29/7% were confirmed clinically or epidemiologically suspect and 31.6% were of unknown etiology [25].

World cheese production is slightly above 19 million tons. Cheese production increased more than 76.3% during the last thirty years, or rather, from approximately 10.8 million tons in 1978 to more than 19 millions in 2008. The expansion of milk-producing regions and production increase throughout recent years provided a highly relevant presence of Brazilian production within the world market of milk-derived exports. Concern is therefore high with regard to the quality of commercialized goods for internal and external consumption.

Family-run agriculture in Brazil has an important share in the milk production chain, with approximately 86% of milk producers. However, the production and management of these milk producers are foregrounded on a homemade basis with scanty technical assistance and high influence of cultural factors that may put to risk consumers' health. Technical and educational orientation through the introduction of healthy manufacturing practices are deemed necessary to minimize contamination risks and food intoxication by the product.

Research in all Brazilian regions, where the production and commercialization of cheese is undertaken mainly by small producers, has demonstrated the risk of toxin infections in the consumption of these products by the population.

The curd-cheese is the most produced and consumed milk-derived product in the northeastern region of Brazil. Several investigations [26] have shown that the handling and carelessness in hygiene within the production system have made it foremost as a contamination source. The manufacturers are transmission vectors of the pathogen Staphylococcus aureus and others that may cause food intoxication. The presence of positive coagulase staphylococcus witnesses the lack of hygiene and sanitary conditions during the production, processing, distribution, storing and commercialization stages of samples of curd-cheese. Sanitary education of the producers and the spreading of processing techniques based on good manufacturing practices are mandatory.

Researches in the state of Mato Grosso, in the Mid-Western region of Brazil, (Loguercio & Aleixo 2001) [27] have shown the poor hygiene and sanitary conditions that characterize the production of fresh Minas cheese. Staphylococcus aureus bacteria rates higher than those permitted by current legislation are rife. The need for more sanitary surveillance and orientation by government authorities is urgent.

Research work in the southeastern region of Brazil [28] (Salotti et al 2006) evaluated the microbiological quality of fresh Minas cheese samples. Results from the hinterlands of the state of São Paulo, Brazil, showed non-compliance to rules established by the Brazilian Agency for Sanitary Vigilance (ANVISA) for 83.4% of homemade products and 66.7% for industrial samples with regard to thermotolerant coliforms. In the case of positive coagulase Staphylococcus, 20% of homemade samples and 10% of industrial products failed to comply with the ANVISA regulations. Microbiological results revealed the potential risk of the product for consumers.

After analyzing samples of fresh Minas cheese in Minas Gerais for coliforms and E. coli, a recent study [29] showed the presence of microorganisms, above the rates allowed by current legislation, in 30% of cheese with certificate; 70% of cheese without certificate and 61.4% of mild cheese. Since E. coli, Proteus, Providencia, Serratia, Klebsiella and Enterobacter were identified within the Enterobacteriaceae isolated in fresh Minas cheese, the risk to public health when the products are consumed is amply demonstrated.

Was reported [30] on the risk in the consumption of fresh Minas cheese by the population of the state of Paraná, southern Brazil. Samples inspected by the Federal Inspection Service of Santa Helena PR Brazil revealed that only 15% were in accord to ANVISA standards. All homemade cheese samples and 70% of inspected ones were not according to legislation. Studies [31] confirmed the above results and reported that 50% of samples of analyzed cheese had thermotolerant coliforms, 100% had positive coagulase Staphyloccocus and 12.5% had Salmonella sp. These samples were inadequate for human consumption since they were not consonant to cheese microbiological standards.

4. MICROBIOLOGICAL CONTAMINANTS OF JERKED BEEF

One of the most traditional products of the northeastern region of Brazil is jerked beef which may be characterized as a nutrition food with high calorie rates and widely accepted by consumers for its peculiar sensorial features. Jerked beef is produced from cuts derived from all parts of cattle carcass, salted and dried, with longer durability when compared to that of fresh meat [32].

Due to different nomenclature in Brazil, such as 'carne-de-sertão', 'carne serenada', 'carne de- viagem', 'carne-mole', 'carne-do-vento', 'cacina' or more simple still, dehydrated meat, jerk beef is often confused with another type of salted beef, albeit industrialized, called 'charque' or dried salted meat [33].

Jerked beef was first used in the northeastern region of Brazil as an alternative to preserve beef surplus which could not be consumed immediately and so that the meat would not deteriorate quickly due to difficulties in its preservation especially among the poor population with no refrigeration equipments. Favorable climate conditions and availability of seawater salt, fresh meat could be preserved by being dehydrated and salted.

Currently the above-mentioned preservation process is less relevant due to the introduction of refrigeration. However, many people from different regions of Brazil, especially from the northeast, became accustomed to the produce's characteristic taste and continued to produce jerked beef will less amounts of salt and frequently without exposure to the sun.

Each Brazilian state developed its own technology and thus produced jerked beef with different characteristics with regard to aspect, taste, color, amount of salt and shelf life. The states of Rio Grande do Norte and Ceará are the greatest producers of jerked beef mainly due to climatic conditions that favor the food's dehydration. In fact, jerked beef passed from a locally consumed product and used in certain food receipts to wider conditions. In fact, it is appreciated throughout Brazil and in several meal preparations. Jerked beef may be found in big city centers such as São Paulo and Rio de Janeiro, in homes and restaurants, outside the restricted circle of northeastern cuisine [34], and in the menu of the poorest worker [35,36].

Owing to the popularization of homemade salting technique, jerked beef production follows typically regional norms. Consequently, it is produced in a highly rudimentary way under inadequate sanitary conditions [37,38]. Analysis of the hygiene conditions in the production and commercialization of jerked beef in the region of Itapetinga BA Brazil may be brought forward as an example of the popularization of the technique. In fact, 73.3% of the shopkeepers interviewed admitted that they themselves produced the jerked beef on sale in their shops. Whereas 63.6% used non-inspected meat, 27.3% used meat inspected by municipal health officers and only 0.1% was inspected by federal health officers. Jerked beef was stored and commercialized in 71% of the shops at room temperature, which favored the multiplication of contaminant microorganisms and flies. These facts bring health risks to consumers and jeopardize the product's physical aspects [39].

Salting technique consists in the removal of water from the meat tissues; decrease in water activity ensues, inhibits microbial development and the speed of unwanted reactions of the final product. When salted beef is conserved without any type of refrigeration, its shelf life is higher than that of fresh meat [40]. However, jerked beef has low sodium chloride (NaCl) rates, between 5 and 6%, high moisture, between 65 and 70% [35,41,42] and water activity of 0.92. It may be characterized as partially dehydrated meat in which water activity is not sufficient decreased to avoid microbial development (and consequently degradation) or the production of microbial toxins [43,44].

Although the literal translation of the jerked beef in Portuguese is 'meat exposed to the sun', it is actually only rarely exposed to the sunrays during the dehydration process. The end product is a semi-dehydrated homemade product with four-day shelf-life at room temperature and up to eight days under refrigeration [43,45,41].

Data on the physical and chemical qualities of jerked beef sold in butcheries and supermarkets in João Pessoa PB Brazil showed that water activity in all samples was relatively high, between 0.898 and 0.967, and that the rates of sodium chloride (NaCl) ranged between 3.73% and 9.79%. Consequently, NaCl employed in the process was insufficient to decrease water activity in the product and thus it did not have a significant inhibitory action in the development of most microorganisms in the beef [46]. Lack of standardization in the quality of jerked beef was also assessed in samples collected at inspected shops. Mean rates of water activity were 0.94±0.02. The same average was obtained for samples collected in shops without any health inspection [47]. Variations in sodium chloride rates were also registered in the samples. Techniques for more efficient conservation are required to decrease such risks since it is a type of food with contamination possibilities throughout the manufacturing process.

With regard to the microbiological contamination of jerked beef, the transformation by which meat in natura is processed into jerked beef requires that technological alterations modify the initial microbiota by which the salting and dehydration process selects more tolerant microorganisms for such conditions [48]. Pathogens that may contaminate jerked beef comprise Clostridium perfringens, Staphilococus aureus, Salmonella, verotoxin-producing Escherichia coli, Campylobacter, Yersinia enterocolítica, Listeria monocytogenes, Aeromonas hydrofila, and other deteriorating bacteria [49]. However, low NaCl rates used in jerked beef is one of the factors that trigger microbiological development since decrease in water activity is insufficient to hinder the development of deterioration-producing bacteria of the genus Pseudomona. It also provides proper conditions for the growth of gram-positive bacteria as those of the genus Staphylococcus [38].

Samples of jerked beef from the north of the state of Minas Gerais, Brazil, showed that the amount of mesophile aerobic bacteria, an index of food hygiene quality, was between 2.0x104 UFC/g and 8.9x108 UFC/g. Psichrotrophic bacteria were found in 93.33% of samples, between 5.4x103 UFC/g and 2.9x106 UFC/g. Results show poor hygiene in the manufacture of jerked beef [50]. Similar results were reported in samples of jerked beef commercialized in João Pessoa where the number of mesophile bacteria ranged between 1.8x105 and 7.5x107 UFC/g, with a clear correlationship between mesophile contamination and hygiene and sanitary standards [42].

High thermotolerant coliform rates, which also demonstrate unsatisfactory hygiene and sanitary conditions during the processing stages in the manufacture of jerked beef, were also registered in most jerked beef samples sold in butcheries and supermarkets in João Pessoa PB Brazil [46]. However, total coliforms in food did not report recent fecal contamination or the occurrence of enteropathogens [51,52]. However, Brazilian sanitary laws did not regulate the presence of this microorganism group in meat.

The commercialization of jerked beef in health inspected or not in the region of João Pessoa PB Brazil has been evaluated and results showed high rates in both groups. Ninety-six samples were analyzed and high contamination by feces-derived microorganisms was reported. Staphylococcus ssp. rates were high in both groups, with a low frequency for S. aureus [47]. Staphylococcus aureus rates were higher than 5logUFC/cm2 in 50% of jerked beef samples commercialized in butcheries and supermarkets in João Pessoa PB Brazil. The above amounts demonstrate high contamination causing gastrointestinal disorders in consumers [53].

Mesophile microorganisms Salmonella sp. and Staphylococcus aureus in jerked beef commercialized at room temperature and under refrigeration in Campina Grande PB Brazil showed no significant difference in S. aureus counts for samples commercialized at room temperature and under refrigeration. Salmonella ssp. was detected in 40% of jerked beef samples commercialized at room temperature and in 30% of samples under refrigeration.

Another source of contamination in the commercialization of jerked beef may be found in supermarkets, open market stalls and butcheries. Data reveal that the utensils used in 75% of these outlets were not exclusively for meat cutting and that the handling of money and food was common practice in 25% of the businesses. Aprons, disposable caps and clean closed shoes were only found in 25% of the shops.

The inadequate washing of hands and other habits such as talking during the handling and commercialization of food were also reported in all commercial enterprises [54]. It has been verified that in João Pessoa, supermarkets had the best hygiene and sanitary profile in jerked beef quality, whereas open markets and stalls in fairs had the worst [42]. In the latter case, meat is exposed without any type of protection and any passerby may handle it at will.

Investigations were carried out with regard to alien matter, such as flies, acarids, larvae, insects, feathers and others, found in jerked beef sold in 20 (90.9%) shops in Diadema SP Brazil, specialized in typical products from the northeastern region of Brazil. Exposure of products without any wrappings is an excellent condition for attacks by insects, especially flies, and rodents, making it improper for human consumption in the wake of health-hazard matter [55].

Almost all jerked beef is manufactured and sold in small shops and specifically prepared for people who appreciate the product. Consequently, lack of sanitary rules for its production, precarious conditions in its commercialization, storage without refrigeration and its exposure without any protection characterize jerked beef in such conditions as haphazard to public health.

5. MICROBIOLOGICAL CONTAMINANTS IN MEAT FILLINGS (SAUSAGES MADE FROM BEEF AND FOWL MEAT, SALAMI)

Animal-derived food conveys a host of microorganisms dangerous to human health. The incidence of toxin infections in Brazil is high, although statistics are rather lacking on the matter. Bacteria causing toxin infections are widely distributed although their main natural habitat is the human or animal intestine tract [14]. The most common bacteria in food contamination are of the genera Escherichia, Salmonella, Shigella, Yersinia, Vibrio, Brucella, Clostridium, Listeria, Campylobacter, Bacillus cereus and Staphylococcus aureus [56]. Sausages, widely used in Europe, is a type of food stuffed with meat from swine, fowls, goats, cattle and fish, seasoned with several types of spicy ingredients. Sausages are a highly popular food in Brazil, easily accessible to all classes of people and consumed throughout the country. Sausages have great acceptance in the southern and southeastern regions due to a more Europeanized culture.

Brazilian swine breeding has a very important role in several sectors of Brazilian economy. It produces jobs and intensifies demand of agricultural products in the industrialization and commercialization of animal-derived products. Besides providing excellent animal protein to the population, the meat industry exports meat and important economical assets are aggregated [57].

Data by the Brazilian Association of Production and Exportation Industry of Pork (ABIPECS) showed that approximately 65% of the Brazilian pork production is directed towards the internal market through industrialized products. Among the processed products, the fresh Tuscan-type sausage, made exclusively from pork, uses the less important animal parts as food, with great acceptance among the population.

Pork and its derived products undergo bacterial alterations owing to several factors such as animal health and fecal contamination by Escherichia coli highly relevant worldwide as a microorganism hazardous to animal and population health involving hygiene and sanitary issues [57]. The same author evaluated the occurrence of E.coli in swine in the abattoirs of Rio de Janeiro, Brazil, from which the Tuscun-type sausages were made. Different parts of the animal used in the stuffing process were examined and concluded that, depending on the meat and the manufacturing process, sausages were not fit for consumption.

Toxoplasma gondhii in fresh pork sausages commercialized in Botucatu SP Brazil was evaluated by researches [58]. Pork represents one of the main sources of infection by T. gondii in humans. Swine were the most important animals in the process of toxoplasmosis transmission [59,60,58,61]. Mendonça's data did not show any evidence of T. gondii in the samples, perhaps due to salt, used in the manufacturing process, which eliminated the microorganism.

The occurrence of food infection by pork sausages contaminated with Salmonella sp. has been suggested [62]. Brazilian sanitary laws [63] make it mandatory that the microorganism should be lacking in 25% so that human intoxication may occur. However, such possibility may vary since it depends on serotype and the person's health conditions and tolerance. Mürmann's results [62] showed that 24% of pork sausages samples were contaminated by Salmonella enterica.

Contamination by Salmonella sp in pork may occur in pens through contact with feces, lack of hygiene and sanitation in the installations and by other animals during the transport, waiting or pre-finishing period. A high increase of S. enteriditis in food toxin infections in humans and in aviary products has been reported in Brazil since the 1990 [64].

Fecal coliforms, positive coagulase staphylococcus, Salmonella spp and Campylobacter spp in fresh sausages were evaluated [65]. When the hygiene and sanitary quality among the different types of fresh sausages was compared, pork sausages had the worst scores with regard to risks in public health, as ruled by the RDC n.12 of Anvisa [63].

The authors also registered that most samples were not in accordance to microbiological standards and thus hazardous to consumer's health. Another datum refers to the absence of Campylobacter spp in the samples, perhaps due to sodium chloride concentrations over 1.5% that may have inhibited these microorganisms.

Was analyzed [66] the presence of Listeria spp, principally L. monocytogenes, during the manufacture of fresh mixed-meat sausages in three abattoirs, supervised by state health authorities, in Pelotas RS Brazil. Results showed that all samples from the three abattoirs were contaminated by Listeria spp, of which the most frequent species was L. innocua (97.6), followed by L. monocytogenes (29.3%) and L. welshimeri (24.4%).

When the hygiene and sanitary conditions in the manufacture of fresh sausages in the northwestern region of the state of Paraná, Brazil, were analyzed [67] data failed to show any microbiological contamination that would jeopardize the health of the consumer. The manufacture of these samples followed strict handling and processing procedures.

On the other hand, another authors [68] studied the prevalence of antimicrobial resistance by serotypes of Salmonella isolated from fresh pork sausages and found significant quantities of the above in samples collected in the southern state of Santa Catarina, Brazil. These serotypes resisted the antimicrobial products sulfonamide and tetracycline (81%); ampicillin (50%) and chloramphenicol (31.25%). Was evaluated the microbiological quality of fresh sausages in two towns of the state of Minas Gerais, Brazil [69]. Results confirmed positive coagulase Staphylococcus in 35% of samples which made them improper for human consumption. The same author also demonstrated that 35% of samples were contaminated by thermotolerant fecal coliforms above the maximum limits.

The consumption of chicken meat and its derivates has recently increased considerably in Brazil due to price decrease, good quality and practical cuttings provided [70]. Per capita consumption increased from 10 kg to 35.4 kg, only slightly lower than beef consumption (União Brasileira de Avicultura) [71]. The products' quality is highly important and a great concern to health authorities, food industry and consumers. Chickens bred for human consumption may host several pathogenic microorganisms such as Campylobacter jejuni, Salmonella sp and E. coli [72,73].

Rall investigated [70] the hygiene and sanitary conditions of chicken meat and several types of sausages commercialized in the interior of the state of São Paulo, Brazil, by determining the Most Probable Number of coliforms at 45oC. The same authors also analyzed the presence of Samonella sp by the traditional method and by PCR. Data showed that 40% of the 75 sausage samples analyzed were improper for human consumption due to excess in coliforms and 7 samples (9.3%) were positive for Salmonella sp. (9.3%). Research by PCR increased to 56% Salmonella-positive samples. When the frequency rate of Salmonella was added to the microbiological limits for coliforms, it might be concluded that 86.7% of sausages were improper for human consumption.

In their research in the northwestern region of the state of São Paulo, Brazil, others authors [74] found contamination by Salmonella in 16% of chicken sausages samples. The most relevant item in the above result may be the handling of the product during processing, coupled to the exposure of the meat to several contamination sources or to already contaminated chickens that provided the contamination of the final product.

The above authors researched the microbiological quality of industrialized avian products and their derivates in another region of the state of São Paulo. Research determined the presence of Campylobacter jejuni and Salmonella sp. Sausages samples analyzed were 42.8% positive for C. jejuni and 28.5% for Salmonella sp.

The presence of microorganisms in the above research works suggests the need for greater care during the handling and preparation of sausages that may be eaten in natura, without any heating treatment that would reduce the number of microorganisms causing toxin infections [75].

Vienna sausage may be defined as an industrialized meat-stuffed product obtained from the emulsion of animal meat to which are added a variety of ingredients and condiments, filling a natural or artificial casing, and submitted to proper thermal process [76]. Vienna sausages are highly popular in Brazil due to their low costs and for the manufacturing of the ubiquitous hot dog.

The physical and chemical characteristics of Vienna sausages should contain a maximum of 65% moisture, 30% fat, 2% starch, 7% total carbohydrates, 12% protein. Fresh sausages should be under permanent refrigeration (0oC to 5oC) from manufacture until consumption, with expiry period after 48 hours [77].

Vienna sausages samples of the hot-dog type were analyzed in Niterói and Rio de Janeiro RJ Brazil to detect thermotolerant coliforms, positive coagulase Staphylococcus, Clostridium spp and Salmonella spp by conventional methods with the necessary modifications [78]. When compared to health norms, results showed that 33% of samples were inadequate for consumption due to the presence of their isolated microorganisms.

Salami is another highly appreciated product in southern Brazil. Its homemade manufacture started in the early 20th century with an enormous variety of industrialized types that differed in composition, casing, size of meat and fats, spices, smoking process and maturation period prior to commercialization. Researchers revaluated the various characteristics [79] of salamis produced by small- and medium-sized agro-industries in the southern state of Santa Catarina, Brazil. Bacteria Staphylococcus aureus, Salmonella spp, Listeria monocystogenes and E.coli were researched in the products. Although results did not identify contamination by Salmonella spp, the E. coli and S. aureus counts were significant, but within the reliability parameters.

Was analyzed the quality [80] of salami in the interior of the state of São Paulo, Brazil, and verified that, despite samples with E. coli and fecal coliforms, all samples were within health standards. Nevertheless, 60% of samples were contaminated by Staphylococcus aureus and 22% were unhealthy for consumption.

6. CONCISE SUMMARY

Due to its significance for public health, the proper management of meat and dairy products need more attention, diligence, and oversight from the relevant health authorities. Given the significant cultural diversity in food production in Brazil, the active involvement of all sectors within the food supply chain is essential to ensure the health and safety of products, hence reducing foodborne illnesses.

REFERENCES

1. Arcuri EF, Brito MAVP, Brito JRF, Pinto SM, Angelo FF, Souza GN, Qualidade Micro biológica do leite refrigerado nas fazendas. Arquivo Brasileiro Medicina Veterinária Zootecnia, Belo Horizonte, 2006; 58(3): 440-446.
2. Lange CC, Brito, JRF. Inluência da qualidade do leite na manufatura e vida de prateleira dos produtos lácteos: papel das altas contagens microbianas. In: Brito, J.R.F; Portugal, JA (Eds.) Diagnóstico da qualidade do leite, impacto para a indústria e a questão dos resíduos de antibióticos. Empresa Brasileira de Pesquisa Agropecuária (Embrapa), Juiz de Fora, p. 117-138, 2003.
3. Gomes HÁ, Gallo CR. Ocorrência de Staphylococcus aureus e produção de enterotoxinas por linhagens isoladas a partir de leite cru, leite pasteurizado tipo C e queijo Minas frescal comercializados em Piracicaba,SP. Revista Ciência e Tecnologia de Alimentos, 1995; 15(2): 158-161.
4. Gonçalves CA, Vieira LC. Obtenção e higienização do leite in natura. Empresa Brasileira de Pesquisa Agropecuária (EMBRAPA). Amazônia Oriental, Belém: Documento 141. 2002. 28p.
5. Olivieri D. de A. Avaliação da qualidade microbiológica de amostras de mercado de queijo mussarela, elaborado a partir do leite de búfala (Bubalus bubalis). 61 p. Disser tação (Mestrado) - Escola Superior de Agricultura Luiz de Queiroz, Piracicaba, SP, 2004.

6. Huhn S, Hajdenwurcel JR, Moraes JM de, Vargas OL. Qualidade microbiológica do leite cru obtido por meio de ordenha manual e mecânica e ao chegar à plataforma.Re vista do Instituto de Laticínios Cândido Tostes, 1980; 35(209): 3-8.

7. Brasil. Ministério da Agricultura, do Abastecimento e da reforma Agrária. Portaria n. 146, de 07 de março de 1996. Diário Oficial da União, Seção I, Brasília, DF, p 3977-3886, 1996.

8. Vieira DAS, Neto JPM. Elaboração de queijos frescais em pequena escala. Informe Agropecuário, 1982; 8(88): 28-29.

9. Loguercio AP, Aleixo JAG. Microbiologia de queijo tipo Minas Frescal produzido artesanalmente. Ciencia Rural, 2001;31(6): 1063-1067.

10. Hoffman FL, Cruz CHG, Vinturim TM. Qualidade microbiológica de queijos comer cializados na região de São José do Rio Preto, SP. Revista do Instituto de Laticínios. Cândido Tostes,1995; 50: 42-47.

11. Furtado MM, Mosquim MCA, Fernandes AR, Silva CAB da. Laticínios diversificados. In: Silva CAB da, Fernandes AR. Projetos de empreendimentos agroindustriais: pro dutos de origem animal. Viçosa: UFV, 2003. 308p.

12. Almeida Filho ES. Características microbiológicas do queijo Minas frescal, produzido artesanalmente e comercializado no Município de Poços de Caldas, MG. Dissertação (Mestrado) - Faculdade de Ciências Agrárias e Veterinárias, Campus de Jaboticabal da UNESP; 1999.

13. Pinto PSA, Germano MIS, Germano PML. Queijos minas: problemas emergentes de vigilância sanitária. Revista Higiene Alimentar,1996; 10(44): 22-27.

14. Cunha CP, Nascimento MGF, Jesus VLT, Nascimento ER, Corbia ACG. Queijo tipo minas frescal com e sem serviço de Inspeção Federal - Contaminação por coliformes fecais e Escherichia coli. V Congresso Brasileiro de Higienistas de Alimentos. Foz de Iguaçu, PR, 1999. Revista Higiene Alimentar,1999; 13(61): 34-35.

15. Franco, GMB, Landgraf, M. Microbiologia dos alimentos. São Paulo: Atheneu; 1996.

16. Santos, TBA, Balioni, GA, Soares, MMSR, Ribeiro, MC. Condições higiênico-sanitárias de alfaces antes e após tratamento com agentes antibacteriano. Revista Higiene Ali mentar, 2004; 18(121): 85- 88.

17. Nascimento, MS, Silva, N, Catanozi, MPLM. Avaliação microbiológica de frutas e hor taliças frescas, comercializadas no município de Campinas, SP. Revista Higiene Ali mentar, 2003;17(114): 73-76.

18. CENEPI/FUNASA/MS. Manual integrado de prevenção e controle de doenças transmitidas por alimentos, 2001.

19. Lisita, MO. Evolução da população bacteriana na linha de produção do queijo minas frescal em uma indústria de lacticínios. 61 p. Dissertação (Mestrado) - Escola Superior de Agricultura Luiz de Queiroz, Piracicaba, SP, 2005

20. Silva, JEAE. Manual de controle higiênico sanitário de alimentos. São Paulo: Varela; 1999

21. Trabulsi, LR. Microbiologia. São Paulo: Atheneu; 1999.

22. Correia, M, Roncada, M.J. Características microscópicas de queijo prato, mussarela e mineiro comercializados em feiras livres da Cidade de São Paulo. Revista de Saúde Pública, junho 1997; 3 (31).

23. Franco, RM, Almaida, LEF. Avaliação microbiológica de queijo ralado, tipo parmesão, comercializado em Niterói, RJ. Revista Higiene Alimentar, 1992; 6(21): 33-36.

24. Brasil, Agência Nacional de Vigilância Sanitária (ANVISA). Resolução RDC nº 12 de 02 de Janeiro de 2001.

25. Silva, LF. Procedimento operacional padronizado de higienização como requisito para segurança alimentar em unidade de alimentação. Dissertação (Mestrado em Ciência e Tecnologia de Alimentos) - Universidade Federal de Santa Maria, RS, 2006.

26. Tigrel, DM, Borelly, MAN. Pesquisa de Estafilococos coagulase-positiva em amostras de "queijo coalho" comercializadas por ambulantes na praia de Itapuã (Salvador-BA). Revista Ciências medica e biologia, 2011;.10(2): 162-166.

27. Loguercio, AP, Aleixo, JAG. Microbiologia de queijo tipo minas frescal produzido artesanalmente. Ciência Rural, 2001; 31(6).

28. Salotti, BM, Carvalho, ACFB, Amaral, LA, Vidal-Martins, AMC, Cortez, AL. Qualidade Microbiológica do queijo minas frescal comercializado no município de Jaboticabal, SP, Brasil. Arquivos do Instituto Biologico, 2006;73 (2): 171-175.

29. Okura, M.H. Avaliação microbiológica de queijo tipos minas frescal comercializados na região do triângulo mineiro Tese (Doutorado em Microbiologia) - Tese (doutorado) - Universidade Estadual Paulista, Faculdade de Ciências Agrárias e Veterinárias, 2010.

30. Pinto, FGS, Souza, M, Saling, S, Moura, AC. Qualidade microbiológica de queijo Minas Frescal comercializado no Município de Santa Helena, PR, Brasil. Arquivos do Instituto Biológico, 2011; 78 (2): 191-198.

31. Oliveira, DF, Tonial, CEC. Sazonalidade como fator interferente na composição físico- quimica e avaliação microbiológica de queijos coloniais. Arquivo Brasileiro de medic ina veterinária e zootecnologia, 2012; 64(2): 521-523.

32. Vasconcelos O. Por cima da carne seca. Revista Globo Rural, 1986; 1 (5): 15-20.

33. Lira GM, Shimokomaki M. Parâmetros de qualidade da carne de sol e dos charques.Higiene Alimentar, São Paulo,1998; 44 (13): 66-69.

34. Serviço de Informação da Carne [SIC]. Charque, carne de sol, carne seca. Desenvolvido pelo Comitê Técnico do SIC. São Paulo. Disponível em: http://www.sic.org.br/char que.asp (Acesso 14 de agost 2012).

35. Nóbrega DM, Schneider I S. Contribuição ao estudo da carne de sol visando melhorar sua conservação. Revista Higiene Alimentar. 1983; 2(3): 150-4.

36. Gouvêa JAG, Gouvêa AAL. Tecnologia de fabricação da carne de sol. Bahia: Rede de Tecnologia da Bahia– RETEC/BA, 2007. 23 p. Dossiê Técnico.

37. Menucci TA. Avaliação das condições higiênico-sanitárias da carne de sol comerciali zada em "casas do norte" no município de Diadema- SP. 2009. 121 f. Dissertação (Mes trado em Saúde Pública) – Faculdade de Saúde Pública, Universidade de São Paulo, São Paulo, 2009.

38. Azevedo PRA, Morais MVT. A tecnologia da produção da carne de sol e suas simpli ficações nos aspectos higiênico-sanitários. Revista Nacional da Carne, São Paulo, 2005; 29(98): 12-13.

39. Ramos ALS, Ramos EM, VIANA EJ. Avaliação das condições higiênicas na produção e comercialização da carne de sol na região de Itapetinga, BA. Revista Higiene Ali mentar, 2007; 21(150): 371-374.

40. Picchi, V, Cia G. Fabricação do charque. Boletim do Centro de Tecnologia de Carnes, 1980; 5: 11-30.

41. Shimokomaki M, Olivo R. Suplementação de vitamina e melhora a qualidade de carnes e derivados. In: Shimokomaki, M. et al. (Ed.). Atualidades em ciência e tecnologia de carnes. São Paulo: Varela, 2006. cap. 11, p. 115-121

42. Farias SMOC. Qualidade da carne de sol comercializada na cidade de João Pessoa-PB, Dissertação de Mestrado- UFPB/CT, 142f. 2010.

43. Lira GM, Shimokomaki M. Parâmetros de qualidade da carne de sol e dos charques. Revista Higiene Alimentar. 1998; 12(58): 33-5.

44. Felicio PE. Carne de sol – Produto artesanal, de consumo regional, tem potencial para ser fabricado e comercializado no país todo. ABCZ, 2002.

45. Costa EL, Silva JA. Avaliação Microbiológica da carne de sol elaborada com baixos teores de cloreto de sódio. Revista Higiene Alimentar 2001; 21(2): 149-53.

46. Costa EL, Silva JA. Qualidade sanitária da carne de sol comercializada em açougues e supermercados de João Pessoa – PB. Bol. CEPPA. Curitiba. 1999; 17(2): 137-44

47. Costa EL, Silva JA. Avaliação Microbiológica da carne de sol elaborada com baixos teores de cloreto de sódio. Revista Higiene Alimentar. 2001; 21(2): 149-53.

48. Maca JV, Miller RK, Acuff, GR. Microbiological, sensory and chemical characteristics of vacuum-packaged ground beef patties treated withs salts of organic acids. Journal of Food Science, Chicago,1997; 62(3): 591-596.

49. Gill CO. Microbiological contamination of meat during slaughter and butchering of cattle, sheep and pigs. In: The microbiology of meat and poultry. Londres: Blackie Aca demic & Professional, 1998. cap. 4, p. 119-157

50. Cruz ALM. Produção, comercialização, consumo, qualidade microbiológica e características físico-químicas da carne de sol do Norte de Minas Gerais, Dissertação de Mestrado,Montes Claros, MG: ICA/UFMG, 2010.

51. Franco BDGM, Landgraf, M. Microrganismos patogênicos de importância em alimentos. In: Microbiologia dos alimentos. São Paulo: Atheneu, 2008. cap. 4, p. 33-82.

52. Silva N, Junqueira VCA, Silveira NFA, Tanawaki MH, Dos Santos, R S, Gomes R A R. Manual de Métodos de Análise Microbiológica de Alimentos. 4° edição. São Paulo. Ed. Livraria Varela, 2010.

53. Costa EL, Silva JA. Qualidade sanitária da carne de sol comercializada em açougues e supermercados de João Pessoa – PB. Bol. CEPPA Curitiba 1999; 17(2): 137-44.

54. Miranda PC, Barreto NSE. Avaliação Higiênico-Sanitária de diferentes Estabelecimen tos de Comercialização da Carne-de sol no Município de Cruz Das Almas-Ba. Revista Caatinga, Mossoró, 2012; 25(2): 166-172.

55. Mennucci TA, Marciano MAM, ATUI, MB, Polineto A, Germano PML. Study on con taminant materials within "sun dried meat (jerked beef)" at the "Northern Houses. Revista Instituto Adolfo Lutz 2010; 69(1): 47-54.

56. Pinto A. Doenças de origem microbiana transmitidas pelos alimentos. Millenium 1996; (4): 91-100.

57. Franco R. E.coli: ocorrência em suínos abatidos na grande Rio e sua viabilidade exper imental em linguiça frescal tipo toscana.Tese doutorado. Universidade Federal Flumi nense; 2002.

58. Mendonça A O. Detecção de Toxoplasma gondii em linguiças comercializadas no muni cípio de Botucatu- SP. Tese doutorado. Universidade Estadual Paulista Julio Mesquita Filho; 2003.

59. DurbeyJ P. Refinemente of pepsin digestion method for isolation of Toxoplasma gondii from infected tissues. Veterinary Parasitology 1998; 74: 75-77.

60. Gamble H R, Murrel K D. Detection of parasites in food. Parasitology 1998; (117): 97-111.

61. Tenter A M. Current knowledge on the epidemiology of infections with Toxoplasma. The Tokai Journal of Experimental and Clinical Medicine 1998; 23 (6): 391.

62. Mürmann L. Avaliação do risco de infecção por Salmonella sp em consumidores de linguiça frescal de carne suína em Porto Alegre –RS. Tese doutorado. Universidade Federal do Rio Grande do Sul; 2008.

63. Brasil. Agência Nacional de Vigilância Veterinária. Resolução nº12 de 12 de janeiro 2001. Regulamento técnico sobre os padrões microbiológicos para alimentos. 2001. Disponível: http:/WWW.anvisa.gov.br/Regis/resol/12_oirac. num. (Acesso: 10 agosto 2012).

64. Fuzihara T O, Fernandes S A, Franco B D. Prevalence e dissemination of Salmonella serotypes along the slaughtering process in Brazilian small poultry slaugtherhouses. Journal Food Protection 2000; 63: 1749-1753.

65. Cortez A L L, Carvalho A C F B; Amaral L A, Salotti B M, Vidal-Martins A M C. Coli formes fecais, Estafilococos coagulase positiva (ECP) e Campylobacter ssp em linguiça frescal. Alimentos Nutrição. 2004 15 (3): 215-220.

66. Silva W P, Lima A S, Gandra E A, Araújo M R, Macedo M R P, Duval E H. Listeria spp. no processamento de linguiça frescal em frigoríficos de Pelotas, RS, Brasil. Ciência Rural 2004; 34 (3): 911-916.

67. Corazza M L, Mantovani D, Filho L C, Costa S C. Avaliação higiênico-sanitária de lin guiças tipo frescal após inspeção sanitária realizada por órgãos federal, estadual e mu nicipal na região noroeste do Paraná. Revista Saúde e Pesquisa 2011; 4 (3): 357-362.

68. Spricigo D A, Matsumoto S R, Espindola M L, Vaz E K, Ferraz S M. Prevalência e perfil de resistência antimicrobianos de sorovares de Samonella isolados de linguiças suínas tipo frescal em Lages, SC. Arquivo Brasileiro Medicina Veterinária e Zootecnia 2008; 60 (2): 517-520.

69. Marques S C, Boari C A, Brcko C C, Nascimento A R, Picolli R H. Evaluation of hy gienical-sanitary type frescal commercialized in the cities of Três Corações and Lavras- MG. Ciência Agrotecnologia 2006; 30 (6): 1120-1123.

70. Rall V L M, Prado J G, Candeias J M G, Cardoso K F G, Rall R, Araujo Junior J P. Pesquisa de Salmonella e das condições sanitárias em frangos e linguiças comercializadas na cidade de Botucatu. Brazilian Journal Veterinary Research Animal Science Brazilian. Journal. 2009; 46(3): 167-174.

71. União Brasileira de Avicultura. Relatório anual 2005-2006. Disponível em < http:// www.uba.org.br/ubanews_files/rel_uba_2005_06.pdf< (Acesso em: 23 jul.2012).

72. Lucey B, Feurer C. Greer P, Moloney P, Cryan B, Fanning S. Antimicrobial resistence profiling and DNA amplification Fingerprint (DAF) of thermophlic Campylobacter spp in human, poutry and porcine samples from Cork region of Ireland. Journal of Applied Microbiology 2000; 89 (5): 727-734.

73. Natrajan N, Sheldon B W. Inhibition of Salmonella on poultry skin protein and poly saccharide-based films containing a nisin formulation. Journal of Food Protection 2000; 63 (9): 1268-1272.

74. Carvalho A C F B, Cortez A L L. Samonella spp in carcasses, mechanically deboned meat sausages and chiken meat. Ciência Rural 2005; 35 (6): 1465-1468.

75. Carvalho A C F B, Cortes A L L. Contaminação de produtos avícolas industrialização e seus derivados por Campylobacter jejuni e Salmonella sp. Arquivo Veterinária 2003; 19 (1): 057-062.

76. Brasil, Ministério da Agricultura, Pecuária e Abastecimento. Secretaria de Defesa Agro pecuária - MAPA/SDA. Instrução Normativa Nº 4 de 31 de março de 2000.-Aprova os Regulamentos Técnicos de Identidade e Qualidade de Carne Mecanicamente Separada, de Mortadela, de Lingüiça e de Salsicha -Diário Oficial da União, Brasília, DF, p.6, de 05 de abril de 2000. Seção 1.

77. Ferreira M C, Fraqueza M J, Barreto A S.Avaliação do prazo de vida útil da salsicha fresca. Revista Portuguesa de Ciências Veterinárias 2007; 102 (561): 141-143.

78. Martins L L, Santos J F, Franco R M, Oliveira L A T, Bezz J. Bacteriological study in bovine and chikem hot dog type- sausages sold in vacuumed packing-case and retail commercialized in Rio de Janeiro city and Niterói, RJ/ Brazil supermarkets. Revista. Instituto. Adolfo Lutz 2008; 67 (3): 215-220.

79. D'Agostini F P, Campana P, Degenhart R. Qualidade e identidade de embutidos pro duzidos no baixo Vale do Rio do peixe, Santa Catarina- Brasil. E. Tech Tecnologias para Competitividade Industrial 2009; 2 (2): 1-13.

80. Hoffmann F L, Garcia-Cruz C H, Vinturim T M, Carmello M T. Qualidade microbiológica do salame. B. Ceppa 1997; 15 (1): 57-64.

13 Chapter

PRESENCE OF ORGANOCHLORINE PESTICIDE RESIDUES IN ANIMAL FEED AND ADIPOSE BOVINE TISSUE

1. INTRODUCTION: AN OVERVIEW

Currently, over 800 distinct types of pesticides are employed to manage insects, rodents, fungi, and undesirable vegetation in agricultural production. While the majority of pesticides either dissipate or dissolve in soil, water, and the environment, trace levels of pesticide residues may be transmitted to humans through the food chain, posing potential risks to human health[1] . Pest management in intensive agriculture include the treatment of crops (fruits, vegetables, cereals, etc.) throughout both pre-harvest and post-harvest phases. Rodenticides are utilized in the post-harvest storage phase, while fungicides may be applied at any step of the process, contingent upon the specific crop. These compounds can be transmitted from plants to animals through the food chain. Moreover, the breeding animals and their housing might be treated with pesticide solution to avert pest infestations. Thus, both contamination pathways can result in the bioaccumulation of persistent pesticides in animal-derived food products, including meat, fat, fish, eggs, and milk. [2,3] In recent decades, significant attention has been directed towards this group of substances at the international level, as it became evident that they are disseminated throughout the environment and that critical concentrations have been attained in certain regions, including areas where they have never been manufactured or utilized. Numerous nations prohibited the use of Organochlorine Pesticides (OCPs) in the 1970s and 1980s, while its application persists in various other countries. Organochlorine pesticides (OCPs) are recognized as significant environmental contaminants due to their persistence, capacity for long-range transport, and harmful effects on humans and animals. OCPs are carcinogenic in both animals and humans (International Agency for Research on Cancer, 1987). The immunotoxicity of specific organochlorine pesticides (OCPs) has been established in vitro[4], in vivo[5],, and in animal models, as well as in the immune systems of human fetuses, neonates, and infants. [6,7,8,9].

A growing number of epidemiological studies have investigated blood or adipose levels of OCPs and their metabolites in relation with cancer, neurodevelopmental effects, immunotoxicity and reproductive efficiency [10,11,12]. The main sources of OCPs in the human diet are foods of animal origin and environmental exposure. It has been concluded that humans are exposed to toxic compounds via diet in a much higher degree compared to other exposure routes such as inhalation and dermal exposure. Low volatility and high stability, together with lipophilic behaviour, are responsible critical factor for their persistence in the environment (air, water and soil) and subsequent concentration in fatty tissues through the food chain. Therefore, it's important to identify and to monitor levels of OCPs in foodstuff of animal origin (meat and tissues that contain fat, milk and dairy products, eggs, honey and fish). The main pathway for the OCPs contamination of animal food is the ingestion of the contaminated food and/or water by the animals. [13,14,15] Breeding animals can accumulate persistent organic pollutants from contaminated feed and water, and/or from pesticides application in livestock areas (treatment of cowshed, pigsties, sheepfold etc.).[16,17,18] The utilization of feedstuffs on farms has become essential for animal nutrition in developed nations due to escalating production demands. Animal feed significantly influences the food chain and affects the composition and quality of animal products consumed by humans. Consequently, the regulation of OCP residues in animal feed is crucial, as is the monitoring in adipose tissues.

1.1 Organochlorine Pesticides (OCPs)

Organochlorine pesticides (OCPs) were intensively used in agriculture to protect cultivated plants in mid-twentieth century. 1,1,1-Trichloro-2,2-bis(4-chlorophenyl) ethane (DDT), one of the common OCPs, was used to prevent spreading of malaria and other vector-borne diseases such as dengue, leishmaniasis and Japanese encephalitis through the prevention of growth of mosquito.[19,20] After OCPs were used widely in soil and plants for some years and due to their relative stability and bioaccumulation property, these persistent chemicals can be transferred and magnified to higher trophic level through the food chain. Consequently, OCP residues are present in fatty foods, both foods of animal origin such as meat, eggs and milk, and of plant origin such as vegetable oil, nuts, oat and olives. Besides, these chemicals are widely distributed in the environment, which provides another route of unwanted intake in human. [21,22,23] Nevertheless, human exposure occurs still primarily via low level food contamination. Since their mode of action is by targeting system or enzymes in the pest which may be identical or very similar to system or enzymes in human beings, these OCPs pose risks to human health and the environment. [24,25] Thus, monitoring of OCPs residues in food becomes a routine analysis of pesticides monitoring laboratories. All US government pesticides datasets showed that persistent OCP residues were surprisingly common in certain foods despite being off the market for over 30 years. Residues of dieldrin, in particular, posed substantial risks in certain root crops. About one quarter of samples of organically labelled fresh produce contained pesticides residues, compared with about three quarters of conventional samples. [26,27] Among the contaminated organic vegetable samples, about 60% of them were contaminated with OCPs. After some OCPs were banned for use since the 80s, common daily food items such as eggs, milk, poultry, meat and fish have been used for monitoring the residuals levels of OCPs. As regards food of animal origin, one efficient way to avoid large-scale contamination is to control and monitor the levels of OCPs residues present in animal feeds before being fed to the husbandry animals. [28,29,30]

At the same time, public health safety authorities should constantly monitor the OCPs in animal food commodities as the major source of human background exposure to OCPs is through food of animal origin. Most persistent organic pollutant (POPs) are OCPs, namely, aldrin, endrin, chlordane, DDT and hexachlorobenzene (HCB). They have been banned for agricultural or domestic use in Europe, North America and many countries of South America, in accordance with Stockholm Convention in 1980s. However, some OCPs are still used, e.g. DDT is used to control the growth of mosquito that spread malaria or as antifouling agent in some developing countries. [31,32] Residues of OCPs have been detected in breast milk (including DDT, HCB and HCH isomers) in contaminated areas. Recently, the scope of POPs was extended to include nine plus one chemicals. Among these new POPs, chlordecone, lindane, α-HCH, β-HCH, pentachlorobenzene (PeCB) and endosulfan, also belong to OCPs. [33,34] In order to fulfil the requirements of the Stockholm convention, the participating countries have to develop their own implementation plant to monitor the background level and collate the exposure data. To ensure the pesticide residues are not found in food of feed at levels presenting an unacceptable risk for human consumption, maximum residue levels (MRLs) have therefore been set by the European Commission. [35,36,37] MRLs are the upper legal concentration limits for pesticides in or on food or feed. They are set for a wide range of food commodities of plant and animal origin, and they usually apply to the product as placed in the market. MRLs are not simply set as toxicological threshold levels; they are derived after a comprehensive assessment of the properties of the active substance and the residues behaviour on treated crops. Both the periodic estimation of human exposure to persistent organic pollutants and the establishment by the EU authorities of MRLs in foods have required the development of analytical methods suitable for research purposes and inspection programmes. As an example, the European Union has established maximum contents for these compounds in animal feed which can be as low as 5 μg Kg-1 for some OCPs in fish feed and β-HCH in cattle feed. In the rest of feed materials these values can be as low as 10 μg Kg-1 relative to feedstuff with moisture content of 12%. [38,39,40]

1.2 Extraction methods and clean-up of OCPs

Animal feed as well as animal fat are considered a very complex matrices with large number of components especially lipids. Consequently, the development od sensitive methods for its analysis with elimination of interferent compounds and enough efficiency in term of analyte recovery represents an interesting task. [41,42] The most intricate step in these procedures is represented by the sample extraction and clean-up that should be efficient enough to allow a reliable screening of contaminated samples. The selection of suitable solvent (s) and extraction method is critical for obtaining satisfactory recovery of OCPs from the food matrix. Of course, if co-extracted materials are minimised in the extract, the clean-up procedure would became simpler. Owing to the lipophilicity of OCPs, organic solvent (s) normally can extract OCPs form food efficiently but lipids are also co-extracted. Solid-liquid extraction method was applicable for extracting OCPs from various types of food samples including vegetables, meats and its products, fish, eggs and animal fats. In addition, several standardised methods, including AOAC 970.52, EN 1528 and EN 12393, have employed such solid-liquid or liquid-liquid extraction techniques for the determination of OCPs in both fatty and non-fatty foods. [43,44,45] In some occasions, sonication or Polytron was also applied to improve the extraction efficiency and recoveries.

1.3 Clean-up methods

Matrix constituents can be co-extracted and later co-eluted with analysed components and can consequently interfere with analyte identification and quantification. Moreover, co-extracted compounds, especially lipids, tend to adsorb in GC system such as injection port and column, resulting in poor chromatographic performance. A through clean-up minimised such matrix issues, improves sensitivity, permits more consistent and repeatable results, and extend the capillary column lifetime. Several approaches have been attempted to eliminated co-extracted interferences from extracts, including freezing centrifugation or filtration, liquid-liquid partitioning, gel permeation chromatography (GPC), solid phase extraction (SPE) and solid-phase microextraction (SPME). The simplest approach to remove the fatty co-extracted is by freezing centrifugation. [46,47] The logic behind is that fatty substances (mainly lipids) have lower melting point than the solvent so that frozen lipids can be removed by centrifugation or filtering while OCPs remain dissolved in the solvent. Different freezing temperatures ranged from -24 °C to -70 °C have been used. However, the solubility of lipids in solvent not only depends on the temperature but also the solubility product. Therefore this technique can remove significant amount of lipids for some food matrix but not for every matrix. Certain amount of lipids would remain in the solvent after the freezing centrifugation step and hence further cleanup is required. Using materials with large surfaces area for absorption of lipids have been employed since early 1970s. These materials include, Florisil, Lipid Removal Agent (LRA) media from Supelco, micro Cel E and Calflo E from Johns-Manville. Micro Cel E and Calflo E and LRA are synthetic calcium silicate while Florisil is a magnesium silicate with high specific surface area. [48,49] They can be applied to remove lipids either in sample preparation, solid phase extraction step or during sample clean-up step, with minimal effect on non-lipid chemicals. When food sample is mixed with these lipids absorbing materials, edible fat could be removed. Therefore it is common to conduct a clean-up step by solid phase extraction (SPE) nowadays. Both, conventional glass column packed with sorbent(s) and ready-to-use cartridges have been utilised and the common used phases are silica, Florisil, alumina and C18-bounded silica. Doong and Lee compared the cleaning efficiency of ready-to-use cartridge filled with three different adsorbents for shellfish extract. [50,51,52] Their results demonstrated that out of 14 OCPs tested, two were retained in the C18-cartridge. As for alumina and Florisil SPE, though all 14 pesticides tested could be recovered, Florisil provide better results in term of recoveries, repeatability and removal of interfering substances. Similarly, Hong et al., also showed that Florisil had better cleaning efficiency of fatty acids in fish extract when compared with C18. Besides, recoveries of some OCPs were poor with hexane as eluent and these more polar OCPs could be eluted out from the column with acetone. Bazlic et al., reported also that the quality of Florisil was important in avoiding possible interference and misinterpretation of results. Even though GC-MS was employed as the detection system, poor quality Florisil could introduce false positive results for lindane and dieldrin. [53,54] To sum up, the combination of sorbent(s) and eluting solvent(s) have to be chosen very carefully. Otherwise, some OCPs or their metabolites/derivatives would be lost during the clean-up step. [55,56,57] These OCPs could either break down or adhere to the sorbent material, leading to low or even no recovery. Finding of the optima clean-up conditions is an art itself. As the targeted OCPs might cover a wide range of polarities, it is quite difficult to find the best combination of SPE column material and eluting solvent, which permits recovering the polar OCPs (but leaving the polar interferents behind on the column), as well as recovering the non-polar OCPs (without eluting any residual oil present in the extract from the column).

1.4 Detection techniques of OCPs

A number of different selective detectors can be coupled with GC for analyzing OCPs, including electron capture detector (ECD), halogen specific detector (XSD), electrolytic conductivity detector (ELCD) and atomic emission detector (AED). GC-ECD is the most commonly used detection method with low detection limits. It is particularly useful for detecting halogen containing molecules. However, other organic molecules, such as aromatic compounds, would also give positive signal. Users have to confirm the presence of OCPs by another confirmative technique. Even though the above-mentioned selective detector can be used for quantification, it is unlike to fulfil the European Commission's stringest requirements as set for pesticides analysis. Confirmation with GC-hyphenated with mass spectrometric (MS) detector is normally required. Single quadrupole MS detector running in electron ionisation (EI) mode with target analytes monitored by selective ion monitoring (SIM) becomes a routine monitoring tool for OCPs nowadays. Since some OCPs are electronegative in nature, GC-MS detector under negative chemical ionisation mode with methane as reagent gas could provide better sensitivity. [58,59,60,61] To further increase confidence in confirmative analysis, GC coupled with tandem Ms is one of suitable techniques. Besides providing a more definitive detection tool, tandem MS also decrease matrix interferences, improves selectivity and achieves higher signal-to-noise ratio and subsequently improves the detection limit. Both tandem-in-time (ion-trap) and tandem-in-space (triple quadrupoles) detector have been applied for OCPs residues analysis in different matrices. The determination of pesticides residues in the environment and in food is necessary for ensuring that human exposure to contaminants, especially by dietary intake, does not exceed acceptable level for health. Consequently, robust analytical methods have to be validated for carrying out both research and monitoring programmes, and thus for defining limitations and supporting enforcement of regulations. In this field, reproducible analytical methods are required to allow the effective separation, selective identification and accurate quantification of pesticides analyses at low levels in food-stuff including food of animal origin.

2. AIM OF THE RESEARCH

The aims of the present work were:

- To develop and optimise a simple extraction and clean-up method to quantify non-polar chlorinated compounds in high lipid containing samples (animal feed and subcutaneous fat bovine tissue).
- To validate a multiresidues method for the simultaneous determination of 20 OCPs by using GC-MS/MS in term of repeatability, precision, limit of detection (LOD), limit of quantification (LOQ) etc. The coupling of this detection mode is very useful for the analysis of these complex samples allowing the separation, identification, quantification and confirmation of a large number of pesticides at trace level.
- To monitor the OCPs level in animal feed samples used in bovine farm.
- To monitor the OCPs level in subcutaneous fat bovine tissue to asses and to verify the concentration phenomena of these persistent pollutants.

3. EXPERIMENTAL

3.1 Feed and subcutaneous bovine fat samples

25 feed samples used for bovine with different composition were obtained from intensive livestock farming. An example of feed mixture was shown in figure 1. 35 fat samples were obtained from bovine for slaughter (18-24 month age) and presented in figure 2.

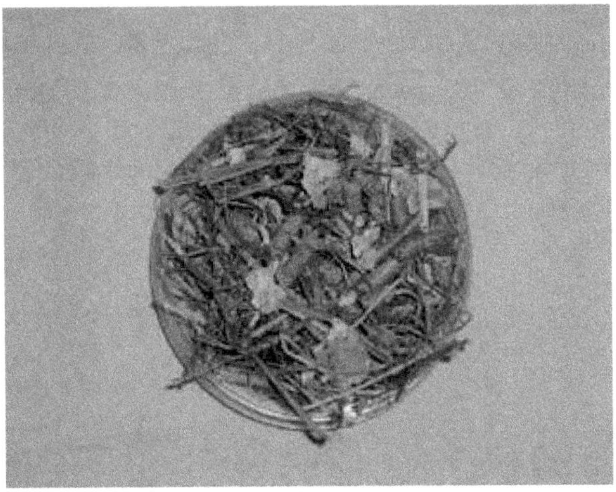

Figure 1. Feed sample mixture

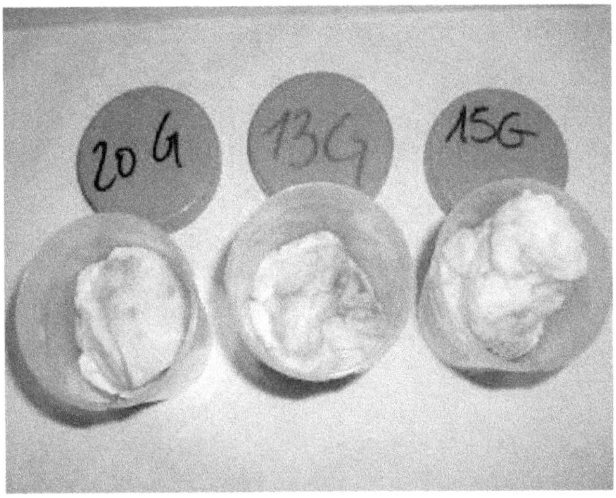

Figure 2. Subcutaneous fatty tissue sample

3.2 Chemicals and reagents

All OCPs were purchased from Supelco Inc.: mix 32094 and 32412 (Bellefonte, PA, USA). The figure 3 displays the structures of these OCPs considered in this study. Purities of pesticides standards were greater than 99%. Working standard solution was prepared at concentration of 0.1-5 µg mL-1 by volume, dilution with acetone and hexane. Organic solvents (hexane, acetone and acetonitrile) were of pesticide residue analysis grade (Sigma Aldrich, USA). All glassware was cleaned with laboratory reagent, sequentially rinsed with distilled water, acetone and methanol and finally baked in a oven at 300 °C. Distilled water was obtained with a Milli-Q system (Millipore, Bedford, MA, USA). For SPE, Florisil 5 g was purchased from Supelco.

3.3 Equipments

Ultrasonic bath (Branson) was used for the extraction of chlorinated pesticides form feed and fat samples. The generator of ultrasonic bath has an output of 150 W and a frequency of 35 kHz. Rotary evaporator (Buchi, Swiss) was used for the concentration of organic solvent. High intensity planetary mill Retsch (model MM 400, Retsch, GmbH, Retsch-Allee, Haan) was used to obtain representative aliquots of feed samples powder.

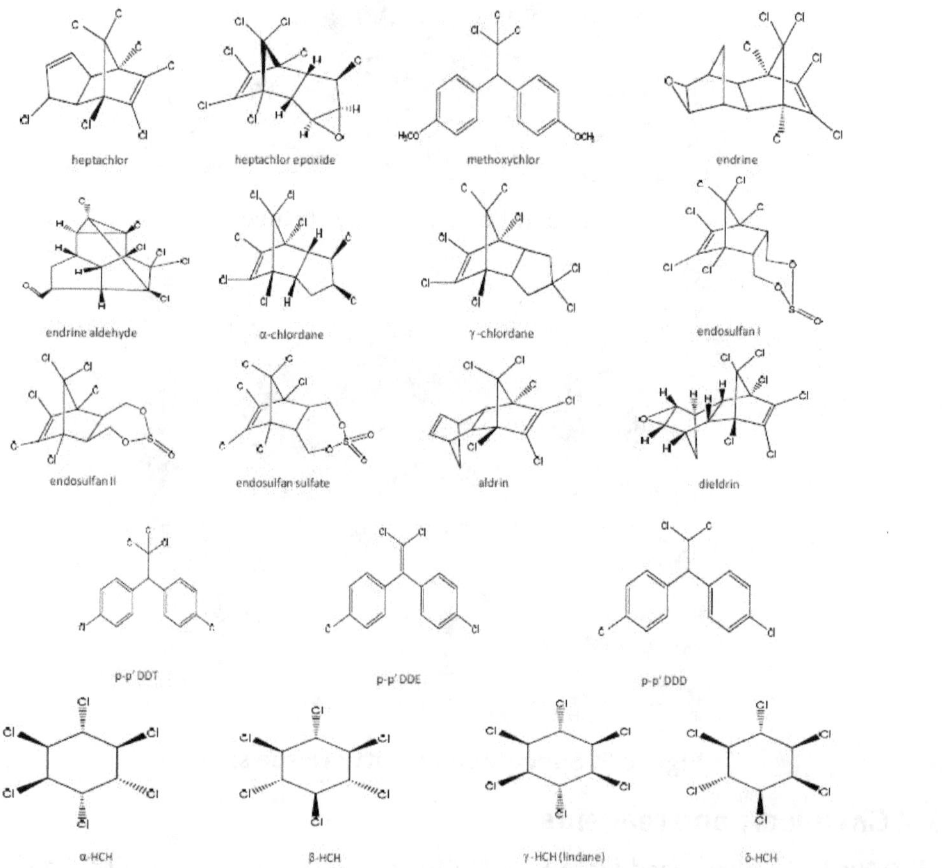

Figure 3. Chemical structures of chlorinated pesticides investigated in this study (19 OCPs)

3.4 Sample extraction, delipidation and clean-up procedure

3.4.1 Superfine Grinding (SFG) of feed sample

In order to obtain a representative feed sample a superfine powder was prepared from feed using mechanical grinding-activation in an energy intensive vibrational mill. 50 g of different feed sample were ground in a high intensity planetary mill. The mill was vibrating at a frequency of 25 Hz for 4 min using two 50 mL jars with 20 mm stainless steel balls. Pre cooling of jars were carried out with liquid nitrogen in order to prevent temperature increasing during the grinding process. The speed differences between balls and jar resulted in the interaction of frictional and impact forces, releasing high dynamic energies. The interplay of all these forces resulted in the very effective energy input of planetary ball mills. The application of mechanochemistry deal with the physical changes of substances in all state of aggregation, for instance occurring with the combined action of pressure and shear in energy- intensive grinding mills. Mechanochemical technology has been developed and applied in different fields (synthesis of superfine powder, surface modification and drug modification) and could represent a novel tool of research. [62,63,64] The procedure is presented in Fig.4.

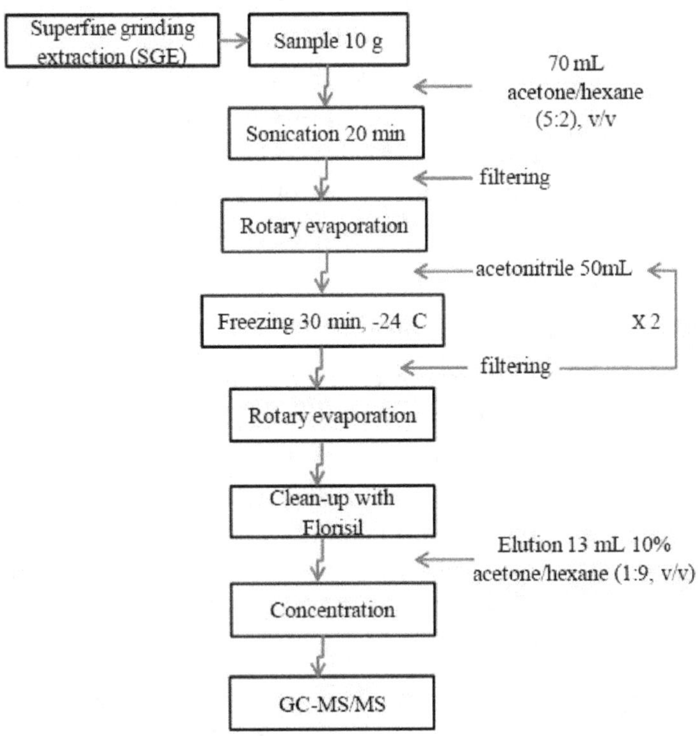

Figure 4. Analytical procedure for the extraction and purification of OCPs from feed and subcutaneous fatty tissue samples

3.4.2 Samples extraction

10 g of subcutaneous fat tissue (homogenised in a cooled mixer) or feed sample finely grounded and prepared with the procedure described above (SFG) were extracted by ultrasonic agitation with a mixed solvent of 70 mL of acetone-n-hexane (5:2, v/v) for 20 min. Extract was filtered to remove traces of water with filter paper containing 5 g of sodium sulphate, and then transferred into a 250 mL round flask. The extraction was repeated one more time. Extracted solvent was dried and redissolved in 50 mL of acetonitrile that has low solubility for lipids. Acetonitrile extract was stored in the freezer at -24 °C for 30 min to freeze lipids. Most of the lipids were precipitated as pale yellow, condensed lump on glassware surface. Cold extract at -24 °C was immediately filtered with filter paper to remove frozen lipids. The precipitated lipid on glassware surface was redissolved in 50 mL of acetonitrile to perform filtration again by same procedure. The filtered extracts were combined and concentrated to a final volume of 1 mL by a rotary evaporator to follow Florisil-SPE clean-up.

3.4.3 Sample clean-up

The SPE cartridge was cleaned with 12 mL of n-hexane and air dried by positive pressure prior sample application. 5 mL of hexane were used to condition the cartridge. After sample loading, the cartridge was air dried for 10 min. Desorption of the OCPs, which had been concentrated on the Florisil sorbent, was carried out using 13 mL of acetone-n-hexane (1:9, v/v) mixture at a flow of 1 mL min-1 and collected in a 50 mL round flask. The eluate was then concentrated at 45 °C under nitrogen stream until just the disappearance of the last drop of solution. Finally, the residue was redissolved in 1 mL hexane Pestanal prior to its injection in GC-MS/MS system.

3.5 GC-MS/MS analysis and detection

A Varian GC 3800 gas chromatograph coupled to a Varian Saturn 2000 ion trap mass spectrometer was used for the analysis and detection of the OCPs. The gas chromatograph was equipped with a Rtx-5 fused-silica capillary column (30 m x 0.25 mm i.d., 0.25 um film thickness) obtained from Restek. Helium (purity 99,99%) was the carrier gas at constant flow of 1 mL min-1. The GC injector temperature was maintained at 280 °C. The oven program temperature was: initial temperature 120 °C increased by 5 °C min -1 to 280 °C and held for 10 min. The ion trap spectrometer was operated in electron ionisation (EI) mode.

The ionization energy was set at 70eV. The detector range was m/z 40-650. The transfer line and trap temperature were 250 °C and 170 °C respectively.

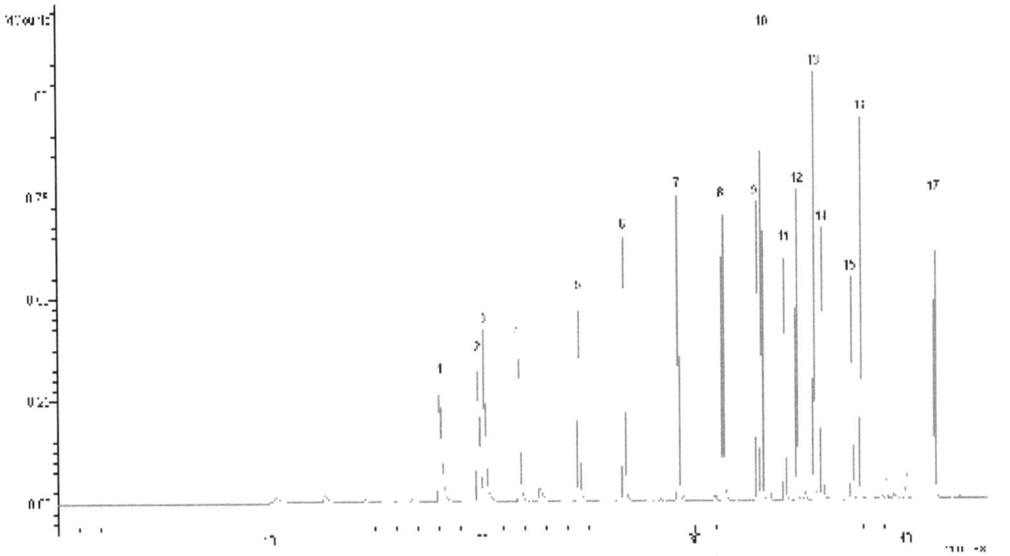

Figure 5. GC-MS chromatogram (TIC mode) of a standard OCPs mixture (MRL 0.5 mg kg-1) used in the present study. 1:α-HCH, 2:β-HCH, 3:γ-HCH, 4:δ-HCH, 5:Heptachlor, 6:Aldrin, 7: Heptachlor epoxide, 8:Endosulfan I, 9:Dieldrin, 10:p-p'DDE, 11:Endrin, 12:Endosulfan II, 13:p-p'DDD, 14:Endrin Aldheyde, 15: Endosulfan Sulphate, 16:p- p'DDT, 17:Methoxychlor

4. RESULTS AND DISUSSION

4.1 Key results about extraction and clean-up method

Two extraction and clean-up methods have been developed, tested and optimised for the extraction of 20 OCPs from animal feed sample and subcutaneous fatty tissue samples from bovine.

Large amounts of lipids were extracted when n-hexane or acetone was used as extraction solvents. In general, complex mixtures of of several types of lipids were co-extracted during the extraction of chlorinated pesticides from biological sample. Triglycerides and sterol esters are the major components in meat fats.

The key point of the extraction method take advantage of significant difference of melting points between lipids (below about 40 C) and chlorinated pesticides (above 260 C), so that lipid components can be easily separated from chlorinated compounds. After extraction, lipids in organic extracts were precipitated as frozen at -24 C in the freezer, while chlorinated compounds were still dissolved in cold organic solvents. Thus frozen-lipids can be removed just by filtering extracts. During overall process, approximately 90% of lipids were eliminated without any significant loss of pesticides. After freezing-lipid filtration, the remaining interferences were successfully removed by a solid-phase (SPE) Florisil cartridge.

Sample clean-up was necessary for the removal of polar coextracted substances. Florisil cartridges have been employed for that purpose since that adsorbent has proved to be very efficient for the clean-up of food samples.

4.2 Optimisation of MS/MS transitions

From full scan spectra, the most intense higher mass precursor ions were selected for development of MRM method. For the most of the analytes these were the base peak ions in the mass spectra, but in some cases higher mass ions of lower intensity were selected to minimise the possibility of matrix interferences. Precursor ions were examined using different collision energies (automated method development) and the most intense product ions were selected for each precursor ion. The products ions for all OCPs determined in this study are summarise in table 1.

For quantification of the target analytes linear calibration curves for all pesticides over six calibration levels (0.005 mg kg-1-1.5 mg kg -1) using a feed and fat blank samples were prepared taking also in consideration the MRLs levels for each compounds. In quantitative analysis one of the main problems is the suppression/ enhancement of the analyte response caused by sample matrix components. Calibration curves were performed by using matrix- matched (in each matrix) because the feed and fat samples contain many compounds that are co-extracted in the extraction organic solvent. The use of Florisil-SPE tries to avoid matrix effect using a clean-up step, but this not eliminates completely the problem. A matrix effect on the analytical signal due to the matrix was noticed for most pesticides.

Table 1. Summary of precursor ions and products ions selected for analysis of OCPs n EI mode and linearity for fat and feed sample calibration curves.

OCPs	R.T. (min)	Precursor ion (m/z)	Product ions (m/z)	Excitation voltage (V)	Linearity fat (r2)	Linearity Feed (r2)
α-BHC	18.01	181	109, 142	1.0	0,9997	0,9974
Hexachlorobenzene	18.41	286	214, 249	1.0	0,9974	0,9951
β-BHC	19.82	181	109, 145	1.0	0,9980	0,9950
γ-BHC	20.12	181	109, 145	1.0	0,9984	0,9985
δ-BHC	21.73	181	109, 145	1.0	0,9994	-
Heptachlor	24.54	272	100, 237	0.4	0,9987	0,9984
Aldrin	26.63	293	220, 255	0.8	0,9992	0,9951
Heptachlor epoxide	29.21	353	263, 334	0.7	0,9982	0,9977
γ-Clordane	30.66	375	266, 301	0.8	0,9944	0,9945
Endosulfan I	31.37	241	170, 260	0.9	0,9993	0,9978

α-Clordane	31.61	375	266, 301	0.8	0,9935	0,9910
Dieldrin	32.96	263	193, 228	0.7	0,9988	0,9979
p-p' DDE	33.16	318	246, 283	0.7	0,9974	0,9987
Endrin	34.23	263	193, 228	0.7	0,9982	0,9975
Endosulfan II	34.84	241	170, 260	0.9	0,9951	0,9973
p-p' DDD	35.66	235	165, 199	0.6	0,9958	0,9976
Endrin aldheyde	36.03	345	243, 279	0.7	0,9968	-
Endosulfan sulphate	37.43	387	251, 289	0.6	0,9988	0,9925
p-p' DDT	37.84	235	165, 199	0.6	0,9996	0,9971
Methoxychlor	41.04	227	196, 212	0.7	0,9982	0,9991

The linearity of the curves was studied for each pesticide considering the area of the peak relative to the internal standard. The calibration data are given in table 2, showing a good linearity of the response for all pesticides at concentration within the interval tested.

LOD and LOQ were evaluated taking into account the baseline noise variations in the chromatogram obtained from the analysis of blank feed and blank fat samples (n=10). The LOD and LOQ were defined as the concentration of the analyte that produced a signal-to-noise ratio of 3 times and 10 times the standard deviation respectively above the blank signal. Table 2 shows the values in mg kg-1 of feed and fat sample calculated with blank sample extracts. The values are similar to those obtained by other authors for the LOD and LOQ in feed animal samples. LOD and LOQ values for subcutaneous fat sample are not present in literature. Our results are very similar to that obtained in fish muscle and meat.

Table 2. MRLs, limits of detection (LOD) and limits of quantification (LOQ) for OCPs in fat and feed samples.

OCPs	Subcutaneous fat tissue			Animal feed		
	MRL* (mg kg^{-1})	LOD (mg kg^{-1})	LOQ (mg kg^{-1})	MRL** (mg kg^{-1})	LOD (mg kg^{-1})	LOQ (mg kg^{-1})
α-BHC	0.5	0.007	0.024	0.02	0.002	0.016
Hexachlorobenzene	0.2	0.007	0.026	0.01	0.003	0.011
β-BHC	0.1	0.012	0.041	0.01	0.003	0.010
γ-BHC	0.02	0.001	0.006	0.2	0.012	0.04
δ-BHC	0.5	0.007	0.023	0.02	-	-
Heptachlor	0.2	0.004	0.010	0.01	0.001	0.005
Aldrin	0.5	0.003	0.010	0.01	0.004	0.015

Heptachlor epoxide	0.2	0.002	0.008	0.01	0.002	0.009
γ-Clordane	0.05	0.005	0.019	0.02	0.003	0.013
Endosulfan I	0.05	0.003	0.012	0.1	0.007	0.024
α-Clordane	0.05	0.005	0.019	0.02	0.005	0.017
Dieldrin	0.2	0.002	0.008	0.01	0.002	0.007
p-p' DDE	1	0.001	0.005	0.05	0.003	0.012
Endrin	0.05	0.002	0.007	0.01	0.004	0.013
Endosulfan II	0.05	0.003	0.010	0.1	0.004	0.016
p-p' DDD	1	0.002	0.008	0.05	0.002	0.008
Endrin aldehyde	0.05	0.003	0.011	0.02	-	-
Endosulfan sulphate	0.05	0.002	0.006	0.1	0.007	0.023
p-p' DDT	1	0.004	0.001	0.05	0.006	0.02
Methoxychlor	0.01	0.002	0.008	0.5	0.002	0.007
*= MRLs of EU regulation guidelines (CE 32/2002); ** = MRLs of EU regulation guidelines (CE 396/2005)						

4.3 Occurrence of OCPs in animal feed samples and subcutaneous fat samples

The OCPs residues may concentrate in the adipose tissue and in blood serum of animals leading to environmental persistence, bioconcentration and biomagnifications through the food chain. Pesticides contamination of meat as well as chicken resulting from feeding a diet containing a low concentration of pesticides is a well established fact. [63,64] OCPs residues in feed may be ingested bi herbivores and eventually find their way into the animal body which ultimately results in contamination of milk, meat eggs, etc. consumed by human being. [65,66]

The most pesticides detected in animal feed were p-p' DDT, heptachlor followed by lindane, methoxychlor and aldrin. The frequency of detection is presented in figure. 6.

In subcutaneous fat sample the most detected OCPs were heptachlor, hexachlorobenzene detected in all samples followed by p-p' DDE, p-p'DDT, methoxychlor, lindane and p-p' DDD as shown in figure 7. Aldrin was detected both in feed samples and animal fat. The presence of aldrin in meat indicates the need for concern from the public health point of view because of its much higher toxicity than other OCPs. [67,68] These results are in accordance with other author that found HCHS and DDTs the most compounds detected in meat samples. In general, it was observed that the p-p' isomers of DDE, DDT and DDDwere detected in samples. All detected pesticides in feed samples and fat samples did not exceed the MRLs established by the European Union for each compounds (Fig 8, 9). The concentration of detected pesticides in the samples are summarised in table 3.

Table 3. Mean organochlorine residues levels (μg kg-1) in subcutaneous fat and feed samples

CPs	Fat samples mean	sd	Feed samples mean	sd
	(n=35)	(±)	(n=25)	(±)
Σ-Heptachlor	4.11	1.15	2.16	1.02
Σ-DDT	38.68	6.60	4.12	1.79
Σ-Aldrin	8.46	6.01	4.53	1.12
Σ-Endosulfan	9.30	1.36	nd	-
α-HCH	1.32	0.07	nd	-
β-HCH	3.07	0.69	nd	-
δ-HCH	5.67	1.51	nd	-
γ-HCH	11.27	1.21	5.17	1.29
Endrin	16.91	2.82	4.15	0.63
Endrin aldheyde	6.89	1.60	12.99	1.57
Methoxychlor	3.78	1.08	nd	-
Hexachlorobenzene	11.73	1.20	nd	-

nd= not detected; sd=standard deviation

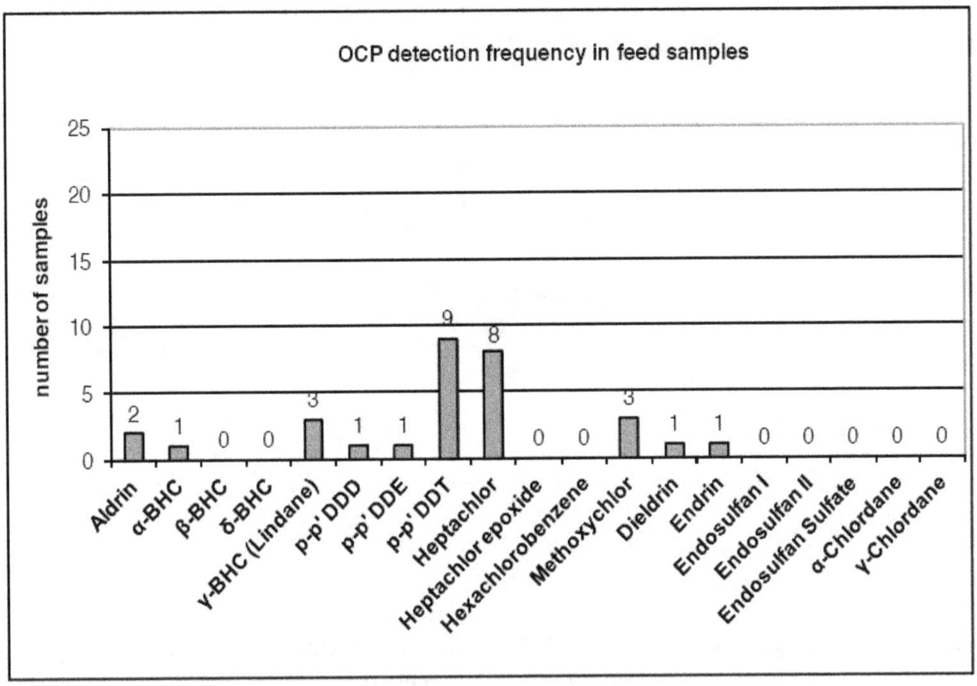

Figure 6. OCPs detection frequency in feed samples

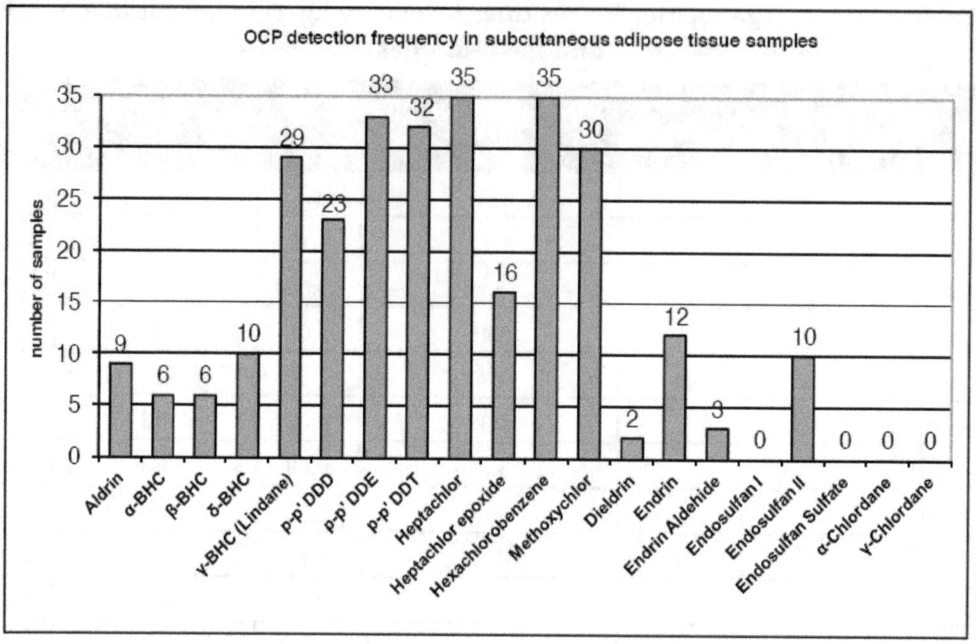

Figure 7. OCPs detection frequency in subcutaneous fat samples

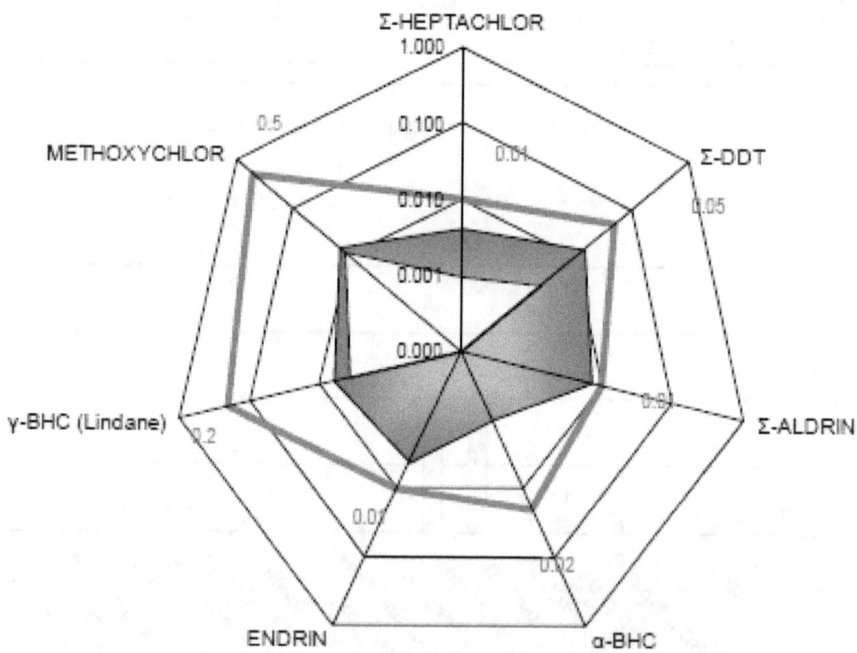

Figure 8. Radar plot of detected OCPs content in (mg kg -1) feed samples in relation to MRLs (red line)

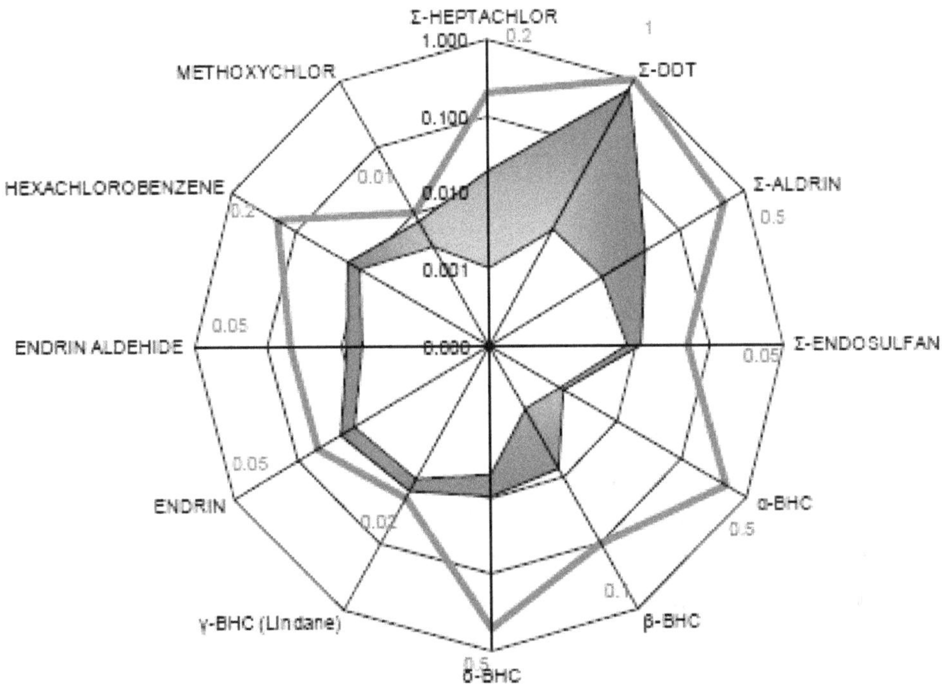

Figure 9. Radar plot of detected OCPs content (mg kg -1) in subcutaneous fat tissue samples in relation to MRLs (red line)

5. CONCISE SUMMARY

In conclusion, procedures for fast extraction, freezing lipid filtering, and GC-MS/MS measurement were devised and employed to quantify chlorinated pesticide levels in animal feed samples and subcutaneous fatty tissue, aiming to evaluate the potential concentration phenomena of these persistent chemicals. The freezing lipid filtration, in conjunction with the Flori-sil-SPE cartridge, facilitated the effective elimination of lipids from feed and fat samples while preserving pesticide integrity. Consequently, the method provides a swift and reliable screening tool with high sensitivity for the identification of organochlorine pesticides utilizing GC-MS/MS technology.

The subcutaneous adipose bovine tissue has been identified as a target organ capable of accumulating lipophilic pesticides, such as organochlorine pollutants. The feed may serve as a potential source of contamination of OCPs via the food chain. Consequently, the assessment of pesticide residues in feed and food is essential to ensure that human exposure to pollutants, particularly through dietary intake, remains within acceptable health limits. One analytical problem in food safety is to deliver accurate results in accordance with regulatory criteria.

REFERENCES

1. Benbrook C.M. Organochlorine Residues Pose Surprisingly High Dietary Risks. J. Epidemiol. Community Health 2002; 56 822.
2. Lehotay S.J., Mastovska K., Yun S.J. Evaluation of Two Fast And Easy Methods For Pesticide Residue Analysis In Fatty Food Matrices. J. Aoac Int. 2005; 88 630.
3. Qiu X., Zhu T., Yao B., Hu, J. Hu, S. Contribution of Dicofol To The Current Ddt Pollution In China. Environ.Sci. Technol. 2005; 39 4385.
4. Bilrha, H., Roy, R., Moreau, B., Belles-Isles, M., Dewailly, E., Ayotte, P.,. In vitro activation of cord blood mononuclear cells and cytokine production in a remote coastal population exposed to organochlorines and methyl mercury. Environ. Health Perspect. 2003; 111.
5. Lavoie E. T. and Grasman, K. A. Effects of in ovo exposure to PCBs 126 and 77 on mortality, deformities and post-hatch immune function in chickens, J. Toxicol. Environ. Health, Part A 2007, 70 547.
6. Bernhoft, A. Skaare J. U. and Wiig O. Possible immunotoxic effects of organochlorines in polar bears (Ursus maritimus) at Svalbard, J. Toxicol. Environ. Health, Part A, 2000; 59 561.
7. Repetto R. and Baliga, S. S. Pesticides and immunosuppression: the risks to public health, Health Policy Plan., 1997; 12 97.
8. Ahlborg, U. G. Lipworth L.and Titus-Ernstoff, L. Organochlorine compounds in relation to breast cancer, endometrial cancer, and endometriosis: an assessment of the biological and epidemiological evidence, Crit. Rev. Toxicol., 1995; 25 463.
9. Hardell L, Liljegren G, Lindstrom G. Increased concentrations of chlordane in adipose tissue from non-Hodgkin's lymphoma patients compared with controls without a malignant disease. Int J Oncol 1996; 9.
10. Morgan D P, Roan C C.. Absorption, storage, and metabolic conversion of ingested DDT and DDT metabolites in man. Arch Environ Health 1971; 22 301.
11. Cocco P, Brennan P, Ibba A, de Sanjose LS, Maynadie M, Nieters A, et al. Plasma polychlorobiphenyl and organochlorine pesticide level and risk of major lymphoma subtypes. Occup Environ Med 2008; 65 132.
12. Bocquene G., Franco A. Pesticide Contamination Of The Coastline Of Martinique. Mar. Pollut. Bull. 2005; 51 612.
13. Gilbert-Lopez B., Garcia-Reyes J.F., Molina-Díaz A. Sample Treatment And Determination Of Pesticide Residues In Fatty Vegetable Matrices: A Review. Talanta 2009; 79 109.

14. Ledoux, M. Analytical Methods Applied To The Determination Of Pesticide Residues In Foods Of Animal Origin. A Review Of The Past Two Decades. J. Chromatogr. A 2011; 1218 1021.

15. Veierov,D.Aharonson,N.ImprovedCleanupOfLargeLipidSamplesForElectron Capture Gas Chromatographic Quantitation And Gas Chromatographic-Mass Spectrometric Confirmation Of Organochlorine Residues. J. Assoc. Off. Anal. Chem. 1980; 63 202.

16. S.J. Young, L.R. Kamps. Gas-Liquid Chromatographic Determination Of Nonpolar Organochlorine Pesticide Residues In A Crude Vegetable Oil And Its Refinery By- Products. J. Assoc. Off. Anal. Chem. 1982; 65 916.

17. Luke, M.A. Forberg J.E, Masumoto, H.T. Extraction And Cleanup Of Organochlorine, Organophosphate, Organonitrogen, And Hydrocarbon Pesticides In Produce For Determination By Gas-Liquid-Chromatography. J. Assoc. Off. Agric. Chem. 1975; 58 1020.

18. Stefanelli, P. Santilio, A. Cataldi, L. Dommarco, R. Multiresidue Analysis Of Organochlorine And Pyrethroid Pesticides In Ground Beef Meat By Gas Chromatography- Mass Spectrometry. J. Environ. Sci. Health B 2009; 44 350.

19. Sannino, A. Mambriani, P. Bandini, M. Bolzoni, L. Multiresidue Method For Determination Of Organochlorine Insecticides And Polychlorinated Biphenyl Congeners In Fatty Processed Foods. J. Aoac Int. 1996; 79 1434.

20. Lazaro, R. Herrera, A. Arino, A.A. Conchello, M.P.S. Bayarri, J. Organochlorine Pesticide Residues In Total Diet Samples From Arago´N (Northeastern Spain). Agric. Food Chem. 1996; 44 2742.

21. Patel, K. Fussell R.J., Hetmanski, M. Goodall, D.M. Keely, B.J. Evaluation Of Gas Chromatography-Tandem Quadrupole Mass Spectrometry For The Determination Of Organochlorine Pesticides In Fats And Oils. J. Chromatogr. A. 2005; 1068 289.

22. Schenck, F.J. Donoghue, D.J. Determination Of Organochlorine And Organophosphorus Pesticide Residues In Eggs Using A Solid Phase Extraction Cleanup. Agric. Food Chem. 2000; 48 6412.

23. Muralidharan, S. Dhananjayan, V. Risebrough, R. Prakash, V. Jayakumar, R. Bloom, P.H. Persistent Organochlorine Pesticide Residues In Tissues And Eggs Of White- Backed Vulture, Gyps Bengalensis From Different Locations In India. Bull. Environ. Contam. Toxicol. 2008; 81 561.

24. Ghidini, S. Zanardi E., Battaglia A., Varisco G., Ferretti, E. Campanini, G. Chizzolini R. Comparison Of Contaminant And Residue Levels In Organic And Conventional Milk And Meat Products From Northern Italy. Food Addit. Contam. 2005; 22 9.

25. Bennett, D.A. Chung, A.C. Lee S.M. Multiresidue Method For Analysis Of Pesticides In Liquid Whole Milk. J. Aoac Int. 1997; 80 1065.

26. Hong, J. Kima H.Y., Kim, D.G. Seo, J. Kimb, K.J.Rapid Determination Of Chlorinated Pesticides In Fish By Freezing-Lipid Filtration, Solid-Phase Extraction And Gas Chromatography-Mass Spectrometry. J. Chromatogr. A 2004; 1038 27.

27. Garridofrenich, A. Martinezvidal, J. Cruzsicilia, A. Gonzalezrodriguez, M. Plazabolanos P. Multiresidue Analysis Of Organochlorine And Organophosphorus Pesticides In Muscle Of Chicken, Pork And Lamb By Gas Chromatography–Triple Quadrupole Mass Spectrometry. Anal. Chim. Acta 2006; 558 42.

28. Fillion, J. Sauve, F. Selwyn, J. Multiresidue Method For The Determination Of Residues Of 251 Pesticides In Fruits And Vegetables By Gas Chromatography/ Mass Spectrometry And Liquid Chromatography With Fluorescence Detection. Aoac Int. 2000; 83 698.

29. Lehotay, S.J. Analysis Of Pesticide Residues In Mixed Fruit And Vegetable Extracts By Direct Sample Introduction/Gas Chromatography/Tandem Mass Spectrometry. J. Aoac Int. 2000; 83 680.

30. Cunha, S.C. Lehotay, S.J. Mastovska, K. Fernandes, J.O. Beatriz, M. Oliveira, P.P. Evaluation Of The Quechers Sample Preparation Approach For The Analysis Of Pesticide Residues In Olives. J. Sep. Sci. 2007; 30 620.

31. Organochlorine Pesticide Residuals In Chickens And Eggs At A Poultry Farm In Bei jing, China; Tao, S. Liu, W.X. Li, X.Q. Zhou, D.X. Li, X. Yang, Y.F. Yue,D.P., Coveney,R.M. Environ. Pollut. 2009; 157 497.

32. Distribution Of Organochlorine Pesticides And Alpha-Hch Enantiomers In Pork Tissues. Chemosphere; Covaci, A. Gheorghe, A Schepens,. P. Chemosphere 2004; 56 757.

33. Doong, R.A. Lee, C.Y. Determination Of Organochlorine Pesticide Residues In Foods Using Solid-Phase Extraction Clean-Up Cartridges. Analyst 1999; 124 1287.

34. Muralidharan, S. Dhananjayan, V. Jayanthi. Organochlorine Pesticides In Commercial Marine Fishes Of Coimbatore, India And Their Suitability For Human Consumption. P. Environ. Res. 2009; 109 15.

35. Campos, A. Lino, C.M. Cardoso, S.M. Silveira, M.I. Organochlorine Pesticide Residues In European Sardine, Horse Mackerel And Atlantic Mackerel From Portugal. Food Addit. Contam. 2005; 22 642.

36. Fidalgo-Used, N. Centineo, G. Blanco-Gonzalez, E. Sanz-Medel, A. Solid-Phase Microextraction As A Clean-Up And Preconcentration Procedure For Organochlorine Pesticides Determination In Fish Tissue By Gas Chromatography With Electron Capture Detection. J. Chromatogr. A 2003; 1017 35.
37. Hopper, M.L. Automated One-Step Supercritical Fluid Extraction And Clean-Up System For The Analysis Of Pesticide Residues In Fatty Matrices. J. Chromatogr. A 1999; 840 93.
38. Argauer, R.J. Eller, K.I. Pfeil, R.M Brown, R.T. Determining Ten Synthetic Pyrethroids In Lettuce And Ground Meat By Using Ion Trap Mass Spectrometry And Electron-Capture Gas Chromatography. J. Agric. Food Chem. 1997; 45180.
39. Ashraf-Khorassani, M. Taylor, L.T. Schweighardt, F.K. Development Of A Method For Extraction Of Orgaochlorine Pesticides From Rendered Chicken Fat Via Supercritical Fluoroform. J. Agric. Food Chem. 1996; 44 3540.
40. Saito, K. A. Sjodin, C.D. Sandau, M.D. Davis, H. Nakazawa, Y. Matsuki, D.G. Patterson Jr., Development Of A Accelerated Solvent Extraction And Gel Permeation Chromatography Analytical Method For Measuring Persistent Organohalogen Compounds In Adipose And Organ Tissue Analysis Chemosphere 2004; 57 373.
41. Suchan, P. Pulkrabova, J. Hajslova, J. Kocourek, V. Pressurized Liquid Extraction In Determination Of Polychlorinated Biphenyls And Organochlorine Pesticides In Fish Samples. Anal. Chim. Acta 2004 520; 193.
42. Weichbrodt, M. Vetter, W. Luckas, B. Microwave-Assisted Extraction And Accelerated Solvent Extraction With Ethyl Acetate-Cyclohexane Before Determination Of Organochlorines In Fish Tissue By Gas Chromatography With Electron-Capture Detection. J. Aoac Int. 2000; 831334.
43. Vetter, W. Weichbrodt, M. Hummert, K. Glotz, D. Luckas, B. Combined Microwave- Assisted Extraction And Gel Permeation Chromatography For The Determination Of Chlorinated Hydrocarbons In Seal Blubber And Cod Livers; Chemosphere 1998; 37 2439.
44. Wilkowska, A.M. Biziuk, M. Rapid Method For The Determination Of Organochlorine Pesticides And Pcbs In Fish Muscle Samples By Microwave-Assisted Extraction And Analysis Of Extracts By Gc-Ecd. J. Aoac Int. 2010; 93 1987.
45. Barriada-Pereira, M. Iglesias-Garcia, I. Gonzalez-Castro, M.J. Muniategui-Lorenzo, S. Lopez-Mahia, Prada-Rodriguez P.D., Pressurized Liquid Extraction And Microwave- Assisted Extraction In The Determination Of Organochlorine Pesticides In Fish Muscle Samples. J. Aoac Int. 2008; 91 174.

46. Karasova, G. Brandsteterova, E. Lachova, Matrix Solid Phase Dispersion As An Effective Preparation Method For Food Samples And Plants Before Hplc Analysis. M. Czech J. Food Sci. 2003; 21 219.

47. Yagüe, C Bayarri,. S. Lazaro, R. Conchello, P. Arino, A. Herrera, A. Multiresidue Determination Of Organochlorine Pesticides And Polychlorinated Biphenyls In Milk By Gas Chromatography With Electron-Capture Detection After Extraction By Matrix Solid-Phase Dispersion. J. Aoac Int. 2001; 84 1561.

48. Yagüe, C. Herrera, A. Arino, A. Lazaro, R. Bayarri, S.. Conchello, P. Rapid Method For Trace Determination Of Organochlorine Pesticides And Polychlorinated Biphenyls In Yogurt. J. Aoac Int. 2002; 85 1181.

49. Schenck, F.J. Wagner, R. Screening-Procedure For Organochlorine And Organophosphorus Pesticide-Residues In Milk Using Matrix Solid-Phase Dispersion (Mspd) Extraction And Gas-Chromatographic Determination. Food Addit. Contam. 1995; 12 535.

50. Schenck, F.J. Calderon, L. Saudarg, D.E. Florisil Solid-Phase Extraction Cartridges For Cleanup Of Organochlorine Pesticide Residues In Foods. J. Aoac Int. 1996; 791454.

51. Lott, H.M. Barker, S.A. Matrix Solid-Phase Dispersion Extraction And Gas-Chroma tographic Screening Of 14 Chlorinated Pesticides In Oysters (Crassostrea-Virginica). J. Aoac Int. 1993; 76 67.

52. Long, A.R. Soliman, M.M. Barker, S.A. Matrix Solid-Phase Dispersion (Mspd) Extraction And Gas-Chromatographic Screening Of 9 Chlorinated Pesticides In Beef Fat. J. Assoc. Off. Anal. Chem. 1991; 74 493.

53. Long, A.R. Crouch, M.D. Barker, S.A. Multiresidue Matrix Solid-Phase Dispersion (Mspd) Extraction And Gas-Chromatographic Screening Of 9 Chlorinated Pesticides In Catfish (Ictalurus-Punctatus) Muscle-Tissue. J. Assoc. Off. Anal. Chem. 1991; 74 667.

54. Valsamaki, V.I. Boti, V.I. Sakkas, V.A. Albanis, T.A. Determination Of Organochlorine Pesticides And Polychlorinated Biphenyls In Chicken Eggs By Matrix Solid Phase Dispersion. Anal. Chim. Acta 2006 195; 573–574.

55. Rogers, W.M. The Use Of A Solid Support For The Extraction Of Chlorinated Pesticides From Large Quantities Of Fats And Oils. J. Assoc. Off. Anal. Chem. 1972; 55 1053.

56. Porter, M.L. Burke, J.A. An Isolation And Cleanup Procedure For Low Levels Of Organochlorine Pesticide Residues In Fats And Oils. J. Assoc. Off. Anal. Chem. 1973; 56 733.

57. Bong, R.L. Determination Of Hexachlorobenzene And Mirex In Fatty Products. J. As soc. Off. Anal. Chem. 1975; 58 557.

58. Goodspeed, D.P. Chestnut, L.I. Determining Organohalides In Animal Fats Using Gel-Permeation Chromatographic Cleanup - Repeatability Study. J. Assoc. Off. Anal. Chem. 1991; 74 388.

59. Bazulic, D. Sapunar-Postruznik, J. Bilic, Arh. S. Significance Of The Quality Of Florisil In Organochlorine Pesticide Analysis; Hig. Rada Toksikol. 49 (1998) 319.

60. Beyer, A. Biziuk, M. Comparison Of Efficiency Of Different Sorbents Used During Clean-Up Of Extracts For Determination Of Polychlorinated Biphenyls And Pesticide Residues In Low-Fat Food. Food Res. Int. 2010; 43 831.

61. Determination Of Pesticides In Composite Dietary Samples By Gas Chromatography/Mass Spectrometry In The Selected Ion Monitoring Mode By Using A Temperature-Programmable Large Volume Injector With Preseparation Column. Rosenblum, L. Hieber, T. Morgan, J. J. Aoac Int. 2001; 84 891.

62. Xie J, Shi L, Zhu X, Wang P, Zhao Y and Su W, Mechanochemical-assisted efficient extraction of ruitn from Hibiscus mutabilis L. Innovative Food Science and Emerging Technologies. 2011; 12: 1446.

63. Liu Y, Jin LJ, Li XY and Xu YP, Application of Mechanochemical Pretreatment to Aqueous Extraction of Isofraxidin from Eleutherococcus Senticosus. Ind. Eng. Chem. Res.: 2007; 46 6584.

64. Zhu XY, Lin HM, Chen X, Xie J and Wang P, Mechanochemical-assisted Extraction and antioxidant Activities of Kaempferol Glycosides from Camellia oleifera Abel. Meal. J. Agr. Food Chem. 2011; 3986.

65. Fillion, J. Hindle, R. Lacroix, M. Selwyn, J. Multiresidue Determination Of Pesticides In Fruit And Vegetables By Gas Chromatography Mass-Selective Detection And Liquid Chromatography With Fluorescence Detection. J. Aoac Int. 1995; 78 1252.

66. Guan, H. Brewer, W.E. Morgan, S.L. New Approach To Multiresidue Pesticide Determination In Foods With High Fat Content Using Disposable Pipette Extraction (Dpx) And Gas Chromatography-Mass Spectrometry (Gc-Ms). J. Agric. Food Chem. 2009; 57 10531.

67. Gillespie, A.M. Walters, S.M. Semi-Preparative Reverse Phase Hplc Fractionation Of Pesticides From Edible Fats And Oils. J. Liq. Chromatogr. 1989; 12 1687.

68. Van Der Hoff G.R., Van Beuzekom, A.C. Brinkman, U.A. Baumann, R.A. Van Zoonen, P. Determination Of Organochlorine Compounds In Fatty Matrices - Application Of Rapid Off-Line Normal-Phase Liquid Chromatographic Clean-Up. J. Chromatogr. A 1996; 754 487.

SECTION 4
FOOD SAFETY

14 Chapter ENZYMES IN BAKERY: CURRENT AND FUTURE TRENDS

1. INTRODUCTION: AN OVERVIEW

The utilization of enzymes predates the recognition of their catalytic properties and the understanding of their chemical composition. The inaugural entirely enzymatic industrial process was established in the 1960s [1]. The processing of starch occurs in two stages: first, the liquefaction of the polysaccharide through bacterial α-amylase, and second, the saccharification facilitated by fungal glucoamylase.

Post-World War II, the utilization of enzymes increased owing to advancements in industrial microbiology and biochemical engineering [1]. Currently, enzymes are utilized throughout various sectors, including food, feed, detergents, textiles, laundry, tanning, medicines, cosmetics, and fine chemicals. Industrial applications comprise more than 80% of the worldwide enzyme market2]. Over 50% of the enzymes currently available in the market are derived from genetically modified organisms, utilizing genetic and protein engineering techniques. Food enzymes are the most prevalent and continue to constitute the largest segment of the enzyme industry.

Advancements in process technology combined with recombinant approaches during recent decades have significantly enhanced fermentation yields, stability, and the specificity and selectivity of enzymes [3-5]. These techniques advance and are expanding the applications of enzymes in food technology and various other fields.

There are two scenarios regarding the use of enzymes, either the enzymes are used to convert the raw material into the main product, or the enzymes are used as additives to alter a functional characteristic of the product. In the first case, the enzymatic process is undertaken in optimized and controlled conditions to enhance the catalytic potential of the enzyme, whereas in the second situation it is more difficult to assure optimal conditions and to control the enzymatic reaction [1]. An example of the first case is the use of immobilized glucose isomerase for the production of high-fructose syrups (HFS), and an example of the second scenario is the use of fungal proteases in dough making [1,6,7].

Enzymes are a crucial component utilized in the majority of baked goods. Recently, enzymes have gained increased significance in baking due to limitations on the use of chemical additions, particularly in the production of bread and other fermented goods [8].

This research aims to examine the existing applications of enzymes in the baking industry and to investigate potential trends within this sector of the food business.

2. BAKERY ENZYMES MARKET

The development of bread process was an important event in mankind. After the 19th century, with the agricultural mechanization, bread's quality was increased while its price was reduced; thereby white bread became a commodity within almost everyone´s reach [9]. An important aspect that contributed to evolution of the baking market was the introduction of industrial enzymes in the baking process, where bakery enzymes represent a relevant segment of the industry.

Among the main industrial enzyme producers, according to Novozymes S/A report 2011 [10], Novozymes S/A occupies 47% of the market, DuPont 21%, DSM 6% and the rest is occupied by other players. Furthermore, in that year, food and beverage enzymes represented 29% of enzyme business and biobusiness sales by the industry [10].

The world enzyme market is in evolution and a growth of 6.8% per year is expected [11]. The world food and beverage enzymes demand requires attention, because it represented $1,220 million dollars in 2010, around 36.5% of the total world industrial enzyme demand, estimated in $3,345 million dollars. Moreover, the world food and beverages enzymes demand is expected to be responsible for 40.1% of the world industrial enzyme demand in 2020, accounting for $2,520 of $6,280 million dollars of the world industrial enzyme market (Figure 1) [11].

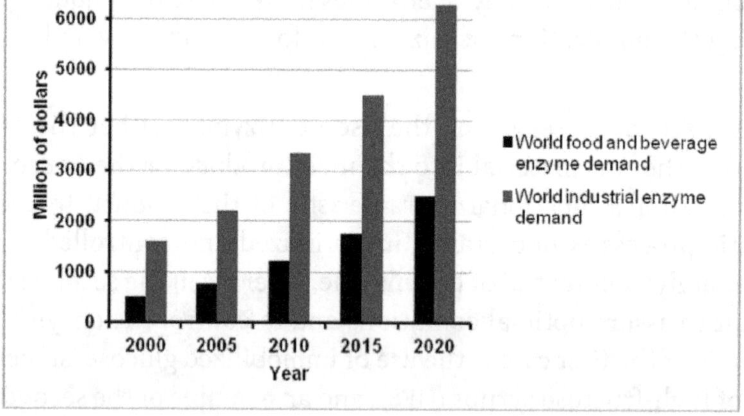

Source: Adapted from The Freedonia Group Inc., World Enzymes to 2015.

Figure 1. Estimated world food and beverage enzyme demand participation on the world industrial enzymes in million dollars from 2000 to 2020.

Table 1 summarizes the world bakery and enzyme demand between 2000 and 2020, segmented according to products. It is possible to observe that the enzymes market for baked goods is expected to increase from 420 million dollars in 2010 to 900 million dollars in 2020, although maintaining its representativeness in this segment, varying from 34.4 in 2010 to 35.7% in 2020 [11].

Table 1. Estimated demand of baked goods, dairy and other food & beverage enzymes in million dollars from 2000 to 2020.

Item	Years				
	2000	2005	2010	2015	2020
World food and beverage enzyme demand	520	760	1220	1770	2520
Baked goods	140	250	420	625	900
Dairy	180	260	360	465	610
Other foods and beverage	200	250	440	680	1010
Source: Adapted from The Freedonia Group Inc., World Enzymes to 2015.					

3. MAIN CONSTITUENTS OF BAKED PRODUCTS

Baking is a common name for the production of baked goods, such as bread, cake, pastries, biscuits, crackers, cookies, pies and tortillas, where wheat flour is both the most essential ingredient and key source of enzyme substrates for the product [12]. Even though based on cereals other than wheat, baked goods such as gluten-free products or rye bread are also considered to be baked products [8]. Baked goods formulations vary significantly depending on the desired final product, and typical ingredients, apart from starch, can include wheat flour (8-16% protein, 71-79% carbohydrate), fats, sugars, eggs, emulsifiers, milk and/or water [13].

Bread is usually made from wheat flour as raw material, which is a mixture of starch, gluten, lipids, non-starch polysaccharides and enzymes. After flour, yeast and water are mixed, complex biochemical and biophysical processes begin, catalyzed by the wheat enzymes and by the yeast, characterizing the dough phase. These processes go on in the baking phase, giving rise to bread. Extra enzymes added to the dough improve control of the baking process, allowing the use of different baking processes, reducing process time, slowing-down staling, compensating for flour variability and substituting chemical additives [14]. Starch is the main component of products such as bread and other bakery goods and is added to different foods, acting as a thickener, water binder, emulsion stabilizer, gelling agent and fat substitute [15]. It is the most abundant constituent and most important reserve polysaccharide of many plants, including cereals, occurring as intracellular, semi-crystalline granules. On a molecular level, its major components are the glucose polymers amylose and amylopectin [16]. Amylose is an essentially linear molecule, consisting of up to 6000 glucose units with α-(1,4)-glycosidic bonds (Figure 2). On the other hand, amylopectin is a highly branched polysaccharide constituted of short α-1,4 linked linear chains of 10–60 glucose units and α-1,6 linked side chains with 15–45 glucose units (Figure 3), containing on average 2 million glucose units [17].

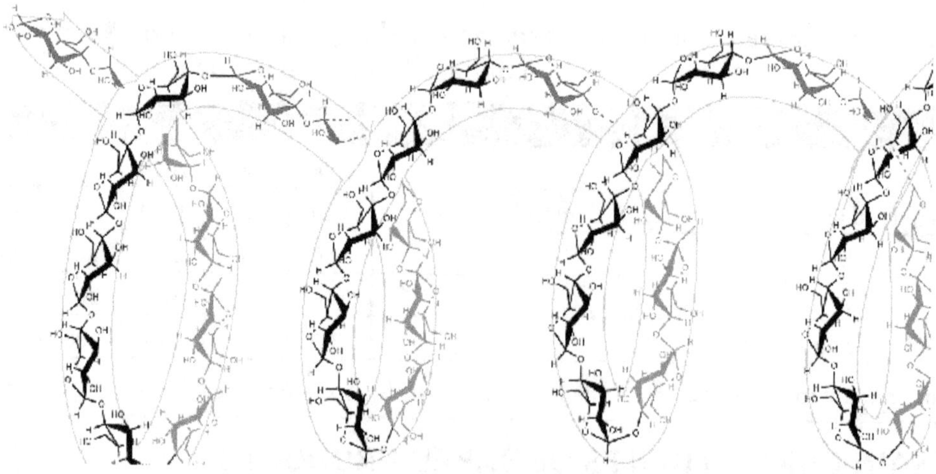

Figure 2. Structure of amylose chain, assumed as a left-handed spiral due to $\alpha(1\rightarrow4)$ glycosidic bonds (n = 500 - 6000 α-D-glucopyranosyl units).

Even though many flour components such as starch, arabinoxylans and lipids affect dough rheological properties [16,18-20], gluten provides dough with extensibility, viscosity, elasticity, cohesiveness and contributes to its water absorption capacity [21]. The unique ability of wheat flour to form visco-elastic dough with gas-holding properties is mostly due to the gluten proteins, the major storage proteins of wheat, which have an essential role in breadmaking [22].

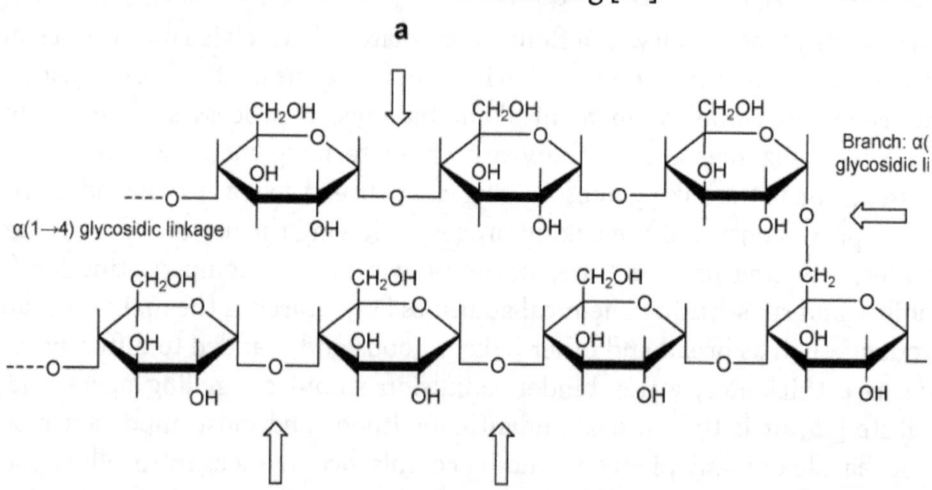

Figure 3. Partial structure of amylopectin with amylolytic enzymes action sites represented by arrows: (a) α-amylases; (b) amyloglucosidases; (c) β-amylases;

(d) isoamylases and pullulanases. Both $\alpha(1\rightarrow4)$ glycosidic linkages between the glucose units in the linear chain and one $\alpha(1\rightarrow6)$ glycosidic linkage to a side chain of the polysaccharide are represented.

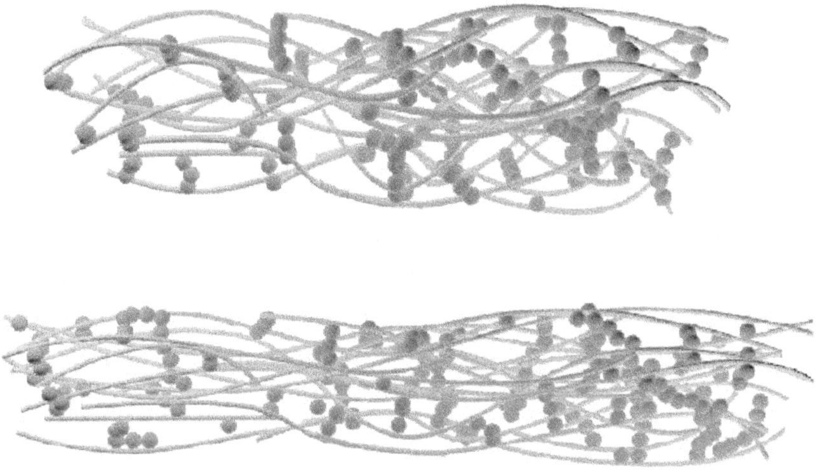

Figure 4. Schematic drawing of gluten proteins structure, where gliadins are represented by spheres and glutenins by filaments. The bulkier structure in the upper part shows the gas retained in the gluten network and consequent dough volume expansion observed in the baking process. The slimmer structure in the lower part represents plasticity, extensibility and viscous properties of the gluten matrix.

Gluten proteins can be divided into monomeric gliadins and polymeric glutenins, based on solubility in 70% aqueous ethanol solutions [23]. Gliadins are globular proteins with molecular weights ranging from 30,000 to 80,000, and are further classified into three groups: α-, γ- and ω-gliadins [24,25]. Except for the ω-gliadins which lack cysteine residues, gliadins contain intramolecular disulfide bonds [21]. Glutenins consist of a heterogeneous mixture of linear polymers with a broad molecular weight range from ca. 80,000 up into the millions [22], made up of disulfide cross-linked glutenin subunits which are biochemically related to the gliadins. The intermolecular disulfide bonds stabilize the glutenin polymers [21]. Gliadins mainly impart the plasticity, extensibility and viscous properties to wheat flour dough whereas glutenins are mostly responsible for the elasticity and cohesive strength of dough (Figure 4) [21,22]. Aspects such as the glutenin polymer structure, size distribution and subunit composition, as well as the gliadin/ glutenin ratio are important to determine gluten quality and, consequently, the breadmaking potential of wheat flour [25-29].

Cereal non-starch polysaccharides are dietary fibre constituents, mostly composed of arabinoxylans, β-glucan and arabinogalactan-peptides. Arabinoxylans make up the largest non- starch polysaccharide fraction of cell walls of many cereals, such as wheat and rye [22,30]. They are polydisperse polyssacharides with similar structural properties, which are present in water-extractable (WE-AX) and water-unextractable (WN-AX) forms [16]. Arabinoxylans consist of a β-1,4 linked D-xylopyranosyl backbone substituted with α-L-arabinofuranose residues at the C(O)3 and/or C(O)2 positions [31-33]. Arabinose residues can be further coupled at the C(O)5 to ferulic acid through an ester linkage [34] (Figure 5). Even though minor flour constituents, arabinoxylans have the capacity to significantly affect the properties of dough and the final baked product [18]. Arabinoxylans and arabinogalactans possess important functional properties for the cereal industry. They can improve dough development and dough stability, by enhancing the water absorption capacity of the dough. These polysaccharides also confer viscosity and may increase gas permeability by contributing to the elasticity of the protein film around them. Additionally, during breadmaking they improve loaf volume, crumb firmness, reduce retrogradation and therefore, enhance the shelf life and storage stability of bread [35].

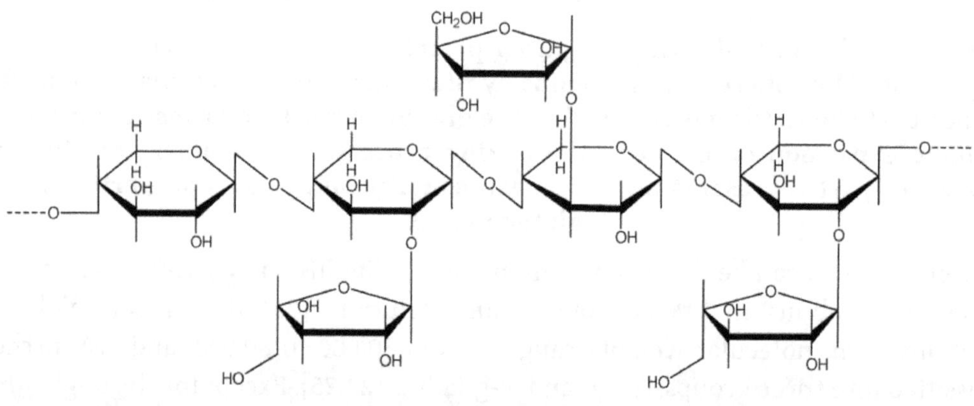

Figure 5. Partial structure of an arabinoxylan: a linear main chain formed by xylan (a pentosan consisting of D-xylose units connected by β(1→4) linkages), randomly attached to L-arabinofuranose residues by α(1→3) or α(1→2) linkages.

In addition to starch, gluten proteins and wheat flour non-starch polysaccharides such as arabinoxylans, lipids and enzymes can considerably improve the breadmaking performance [16,18,22,36,37]. Lipids are important components in breadmaking because they provide a variety of beneficial properties during processing and storage. In bread, lipids come from multiple ingredients, largely wheat flour, shortening and surfactants in a typical bread formula [38]. Wheat flour contains about 2% lipids [23], which occur free and bound to other wheat constituents. They are classified as starch lipids and free and bound non-starch lipids, based on their solubility in solvents of different polarities [39]. The bound non-starch lipids are mainly associated with flour protein and consist predominantly of non-polar lipids, while free non-starch lipids comprise mostly polar glyco- and phospholipids [40].

4. BAKING PROCESS

Bread processing can be divided into three basic operations mixing, fermentation (resting and proofing) and baking. Through baking the mainly fluid dough or batter is transformed into a predominantly solid baked product. Indirectly, baking alters the sensory properties, improving palatability, and extending the range of tastes, aromas and textures of foods produced from raw materials [41].

Although baking has been practiced for a very long time, the whole process is not completely understood, possibly due to the occurrence of several coupled complex physical [42] and molecular processes [43]. The baking process therefore results in a series of physical, chemical and biochemical changes in the product. These changes include volume expansion, evaporation of water, formation of a porous structure, denaturation of protein, gelatinization of starch, crust formation and browning reactions [44].

Bread consists of an unstable, elastic, solid foam structure, containing a continuous phase made up of an elastic network of cross-linked gluten protein molecules and of leached starch polymer molecules, mainly amylose, uncomplexed and complexed with polar lipid molecules, and also a discontinuous phase of entrapped, gelatinised, swollen, deformed starch granules [45]. The nature and properties of the final product are influenced by physical and mechanical mixing, chemical reactions (including enzyme-catalyzed reactions), and thermal effects (baking time and temperature).

The simplest breadmaking procedure is a straight-dough system where all bread formula ingredients are mixed into developed dough [46]. A second process is the sponge and dough method where mixing of ingredients is performed in two steps. Leavening agent is prepared in the first step, by mixing together the yeast and certain quantity of water and flour. The mixture is left to develop for a few hours and then it is mixed with the other ingredients [42]. A third procedure is the Chorleywood method in which all the ingredients are mixed for a few minutes in an ultrahigh mixer [47].

In conventional breadmaking, the most commonly used leavening agent is the yeast Saccharo myces cerevisiae, although other Saccharomyces species such as S. cariocanus, S. mikatae, S. paradoxus and S. kudriavzevii can be also employed [48]. Furthermore, lactic acid bacteria, mainly Lactobacillus species are used as leavening agents for sourdough bread production [49].

The breadmaking process begins with the formation of dough through mixing of flour, water, yeast, sugar, salt, shortening and other ingredients. Flour particles are hydrated and sheared during mixing, and dough develops when gluten proteins form a continuous cohesive network in which the starch granules are dispersed [40]. Depolymerisation and polymerisation reactions possibly give rise to the gluten network, mostly made up of glutenin [50]. Incorporation of air during dough mixing is extremely important, affecting the final crumb structure because the carbon dioxide produced by yeast during fermentation diffuses to pre-existing air bubbles [40,51]. An optimal gluten network confers dough machinability, good gas retention, high bread volume and fine crumb structure [29]. After resting, the dough is divided into loaf-sized pieces, rounded, moulded, placed on a baking tray, proofed and baked.

The combined effects of heat, moisture and time induce starch gelatinisation and pasting which together with heatsetting of gluten proteins occur during baking, giving rise to the typical solid foam structure of baked bread [22]. The partially crystalline starch is converted into amorphous, transient, gelatinised starch networks. The swollen gelatinised starch granules are deformed, part of the starch polymers leach out of the granules and form a continuous network in the bread crumb [40,52]. Besides accumulation of amylose outside the granules, the presence of an amylose-rich region in the centre of gelatinised starch granules was found after baking [22,52].

During baking the transient gluten network formed in dough is transformed into a continuous, permanent network probably due to modifications in protein surface hydrophobicity, sulfhydryl/disulfide interchanges and formation of new disulfide cross-links [22,38,50,53]. Moreover, heat-induced sulfhydryl-disulfide exchange reactions can lead to incorporation of α- and γ-gliadins into the glutenin network [54]. Gas cell opening occurs, and besides becoming gluten continuous the bread is also gas-continuous [38,40].

Macroscopic changes during baking include further expansion of the dough and crust formation and browning [40]. The oven spring is due to continued production of carbon dioxide by yeast, its expansion by heating and vaporisation of ethanol and water. The bread bakes from the outside to the inside, resulting in a baked crumb [38].

The crust browning is directly related to the reducing sugars (glucose, fructose, maltose, etc.) formed by hydrolysis of starch and complex sugars of the flour, during dough making and leavening. Under heating, the sugars can undergo caramelisation, and/or the reducing sugars can react with the free amino acid groups of proteins in the Maillard reaction [54,55]. Besides, different flavour compounds are produced, giving bread its appealing smell and taste [55].

Additional interactions between biopolymers in the bread crumb occur during cooling. Amylose chains form helices, self-associate and crystallise [22,52,57]. Moreover, amylose may form more inclusion complexes with polar lipids. As a consequence, a permanent and in part crystalline amylose network is formed, providing a soft crumb in fresh bread. The gluten network organized during baking and the amylose network developed while cooling thus account for the plasticity of freshly baked bread [22].

Fresh bread consists mainly of a continuous gluten network, which forms a compressed matrix between the swollen, gelatinised starch granules, and the starch network, consisting of entangled, gelatinised starch polymers [22]. It usually presents an appealing brownish and crunchy crust, a pleasant aroma, fine slicing characteristics, a soft and elastic crumb texture, and a moist mouthfeel [47]. However, when a loaf of bread is removed from the oven after baking, a series of undesirable changes called staling starts, eventually leading to deterioration of quality [46].

Staling implies a relatively short shelf life for fresh bakery products. The loss of freshness is paralleled by an increase in crumb firmness and a decrease in flavour and aroma, leading to loss of consumer acceptance. Loss of moisture and starch retrogradation are accepted as two of the basic mechanisms in the firming of the crumb [58]. This subject has been extensively reviewed and discussed in [16,22,45]. In this context, mechanization, large scale production and increase in consumer demand for consistent product quality and longer shelf life of baked goods have led to the use of a wide range of additives (bread improvers) in the baking industry, which include emulsifiers, soy flour, chemical redox agents and enzymes [29,42,59].

5. ENZYMES USED IN BAKED PRODUCTS

Baking comprises the use of enzymes from three sources: the endogenous enzymes in flour, enzymes associated with the metabolic activity of the dominant microorganisms and exogenous enzymes which are added in the dough [60].

The supplementation of flour and dough with enzyme improvers is a usual practice for flour standardization and also as baking aids. Enzymes are usually added to modify dough rheology, gas retention and crumb softness in bread manufacture, to modify dough rheology in the manufacture of pastry and biscuits, to change product softness in cake making and to reduce acrylamide formation in bakery products [8]. The enzymes can be added individually or in complex mixtures, which may act in a synergistic way in the production of baked goods [60-62], and their levels are usually very low.

5.1 Hydrolases

Enzymes as technological aids are usually added to flour, during the mixing step of the breadmaking process. The enzymes most frequently used in breadmaking are the α-amylases from different origins [63].

5.1.1 Amylases and other starch-converting enzymes

The industrial processing of starch is usually started by α-amylases (α-1,4-glucanohydrolase). Most of the starch-converting enzymes belong to the α-amylase family or family 13 glycosyl hydrolases (GH), based on amino acid sequence and structural similarities [64,65,66,67].

α-Amylases (EC 3.2.1.1) are endoenzymes that catalyze the cleavage of α-1,4-glycosidic bonds in the inner part of the amylose or amylopectin chain. The end products of α-amylase action are oligosaccharides, with an α-configuration and varying lengths, and α-limit dextrins, which are branched oligosaccharides [17]. These enzymes can be obtained from cereal, fungal, bacterial and biotechnologically altered bacterial sources. Differences in the number of binding sites and location of catalytic regions determine substrate specificity of α-amylases, the length of the oligosaccharide fragments released after hydrolysis and, consequently, the carbohydrate profile of the final product. The different forms of α-amylases also have diverse thermal stability profiles [15].

Also part of the GH13 family are the exoenzymes maltogenic α-amylase (glucan 1,4-α-glucanhydrolase, EC 3.2.1.133) and other maltooligosaccharide forming amylases (EC 3.2.1.60, for instance). While maltogenic α-amylase mainly releases maltose from starch, maltooligosaccharide producing amylases give rise to maltotetraose or maltohexaose, among others. On the other hand, debranching enzymes, such as pullulanase (EC 3.2.1.41) and isoamylase (EC 3.2.1.68), grouped as well in the GH13 family, hydrolyse α-(1,6)-bonds removing the side-chains from amylopectin [16,17].

β-Amylases (EC 3.2.1.2) and glucoamylases (EC 3.2.1.3) are encompassed in the GH14 and GH15 families, respectively. Both are exoamylases that employ the inverting mechanism to cleave α-glycosidic bonds at the non-reducing ends of amylose and amylopectin, producing low molecular weight carbohydrates in the β-anomeric form [15,68]. β-Amylases are unable to cleave α-1,6-linkages and the final products consist of maltose and β-limit dextrin. Therefore hydrolysis of amylopectin is incomplete, resulting in only 50-60% conversion to maltose. In the case of amylose, the maximum degree of hydrolysis is 75-90% due to the slightly branched structure of this polysaccharide [15]. On the other hand, glucoamylase has a limited activity on α-1,6-linkages and would possibly be able to catalyse total conversion of starch into β-glucose [16].

Malt and microbial α-amylases have been widely used in the baking industry. The malt preparation led the way for the commercial use of many other enzymes in baking [69]. Fungal α-amylases or malt are usually added to optimize amylase activity of the flour, initially aiming to increase the levels of fermentable and reducing sugars. In view of their lower thermostability, fungal α-amylases are more appropriate than malt amylases for flour standardization. The α- and β-amylases have different but complementary functions during the breadmaking process [70]. The supplemented α-amylases break down damaged starch particles into low molecular weight dextrins during the dough stage, while endogenous β-amylase converts these oligosaccharides into maltose which is used as fermentable sugar by the yeast or sourdough microorganisms [15,16]. The increased levels of reducing sugars lead to the formation of Maillard reaction products, intensifying bread flavour and crust colour. In addition, these enzymes can improve the gas-retention properties of fermented dough and reduce dough viscosity during starch gelatinization, with consequent improvements in product volume and softness [8,22,71].

Certain amylases are able to decrease the firming rate of bread crumb, acting as anti-staling agents. Amylase-containing anti-staling products typically consist of bacterial or fungal α-amylases with intermediate thermostability [16,22]. In this context, one of the most effective anti-staling amylases is the Bacillus stearothermophilus maltogenic α-amylase [22]. The anti-staling action of amylases has been attributed to the modified retrogradation behaviour of the hydrolysed starch [72-74]. Yet, other researchers ascribe the effect to the interference of the low molecular weight dextrins with starch-starch and/or gluten-starch interactions [74-76].

5.1.2 Proteases

Proteases can be subdivided into two major groups according to their site of action: exopeptidases and endopeptidases. Exopeptidases cleave the peptide bond proximal to the amino or carboxy termini of the substrate, whereas endopeptidases cleave peptide bonds distant from the termini of the substrate [77]. Most of the proteolytic activity of wheat and rye flours corresponds to aspartic proteases and carboxypeptidases, which are both active in acid pH. Additionally, aspartic proteases of wheat are partly associated with gluten [78]. Nevertheless, the proteolytic activity of sound, ungerminated grain is normally low [79].

Proteases are used on a large commercial scale in the production of bread, baked goods, crackers and waffles [80]. These enzymes can be added to reduce mixing time, to decrease dough consistency, to assure dough uniformity, to regulate gluten strength in bread, to control bread texture and to improve flavour [16,60]. In addition, proteases have largely replaced bisulfite, which was previously used to control consistency through reduction of gluten protein disulfide bonds, while proteolysis breaks down peptide bonds. In both cases, the final effect is a similar weakening of the gluten network [79].

In bread production, a fungal acid protease is used to modify mixtures containing high gluten content. When proteases are mixed in the blend, it undergoes partial hydrolysis becoming soft and easy to pull and knead [7,60]. Proteases are also frequently added to dough preparations. These enzymes have great impact on dough rheology and the quality of bread possibly due to effects on the gluten network or on gliadin [7].

Proteases are also applied in the manufacture of pastries, biscuits and cookies. They act on the proteins of wheat flour, reducing gluten elasticity and therefore reducing shrinkage of dough or paste after moulding and sheeting [8,81]; for instance, hydrolysis of glutenin proteins, which are responsible for the elasticity of dough, has considerable improving effects on the spread ratio of cookies [81].

5.1.3 Hemicellulases

Hemicellulases are a diverse class of enzymes that hydrolyse hemicelluloses, a group of polysaccharides comprising xylan, xylobiose, arabinoxylan and arabinogalactan [82]. This group includes xylanase or endo-1,4-β-xylanase (4-β-D-xylan xylanohydrolase, EC 3.2.1.8), a glycosidase that catalyses the endohydrolysis of 1,4-β-D-xylosidic linkages in xylan and arabinoxylan.

Xylanase, also designated endoxylanase, was originally termed pentosanase [83]. A wide variety of xylanases have been reported from a plethora of microorganisms including bacteria, archaea and fungi [84]. These enzymes are mainly classified in the glycosyl hydrolase (GH) families 10 and 11 [16,64,65], although putative xylanase activities have been reported in GH families 5, 7, 8 and 43 [84,85]. GH10 xylanases are regarded to have broader substrate specificity and release shorter fragments compared to GH11 xylanases, while the latter enzymes are more susceptible to steric hindrance by arabinose substituents [86,87]. In addition, different endogenous xylanase inhibitors occur in cereals: Triticum aestivum L. xylanase inhibitor (TAXI) [88,89], xylanase inhibitor proteins (XIP-type inhibitors) [90] and TLXI-type (thaumatin-like endoxylanase inhibitors) [91].

The complete hydrolysis of arabinoxylans requires the concerted action of different enzymes. The xylan backbone will be cleaved randomly by endo-1,4-β-xylanases, the main arabinoxylan hydrolysing enzymes, yielding arabinoxylo-oligosaccharides. β-D-xylosidases (EC 3.2.1.37) cleave xylose monomers from the non-reducing end of arabinoxylo-oligosaccharides. The arabinose residues are removed by α-L-arabinofuranosidases (EC 3.2.1.55), while ferulic acid esterases (EC 3.1.1.73) cleave ester linkages between arabinose residues and ferulic acid [30,83].

Xylanases were introduced to the baking segment in the years 1970 and are most often used combined with amylases, lipases and many oxidoreductases to attain specific effects on the rheological properties of dough and organoleptic properties of bread [85]. These enzymes have also been used to improve the quality of biscuits, cakes and other baked products [71].

The most favourable xylanases for breadmaking are those that preferentially act on WU-AX and are poorly active on WE-AX, because they remove the insoluble arabinoxylans which interfere with the formation of the gluten network, giving rise to high molecular weight solubilised arabinoxylans, resulting in increased viscosity and thus enhancing dough stability [92-94]. As a consequence, a more stable, flexible and easy to handle dough is obtained, resulting in improved oven spring, larger loaf volume, as well as a softer crumb with improved structure [43]. Moreover, the addition of xylanases during dough processing is expected to increase the concentration of arabinoxylo-oligosaccharides in bread, which have beneficial effects on human health [95].

The potential of GH family 8 xylanases as technological aids in baking was shown for a psychrophilic enzyme from Pseudoalteromonas haloplanktis and a mesophilic enzyme from Bacillus halodurans. Although both enzymes had a positive effect on loaf volume, psychrophilic GH8 xylanase was apparently much more efficient than the mesophilic enzyme from the same family, because much lower concentrations of the former enzyme were required to produce a similar increase in bread volume. Additionally, a psychrophilic GH10 xylanase from Cryptococcus adeliae was found to be ineffective [85].

Recently, a purified GH11 xylanase from Penicillium occitanis was evaluated as an additive during mixing of wheat flours. Significant improvements of bread characteristics, including higher final moisture content, volume and specific volume, were observed. Enhancements in sensory and textural properties were also obtained [96].

5.1.4 Lipases

Lipases (EC 3.1.13) or triacylglycerol acylhydrolases hydrolyse triacylglycerols (TAG) producing monoacylglycerols (MAG), diacylglycerols (DAG), glycerol and free fatty acids. These enzymes are widely found in nature [97]. Besides TAG lipases there are phospholipases A1 (EC 3.1.1.32), A2 (EC 3.1.1.4), C (EC 3.1.4.3), D (EC 3.1.4.4) and galactolipases (EC 3.1.1.26). Even though they are present in all cereal grains; lipase activity of white flour is usually low enough to avoid rancidity due to hydrolysis of native lipids and of baking fat [71,79,98].

The use of lipases in the baking segment is much more recent in comparison to α-amylases and proteases. The first generation of commercial lipase preparations was introduced to the market in the years 1990 and recently a third generation became available [59]. The latter are protein engineered enzymes, claimed to give a better effect in high speed mixing and no- time dough processes. Moreover, third generation lipases have lower affinity for short chain fatty acids, which reduces the risk for off-flavour formation on account of prolonged storage of the baked goods and the use of butter or milk fat in baked products [12].

Lipases (TAG lipases) of the first generation are 1,3-specific, removing preferentially fatty acids from positions 1 and 3 in TAG. These enzymes can improve dough rheology, increase dough strength and stability, thus improving dough machinability [62,99,100]. In addition, lipases lead to an increase in volume which results in an improved, more uniform crumb structure; hence a softer crumb is obtained [99].

The second generation lipases act simultaneously on TAG, diacylgalactolipids and phospholipids, producing more polar lipids, providing a greater increase in volume, better stability to mechanical stress on the dough, and a fine, uniform bread crumb structure compared to the first generation lipases [43,59,101]. Moreover, a third generation lipase was found to increase expansion of the gluten network, increase the wall thickness and reduce cell density, enhancing volume and crumb structure of high fibre white bread [102].

The surface active properties of the hydrolysis reaction products (MAG, DAG, monoacylgalactolipids and lysophospholipids), along with modifications on the interactions between lipids and gluten proteins caused by the lipases, as well as the effect of these enzymes on the incorporation of air during mixing are possible mechanisms by which they affect bread volume [101]. In this context, the roles of lipids and surfactants in breadmaking have been extensively reviewed elsewhere [38,45].

The addition of lipases has been claimed to retard the rate of staling in baked products [8,103,104]. The effect of these enzymes has been attributed to in situ production of MAG following TAG hydrolysis, although this mechanism is not completely accepted because the amount of MAG would be insufficient to account for the antistaling effect [45,99]. Lipases may also be used for the development of particular flavors in bakery products [100].

The effect of a third generation lipase on the quality of high-fibre enriched brewer's spent grain breads has been evaluated. The enzyme produced beneficial effects during bread making, positively affecting loaf volume, staling rate and crumb structure [102].

A recent study compared three generations of lipase enzymes with the emulsifier, diacetyl tartaric esters of monoglycerides (DATEM), on white wheat flour bread. Lipases and DATEM improved most aspects of bread quality. In shorter fermentation times, DATEM, a second generation (Lipopan F-BG) and a third generation (Lipopan Xtra-BG) lipase were more effective. In longer fermentations, unlike the third generation lipase (Lipopan Xtra-BG), moderate amounts of the second generation lipase (Lipopan F-BG) significantly increased the bread volume [59].

The application of lipase and MAG to produce fiber enriched pan bread using the straight dough method was assessed. The use of lipase dosages up to 50 ppm and MAG up to 2% indicated the possibility of replacement of MAG by lipases in fiber enriched pan bread [105].

Recently, the effects of two lipases and DATEM on the rheological and thermal properties of white and whole wheat flour doughs were compared. Lipases were able to cause modifications in the dough components (gluten proteins and starch). The enzymes improved dough handling properties to a similar or greater extent than DATEM, increasing dough stability, maximum resistance to extension and hardness, and decreasing softening degree and stickiness. The possible role of lipases in delaying starch retrogradation was indicated by the greater extent of formation of amylose-lipid complexes promoted by lipases in comparison to DATEM [106].

5.2 Oxidoreductases

5.2.1 Lipoxygenases

Lipoxygenase (linoleate oxygen oxidoreductase, EC 1.13.11.12) is a non-heme iron-containing dioxygenase, found in a wide variety of plant and animal tissues, which with molecular oxygen catalyses the oxidation of polyunsaturated fatty acids (PUFA) containing a cis,cis-1,4-pentadiene system, such as linoleic or linolenic acid, to form fatty acid hydroperoxides [107,108]. These enzymes are abundant in grain legume seeds (beans and peas) and potato tubers, being minor constituents of wheat flour [107]. Multiple isoforms of lipoxygenases are found in plants; for example a multigene family encodes soybean lipoxygenases, three members of which encode the three major seed isoforms L1, L2 and L3 [109].

The main commercial sources of lipoxygenases are enzyme-active soybean flour and, to lower extent, flour from other beans, such as fava beans [12]. Wheat lipoxygenase catalyses the oxidation of PUFA in the free or MAG forms [110] while soybean or horse bean lipoxygenases also catalyse the oxidation of PUFA present in TAG [111]. The transient alkyl, peroxyl and hydroxyl radicals formed during lipoxygenase catalysed reactions are able to oxidise carotenoid pigments and sulfhydryl groups in peptides and proteins present in the dough, mainly giving rise to hydroxyacids [112].

In fact, the initial application of lipoxygenases in doughs was based on their ability to bleach fat-soluble carotenoid flour pigments, through co-oxidation of carotenoids with PUFA [113,114]. However, since the endogenous lipoxygenase content of wheat flour is insufficient to give enough bleaching effect, enzyme-active soybean or fava bean flour is added [114].

Lipoxygenases are also employed to improve mixing tolerance and dough handling properties [115]. In this case, the effect of these enzymes may be explained by oxidation of thiol groups of gluten proteins which can lead to rearrangement of intra- or inter-chain disulfide bonds [21] and also to formation of tyrosine cross-links [116], with consequent strengthening of the gluten network. As a result, improvement in dough rheology occurs, with increase in dough strength through proofing and baking, finally leading to improved loaf volume.

On the other hand, the action of lipoxygenase can lead to undesirable flavors in bread [79,114]. These flavors are possibly due to some of the breakdown products (ketodienes) formed during the anaerobic reaction [117,112].

5.2.2 Glucose oxidase

Glucose oxidase (β-D-glucose:oxygen: 1-oxidoreductase; EC 1.1.3.4) catalyzes the oxidation of β-D-glucose to D-glucono-δ-lactone and hydrogen peroxide [118,119]. This enzyme has been obtained from different fungal sources, mainly from genus Aspergillus and Penicillium, being Aspergillus niger the most commonly used [120-123].

Glucose oxidase has been used successfully to remove residual glucose and oxygen in foods and beverages aiming to increase their shelf life. The hydrogen peroxide generated by this enzyme presents antimicrobial properties, and is easily removed by catalase utilization, which is an enzyme that catalyzes the conversion of hydrogen peroxide to oxygen and water [12,124-127]. Glucose oxidase can be used as alternative oxidizing agent instead of potassium bromate in breadmaking. Potassium bromate is an oxidizing agent that was traditionally used in baking, and its use was prohibited in many countries after it was recognized as carcinogenic [128,129].

Although the mechanism of action of glucose oxidase is still not completely understood, a possible explanation is that the hydrogen peroxide formed during catalysis promotes, indirectly, the formation of either disulfide bonds or dityrosine crosslinks, or both, in the gluten network [116,130,131]. Therefore, the increase in disulfide crosslinking and/or promotion of gelative oxidation on the gluten matrix confers dough machinability, good gas retention, high bread volume and fine crumb structure [54,132-134]. Addition of increasing glucose oxidase concentrations to wheat flour dough produced significant changes on dough rheology and bread quality; and the extent of the effect was highly dependent on the amount of enzyme and the original wheat flour quality [130]. Furthermore, glucose oxidase was able to recover the breadmaking ability of damaged gluten [135]. Another possibility to explain the improvements on crumb properties, in both bread and croissants, as a result of glucose oxidase catalysed reactions would be the crosslinking of the albumin/globulin fraction with both disulfide and non-disulfide bonds, and the slight occurrence of non-disulfide crosslinking in the gluten proteins [131].

5.3 Other enzymes

Among the enzymes which have attracted attention for use in bakery is asparaginase. Differently from other enzymes, its use is not associated with improved bread volume, crumb softening or reduced staling. Instead, asparaginase is claimed to have a high potential of reducing formation of acrylamide during baking [136-138]. Asparaginase (L-asparagine amidohydrolases, EC 3.5.1.1) catalyses the hydrolysis of asparagine to aspartic acid and ammonium, removing the precursor of acrylamide formation [139]. Acrylamide, classified as a probable human carcinogen, is formed in heated foods via Maillard reaction between asparagine and a carbonyl source [137,138,140,141]. Although asparaginase can be found among living organisms, including animals, plants and microorganisms, filamentous fungi as Aspergillus oryzae and A. niger have been explored for enzyme preparation aiming commercial purposes [142-144]

Table 2. Applications of starch modifying enzymes in baking.

Enzyme (classification)	Substrate in foods	Reaction	Applications in baked products	References
Amylolytic enzymes	*Starch*	*Hydrolysis of linkages*		
α-Amylases (EC 3.2.1.1) or α-(1,4)-glucanhydrolases	Amylose and amylopectin	α(1→4)-D-glycosidic [endo], *liberating α-dextrins*		[16,17,22,48,74,174]
β-Amylases (EC 3.2.1.2)	Amylose and amylopectin	α(1→4)-D-glycosidic [exo], *liberating β-dextrins and β-maltose*	Generation of fermentable compounds; Increase in bread volume; Reduction in fermentation time; Improvement in dough viscosity, rheology and bread softness; Improvement in bread texture; Formation of reducing sugars and subsequent Maillard reaction products, intensifying bread flavor and color; Decrease of bread crumb firming rate; Anti-staling effects.	[16,17,22,175]
Glucoamylase (EC 3.2.1.3) or amyloglucosidase	Amylose and amylopectin	α(1→4)- and α(1→6)-D-glucosidic, *liberating β-glucose*		[16,17,22]
Pullulanase (EC 3.2.1.41)	Amylopectin	α(1→6)-D-glycosidic		[16,17,22]
Isoamylase (EC 3.2.1.68)	Amylopectin	α(1→6)-D-glycosidic		[16,17,22]
Maltogenic α-amylase (EC 3.2.1.133)	Amylose and amylopectin	α(1→4)-D-glycosidic, liberating maltose		[16,17,22,74,175]
Maltooligosaccharides forming amylases (glucan 1,4-α-maltotetraohydrolase) (ex., EC.3.2.1.60)	Amylose and amylopectin	Liberation of maltotetraose or maltohexaose		[16,17,22,175]
Transferases Amylomaltases (EC 2.4.1.25) Amylosucrases (EC 2.4.1.4) Ciclodextrin glycosyltransferases (EC 2.4.1.19)	Amylose, amylopectin and dextrins	Hydrolysis of α(1→4) glycosidic bonds and transference of a reducing group to a non-reducing acceptor (monosaccharide unit)		[17]

Transglutaminases (EC 2.3.2.13) from microbial sources also have potential for application in bakery products. Food proteins can be modified through cross-linking by transglutaminases, resulting in textured products, protecting lysine in food proteins from undesired chemical reactions, encapsulating lipids and lipid-soluble materials, forming heat and water resistant films, improving elasticity and water-holding capacity, modifying solubility and functional properties, and producing food proteins of higher nutritive value [29,145-153].

Laccase (EC 1.10.3.2) is a copper containing enzyme that catalyses the oxidation of a wide variety of phenolic compounds via one-electron removal, generating reactive phenolic radicals [29,154]. This enzyme is very interesting for baking due its ablity to cross-link the esterified ferulic acid on the arabinoxylan fraction of dough, resulting in a strong arabinoxylan network [155]. It was also reported that laccase may improve crumb structure and softness of baked products. Futhermore, increases in strength and stability, as well as reduced stickness of dough, which confers improvement of machinability, have been described [149,155-157].

A summary of the main applications of different classes of enzymes in the baking industry is presented in tables 2, 3 and 4.

Table 3. Applications of cellulases and hemicellulases in baking.

Enzyme (classification)	Substrate in foods	Reaction	Applications in baked products	References
Cellulases and Hemicellulases	*Non-starch components of cereals*	*Hydrolysis of linkages*		
Cellulase (EC 3.2.1.4)	Cellulose and β-glucan	β(1→4)-D-glycosidic [endo]	Removal of insoluble arabinoxylans, contributing to gluten network formation; Increase in dough viscosity, stability, with better moldable form; Improvements on rheological properties of dough; Reduction in fermentation time; Increase of bread volume; Synergistic action of glucanases on xylanolytic attack of cereals structure, providing more soluble dietary fiber in bread products; Production of prebiotic oligosaccharides in bread.	[30,71,83,95]
Lamarinase (EC 3.2.1.6)	β-glucans	β(1→3)- and β(1→4)-D-glycosidic		[71]
Lichenase (EC 3.2.1.73)	β-glucans	β(1→3)- and β(1→4)-D-glycosidic		[30,71,83]
Endo β(1,4)-D-xylanase (EC 3.2.1.8) or endoxylanase	Arabinoxylan	β(1→4)-D-xylosidic bonds		[71,83,85,95,165,176,177]
α-L-Arabinosidase (EC 3.2.1.55)	Arabinoxylan	Terminal α-L-Arabinofuranoside residues		[71,83,95,102]
β-D-Xylosidase (EC 3.2.1.37)	Arabinoxylan	β(1→4)-D-xylosidic bonds (non-reducing end)		[83,95]

Table 4. Applications of proteases, transglutaminases, lipases and esterases, and oxidoreductases in baking.

Enzyme (classification)	Substrate in foods	Reaction	Applications in baked products	References
Proteases (EC 3.4.)	Gluten proteins Gliadin and glutenin	*Hydrolysis of peptide bonds*	Reduction of dough mixing time; Control of dough rheology or viscoelastic properties of gluten strength in bread; Enhance dough extensibility; Increase loaf or bread volumes; Formation of aminoacids and flavors; Crispness feature on bread crust; Production of gluten-free products.	[78,79,101,178, 179,180]
Transglutaminases Protein-glutamine γ-glutamyl-transferase (EC 2.3.2.13)	*Gluten Proteins*	Acyl-transfer reaction between γ-carboxyamide and primary amines	Cross-link between gluten and other peptides, forming a new protein network; Increase volume and improve structure of breads, better retention of gas; Improve bread crumb strength, height increase in puff pastry and croissants volume; Improve dough stability; Improve properties of gluten-free breads; Protect frozen doughs from damage.	[29,145-150]
Lipases and esterases	*Lipids*	*Hydrolysis of ester bonds*		
Lipase (EC 3.1.1.3)	Triacylglycerols	Liberation of free fatty acids	Improvement in bread volume and dough stability; Formation of emulsifiers; Retard staling; Development of flavors.	[59,106,158]
Oxidoreductases	*Various*	*Oxi-reductions*		
Glicose oxidase (EC 1.1.3.4) β-D-glucose:oxygen 1-oxidoreductase	β-D-glucose	Oxidation of β-D-glucose to gluconic acid	Control on browning for Maillard reaction; Improvements in crumb properties.	[130,131,181]
Lipoxygenase (EC 1.13.11.12) linoleate:oxygen 13-oxidoreductase	Polyunsaturated fatty acids	Oxidation of fatty acids	Bleaching of fat-soluble flour pigments; Hydroperoxides formed can oxidize sulfhydryl groups in proteins.	[79,182]
Laccase (EC 1.10.3.2)	Feruloyl esters of arabinoxylans;	Oxidation of phenol groups	Dough strength, stability and reduced stickiness; Increase in volume; Improved crumb structure and softness.	[155,163]

Enzyme (classification)	Substrate in foods	Reaction	Applications in baked products	References
or benzene-diol:oxygen oxidoreductase	sulfhydryl groups in gluten proteins			
Sulfhydryl oxidase	Sulfhydryl groups in proteins	Oxidation of sulfhydryl groups	Help gluten network formation and increase dough stability.	[79,162]

5.4 Use of enzyme combinations

It is common practice to use mixtures of enzymes, some of which are commercially available. The enzymes may act individually or present a synergistic effect. The trend is to choose and control the use of complex mixtures of enzymes which may act in a synergistic way and can exert a better effect (than the individually used) on the different flour components [60]. Recent advances in understanding of the dough forming and overall baking processes at the molecular level have focused attention on improvements that can be achieved by application of more specially tailored enzymes alone or in combinations. Usually, integrated experimental design and optimization followed by chemical analyses, rheological experiments and baking trials are necessary in order to provide answers to the more complicated questions [158].

The use of a combination of enzymatic preparations of amylases, xylanases and lipases has been reported by different authors [60,102,159]. This specific mixture is claimed to increase bread volume and shelf-life. The use of α-amylase and glucose oxidase to replace bromate led to a significant improvement in dough extensibility and bread volume [160]. Addition of commercial enzyme mixtures, containing α-amylase and lipase activities to produce bread samples, using the straight dough method, had a beneficial effect on bread keeping properties and resulted in the formation of a more thermostable amylose-lipid complex compared to the control bread [161]. Amylopectin retrogradation was inhibited by the use of the enzyme combinations and this effect was strongly related to a decrease in crumb-firming rates.

The combined use of different enzymes, classified as gluten degrading (like proteases) or adjuvants, such as amylases and xylanases, with a group of crosslink promoting enzymes, such as transglutaminases and glucose oxidase, was also studied [149]. Better shaped bread could be obtained after the use of gluten degrading or adjuvant enzymes, and association with transglutaminase resulted in improvements on texture and rheological properties. The crumb firmness which can further lead to staling, can result from transglutaminase action, but it may be reversed with opposite amylase, xylanase and protease effects.

In a similar way, combinations of enzymes classified as carbohydrate degrading, including amylases and xylanases (pentosanases), and crosslink promoting enzymes, like transglutaminases and oxidases, including glucose oxidase, laccase [149], lipoxygenase and sulfhydryl oxidase [79,162] were evaluated. The most frequent associations contained xylanases and glucose oxidase, but addition of laccase and transglutaminase was also employed. The hydrogen peroxide formed by glucose oxidase catalysis may interfere in gluten network, via oxidized glutathione reaction, leading to gluten disulfide bonds formation [43], and it also interferes in the formation of a soluble pentosan gel (from xylans) that increased dough consistency [146]. Because both oxidases and xylanases influence the xylan properties, xylanases and oxidases could be used advantageously in combination, resulting in a mesh of gluten and gelified xylans matrix, which increases gas retention, dough stability and bread volume. Laccase is reported to catalyze dimerization of feruloylated esters in feruloylated arabinoxylans in doughs [163,164], forming a xylan network, contributing to increase strength of dough and volume.

Lipoxygenases oxidize polyunsaturated fatty acids during dough mixing. The hydroperoxides formed can oxidize the sulfhydryl groups of gluten proteins and thus be advantageous in the formation of the gluten network of dough. Sulfhydryl oxidase combined to glucose oxidase and xylanases has been used to strengthen weak doughs [79,162].

The use of a combination of commercial preparations of glucolipase, hemicellulase and hexose oxidase in formulations of frozen pre-baked French bread, substituted with whole wheat flour, improved parameters such as proofing time, oven spring and cut opening and cut height [158]. An interaction among the three enzymes was observed for most of the parameters, because the responses of each enzyme to variations in dosing were influenced by the doses of the other two.

6. CONCISE SUMMARY

In addition to the desire for substituting chemical additives with natural alternatives, there is a growing consumer concern and, subsequently, an increased demand for the preservation and/or enhancement of foods with ingredients that confer health benefits. The application of enzymes in baked goods has been documented for the production of dietary fiber-enriched bread [102,165], the formulation of gluten-free products [145], and the enhancement of products with elevated levels of arabinoxylan oligosaccharides possessing prebiotic potential [165].

Multiple factors can be identified for the development of enzyme preparations that deliver the necessary effects or possess suitable features for application under certain process circumstances. Potential strategies encompass the identification of novel enzymes from diverse sources[166], including microorganisms derived from the planet's extensive biodiversity, the production of recombinant proteins from genetically modified organisms, and protein engineering[167].

Psychrophilic enzymes usually have higher optimal activity and stability at lower temperatures than their mesophilic counterparts [168]. Due to the fact that the temperatures most frequently used in dough mixing and proofing are around or below 35 °C, it has been suggested that psychrophilic enzymes would be advantageous candidates for use as additives in the baking industry [83,85]. In this context, researchers have shown that much lower dosages of psychrophylic xylanases than of the mesophilic enzymes could be used to attain maximal bread volumes [85,169,170].

Directed evolution is an effective method in protein engineering for designing and altering enzyme properties [171]. This method can be utilized for a diverse array of proteins, the majority of which are pertinent to biocatalytic activities. In the span of ten years, directed evolution has emerged as a conventional technique in protein engineering, applicable alongside rational protein design and other established methods to fulfill the requirements for industrially viable biocatalysts that endure process conditions including elevated substrate concentrations, high temperatures, and prolonged stability, while also exhibiting the requisite specificity and/or selectivity [172]. A recent work demonstrated the synergistic application of directed evolution and high-throughput screening to enhance the efficacy of a maltogenic α-amylase derived from Bacillus sp. for low pH bread applications. One of the resulting variations exhibited a significant enhancement in heat stability at pH 4.5 and a notable antistaling effect in low pH breads [173].

REFERENCES

1. Illanes A. Introduction. In; Illanes A. (ed.) Enzyme Biocatalysis. Principles and Applications. Springer; 2008. p. 19-56.

2. van Oort, M. Enzymes in food technology – introduction. In: Whitehurst RJ, van Oort M. (eds.) Enzymes in Food Technology, second ed. Chichester: Wiley-Blackwell; 2010. p. 1-17.

3. Falch E.A. Industrial Enzymes – Developments in Production and Application. Bio technology Advances 1991;9(4) 643-658.

4. Poulsen PB, Bucholz K. History of Enzymology with Emphasis on Food Production. In: Whitaker JR, Voragen AGJ, Wong DWS. (eds.) Handbook of Food Enzymology. New York: Marcel Dekker; 2003. Chapter 2.

5. Petersen SB. Protein Engineering: Design and Engineering on the Nano Scale. In: Straathof AJJ, Adlercreutz P. (eds.) Applied Biocatalysis, second ed. Amsterdam: Harwood Academic Publications; 2005. p. 263-310.

6. Butler DP, van der Maarel MJEC, Steeneken PAM. Starch-acting enzymes. In: Eliasson A-C. (ed.) Starch in Food. Structure, Function and Applications. Cambridge: Woodhead Publishing; 2004. Chapter 4.

7. Salleh AB, Razak CNA, Rahman RNZRA, Basri M. Protease: Introduction. In: Salleh AB, Rahman RNZRA, Basri M. (eds.) New Lipases and Proteases. New York: Nova Science Publishers; 2006. p. 23-39.

8. Cauvain S, Young L. Ingredients and their influences. In: Cauvain S, Young L. (eds.) Baked Products. Science, Technology and Practice. Oxford: Blackwell Publishing; 2006. p. 72-98.

9. Dupaigne B. The history of bread. In: Dupaigne, B. (ed) The history of bread. Harry N. Abrams, Inc.; 1999. P.10.

10. Novozymes A/S: The Novozymes Report 2011. (http://report2011.novozymes.com/ (accessed 23 August 2012).

11. The Freedonia Group, Inc.: WORLD FOOD & BEVERAGE ENZYME DEMAND BY MARKET (million dollars) 2000 – 2020: Study #: 2824 - World Enzymes to 2015. http://www.freedoniagroup.com/FractionalDetails.aspx?DocumentId=573141 (accessed 20 August 2012).

12. van Oort, M. (2010) Enzymes in bread making. In: Whitehurst RJ, van Oort M. (eds.) Enzymes in Food Technology, second ed. Chichester: Wiley-Blackwell; 2010. p. 103-143.

13. Taggart P. Starch as an ingredient: manufacture and applications. In: Eliasson A-C. (ed.) Starch in Food. Structure, Function and Applications. Cambridge: Woodhead Publishing; 2004. Chapter 12.

14. Tramper J, Poulsen PB. (2005) Enzymes as Processing Aids and Final Products. In: Straathof AJJ, Adlercreutz P. (eds.) Applied Biocatalysis, second ed. Amsterdam: Harwood Academic Publications; 2005. p. 62-102.

15. Synowiecki J. The Use of Starch Processing Enzymes in the Food Industry. In: Polaina J, MacCabe AP. (eds.) Industrial Enzymes. Sctructure, Function and Applications. Dordrecht: Springer; 2007. p. 19-34.

16. Goesaert H, Brijs K, Veraverbeke WS, Courtin CM, Gebruers K, Delcour JA. Wheat flour constituents: how they impact bread quality, and how to impact their functionality. Trends in Food Science and Technology 2005;16(1-3) 12–30.

17. van der Maarel MJEC, van der Veen B, Uitdehaag JCM, Leemhuis H, Dijkhuizen L. Properties and applications of starch-converting enzymes of the α-amylase family. Journal of Biotechnology 2002;94(2) 137–155.

18. Courtin CM, Delcour JA. Arabinoxylans and endoxylanases in wheat flour bread- making. Journal of Cereal Science 2002;35(3) 225–243.

19. Gandikota S, MacRitchie F. Expansion capacity of doughs: methodology and applications. Journal of Cereal Science 2005;42(2) 157–163.

20. Georgopoulos T, Larsson H, Eliasson AC. Influence of native lipids on the rheological properties of wheat flour dough and gluten. Journal of Texture Studies 2006;37(1) 49–62.

21. Wieser H. Chemistry of gluten proteins. Food Microbiology 2007;24(2) 115–119.

22. Goesaert H, Slade L, Levine H, Delcour JA. Amylases and bread firming – an integrated view. Journal of Cereal Science 2009;50(3) 345–352.

23. Belitz HD, Grosch W, Schieberle P., editors. Food Chemistry, third revised ed. Berlin: Springer-Verlag; 2004.

24. Shewry PR, Tatham AS, Forde J, Kreis M, Miflin BJ. The classification and nomenclature of wheat gluten proteins: a reassessment. Journal of Cereal Science 1986;4(2) 97– 106.

25. Veraverbeke WS, Delcour JA. Wheat protein composition and properties of wheat glutenin in relation to breadmaking functionality. Critical Reviews in Food Science and Nutrition 2002;42(3) 179–208.

26. D'Ovidio R, Masci S. The low-molecular-weight glutenin subunits of wheat gluten. Journal of Cereal Science 2004;39(3) 321–339.

27. Khatkar BS, Bell AE, Schofield JD. The dynamic rheological properties of glutens and gluten subfractions from wheats of good and poor bread-making quality. Journal of Cereal Science 1995;22(1) 29–44.

28. Tronsmo KM, Faergestad EM, Longva A, Schofield JD, Magnus EM. A study of how size distribution of gluten proteins, surface properties of gluten and dough mixing properties relate to baking properties of wheat flours. Journal of Cereal Science 2002;35(2) 201–214.

29. Joye IJ, Lagrain B, Delcour JA. Use of chemical redox agents and exogenous enzymes to modify the protein network during breadmaking – A review. Journal of Cereal Science 2009;50(1) 11–21.

30. Goesaert H, Courtin CM, Delcour JA. Use of enzymes in the production of cereal- based functional foods and food ingredients. In: Arendt EK, Dal Bello F. (eds.) Gluten-Free Cereal Products and Beverages. UK: Elsevier Inc.; 2008. p. 237-265.

31. Perlin AS. Isolation and composition of the soluble pentosans of wheat flour. Cereal Chemistry 1951;28(5) 370–381.

32. Perlin AS. Structure of the soluble pentosans of wheat flours. Cereal Chemistry 1951;28(5) 382–393.

33. Hoffmann RA, Roza M, Maat J, Kamerling JP, Vliegenthart JFG. Structural Characteristics of the Warm-water-soluble Arabinoxylans from the Tailings of the Soft Wheat Variety Kadet. Carbohydrate Polymers 1991;16(3) 275-289.

34. Fausch H, Kundig W, Neukom H. Ferulic acid as a component of a glycoprotein from wheat flour. Nature 1963;199(489) 287.

35. Saeed F, Pasha I, Anjum FM, Sultan, MT. Arabinoxylans and Arabinogalactans: A Comprehensive Treatise. Critical Reviews in Food Science and Nutrition 2011;51(5) 467–476.

36. Peressini D, Sensidoni A. Effect of soluble dietary fibre addition on rheological and breadmaking properties of wheat doughs. Journal of Cereal Science 2009;49(2) 190-201.

37. Stojceska V, Butler F. Investigation of reported correlation coefficients between rheological properties of the wheat bread doughs and baking performance of the corresponding wheat flours. Trends in Food Science and Technology 2012;24(1) 13-18.

38. Pareyt B, Finnie SM, Putseys JA, Delcour JA. Lipids in bread making: Sources, interactions, and impact on bread quality. Journal of Cereal Science 2011;54(3) 266-279.

39. Van Der Borght A, Goesaert H, Veraverbeke WS, Delcour JA. Fractionation of wheat and wheat flour into starch and gluten: overview of the main processes and the factors involved. Journal of Cereal Science 2005;41(3) 221–237.

40. Delcour JA, Hoseney RC., editors. Principles of Cereal Science and Technology, third ed. St. Paul: Association of Cereal Chemists Inc.; 2010.

41. Abdullah, MZ. Quality Evaluation of Bakery Products. In: Sun DW. (ed.) Computer Vision Technology for Food Quality Evaluation. UK: Elsevier Inc.; 2007. p. 481-522.

42. Mondal A, Datta AK. Bread baking – A review. Journal of Food Engineering 2008; 86(4) 465–474.

43. Heldt-Hansen HP. Macromolecular Interactions in Enzyme Applications for Food Products. In: Gaonkar AG, McPherson A. (eds.) Ingredients Interactions: Effects on Food Quality, second ed. Boca Raton: Taylor & Francis; 2006. p. 363-388.

44. Sablani SS, Baik O-D, Marcotte M. Neural networks for predicting thermal conductivity of bakery products. Journal of Food Engineering 2002;52(3) 299–304.

45. Gray JA, Bemiller JN. Bread Staling: Molecular Basis and Control. Comprehensive Reviews in Food Science and Food Safety 2003;2(1) 1-21.

46. Sahlstrom S, Brathen E. Effects of enzyme preparations for baking, mixing time and resting time on bread quality and bread staling. Food Chemistry 1997;58(1-2) 75-80.

47. Giannou V, Kessoglou V, Tzia C. Quality and safety characteristics of bread made from frozen dough. Trends in Food Science and Technology 2003;14(3) 99–108.

48. Sicard D, Legras J-L. Bread, beer and wine: Yeast domestication in the Saccharomyces sensu stricto complex. Comptes Rendus Biologies 2011;334(3) 229–236.

49. Hammes WP, Brandt MJ, Francis KL, Rosenheim J, Seitter MFH, Vogelmann SA. Microbial ecology of cereal fermentations. Trends in Food Science and Technology 2005;16(1-3) 4-11.

50. Weegels PL, Hamer RJ, Schofield JD. Depolymerisation and re-polymerisation of wheat glutenin during dough processing. 2. Changes in composition. Journal of Cereal Science 1997;25(2) 155–163.

51. Belton PS. New approaches to study the molecular basis of the mechanical properties of gluten. Journal of Cereal Science 2005;41(2) 203–211.

52. Hug-Iten, S, Handschin S, Conde-Petit B, Escher F. Changes in starch microstructure on baking and staling of wheat bread. LWT Food Science and Technology 1999;32(5) 255–260.

53. Morel M-H, Redl A, Guilbert S. Mechanism of Heat and Shear Mediated Aggregation of Wheat Gluten Protein upon Mixing. Biomacromolecules 2002;3(3) 488-497.

54. Lagrain B, Thewissen BG, Brijs K, Delcour JA. Impact of redox agents on the extractability of gluten proteins during bread making. Journal of Agricultural and Food Chemistry 2007; 55(13) 5320–5325.

55. Gökmen V, Açar OÇ, Serpen A, Morales FJ. Effect of leavening agents and sugars on the formation of hydroxymethylfurfural in cookies during baking. European Food Research and Technology 2008;226(5) 1031-1037.

56. Hidalgo A, Brandolini A. Evaluation of heat damage, sugars, amylases and colour in breads from einkorn, durum and bread wheat flours. Journal of Cereal Science 2011;54(2) 90-97.

57. Hulleman SHD, Helbert W, Chanzy H. Single crystals of V amylose complexed with glycerol. International Journal of Biological Macromolecules 1996;18(1-2) 115–122.

58. Selomulyo VO, Zhou W. Frozen bread dough: Effects of freezing storage and dough improvers. Journal of Cereal Science 2007;45(1) 1–17.

59. Moayedallaie S, Mirzaei M, Paterson J. Bread improvers: Comparison of a range of lipases with a traditional emulsifier. Food Chemistry 2010;122(3) 495–499.

60. Di Cagno R, De Angelis M, Corsettic A, Lavermicocca P, Arnault P, Tossut P, Gallo G, Gobbetti M. Interactions between sourdough lactic acid bacteria and exogenous enzymes: effects on the microbial kinetics of acidification and dough textural properties. Food Microbiology 2003;20(1) 67–75.

61. Collar C, Martinez JC, Andreu P, Armero E. Effects of enzymes associations on bread dough performance. A response surface analysis. Food Science and Technology International 2000;6(3) 217-226.

62. Martinez-Anaya MA, Jimenez T. Rheological properties of enzyme supplemented doughs. Journal of Texture Studies 1997;28(5) 569–583.

63. Sanz Penella JM, Collar C, Haros M. Effect of wheat bran and enzyme addition on dough functional performance and phytic acid levels in bread. Journal of Cereal Science 2008;48(3) 715–721.

64. Henrissat B. A classification of glycosyl hydrolases based on amino acid sequence similarities. Biochemical Journal 1991;280(2) 309–316.

65. Henrissat B, Bairoch A. Updating the sequence-based classification of glycosyl hy drolases. Biochemical Journal 1996;316(2) 695–696.

66. Svensson B, Tovborg Jensen M, Mori H, Bak-Jensen KS, Bonsager B, Nielsen PK, Kramhoft B, Praetorius-Ibba M, Nohr J, Juge N, Greffe L, Williamson G, Driguez H. Fascinating facets of function and structure of amylolytic enzymes of glycoside hydrolase family 13. Biologia 2002;57(11) 5–19.

67. Machovic M, Janecek S. Amylolytic enzymes: Types, structures and specificities. In: Polaina J, MacCabe AP. (eds.) Industrial Enzymes. Sctructure, Function and Application. Dordrecht: Springer; 2007. p. 3-18.

68. MacGregor EA, Janecek S, Svensson B. Relationship of sequence and structure to specificity in the α-amylase family of enzymes. Biochimica Biophysica Acta 2001;1546(1) 1–20.

69. Gupta R, Gigras P, Mohapatra H, Goswami VK, Chauhan B. Microbial α-amylases: a biotechnological perspective. Process Biochemistry 2003;38(11) 1599-1616.

70. Martin ML, Hoseney RC. A mechanism of bread firming. II role of starch hydrolyzing enzymes. Cereal Chemistry 1991;68(5) 503–509.

71. Poutanen K. Enzymes: An important tool in the improvement of the quality of cereal foods. Trends in Food Science and Technology 1997;8(9) 300-306.

72. Gerrard JA, Every D, Sutton KH, Gilpin MJ. The role of maltodextrins in the staling of bread. Journal of Cereal Science 1997;26(2) 201–209.

73. Hug-Iten S, Escher F, Conde-Petit B. Staling of bread: Role of amylose and amylopectin and influence of starch-degrading enzymes. Cereal Chemistry 2003;80(6) 654–661.

74. Leman P, Goesaert H, Delcour JA. Residual amylopectin structures of amylase-treated wheat starch slurries reflect amylase mode of action. Food Hydrocolloids 2009;23(1) 153–164.

75. Defloor I, Delcour JA. Impact of maltodextrins and antistaling enzymes on the differential scanning calorimetry staling endotherm of baked bread doughs. Journal of Agricultural and Food Chemistry 1999;47(2) 737–741.

76. Min BC, Yoon SH, Kim JW, Lee YW, Kim YB, Park KH. Cloning of novel maltooligosaccharide-producing amylases as antistaling agents for bread. Journal of Agricultural and Food Chemistry 1998;46(2) 779–782.

77. Rao MB, Tanksale AM, Ghatge MS, Deshpande VV. Molecular and biotechnological aspects of microbial proteases. Microbiology and Molecular Biology Reviews 1998;62(3) 597–635.

78. Ganzle MG, Loponen J, Gobbetti M. Proteolysis in sourdough fermentations: mechanisms and potential for improved bread quality. Trends in Food Science and Technology 2008;19(10) 513-521.

79. Linko Y-Y, Javanainen P, Linko S. Biotechnology of bread baking. Trends in Food Science and Technology 1997;8(10) 339-344.

80. Bombara N, Anon MC, Pilosof AMR. Functional Properties of Protease Modified Wheat Flours. LWT Food Science and Technology 1997;30(5) 441–447.

81. Kara M, Sivri D, Koksel H. Effects of high protease-activity flours and commercial proteases on cookie quality. Food Research International 2005;38(5) 479–486.

82. Shallom D, Shoham Y. Microbial hemicellulases. Current Opinion in Microbiology 2003;6(3) 219–228.

83. Collins T, Gerday C, Feller G. Xylanases, xylanase families and extremophilic xylanases. FEMS Microbiology Reviews 2005;29(1) 3–23.

84. Verma D, Satyanarayana T. Molecular approaches for ameliorating microbial xylanases. Bioresource Technology 2012;117 360–367.

85. Collins T, Hoyoux A, Dutron A, Georis J, Genot B, Dauvrin T, Arnaut F, Gerday C, Feller G. Use of glycoside hydrolase family 8 xylanases in baking. Journal of Cereal Science 2006;43(1) 79–84.

86. Biely P, Vrsanska M, Tenkanen M, Kluepfel D. Endobeta-1,4-xylanase families: Differences in catalytic properties. Journal of Biotechnology 1997;57(1-3) 151–166.

87. Bonnin E, Daviet S, Sorensen JF, Sibbesen O, Goldson A, Juge N, Saulnier L. Behaviour of family 10 and 11 xylanases towards arabinoxylans with varying structure. Journal of the Science of Food and Agriculture 2006;86(11) 1618–1622.

88. Debyser W, Peumans WJ, Van Damme EJM, Delcour JA. Triticum aestivum xylanase inhibitor (TAXI), a new class of enzyme inhibitor affecting breadmaking performance. Journal of Cereal Science 1999;30(1) 39–43.

89. Gebruers K, Debyser W, Goesaert H, Proost P, Van Damme J, Delcour JA. Triticum aestivum L. endoxylanase inhibitor (TAXI) consists of two inhibitors, TAXI I and TAXI II, with different specificities. Biochemical Journal 2001;353(2) 239–244.

90. McLauchlan WR, Garcia-Conesa MT, Williamson G, Roza M, Ravestein P, Maat J. A novel class of protein from wheat which inhibits xylanases. Biochemical Journal 1999;338(2) 441–446.

91. Fierens E, Rombouts S, Gebruers K, Goesaert H, Brijs K, Beaugrand J, Volckaert G, Van Campenhout S, Proost P, Courtin CM, Delcour JA. TLXI, a novel type of xylanase inhibitor from wheat (Triticum aestivum) belonging to the thaumatin family. Biochemical Journal 2007;403(3) 583-591.

92. Rouau X, El-Hayek ML, Moreau D. Effect of an enzyme preparation containing pentosanases on the breadmaking quality of flours in relation to changes in pentosan properties. Journal of Cereal Science 1994;19(3) 259–272.

93. Courtin CM, Delcour JA. Relative activity of endoxylanases towards water-extractable and water-unextractable arabinoxylan. Journal of Cereal Science 2001;33(3) 301- 312.

94. Sorensen JF, Kragh KM, Sibbesen O, Delcour J, Goesaert H, Svensson B, Tahir TA, Brufau J, Perez-Vendrell AM, Bellincampi D, D'Ovidio R, Camardella L, Giovane A, Bonnin E, Juge N. Potential role of glycosidase inhibitors in industrial biotechnological applications. Biochimica et Biophysica Acta-Proteins and Proteomics 2004;1696(2) 275- 287.

95. Bhat MK. Cellulases and related enzymes in biotechnology. Biotechnology Advances 2000;18(5) 355-383.

96. Driss D, Bhiri F, Siela M, Bessess S, Chaabouni S, Ghorbel R. Improvement of breadmaking quality by xylanase GH11 from Penicillium occitanis Pol6. Journal of Texture Studies2012(inpress)http://onlinelibrary.wiley.com/doi/10.1111/j. 1745-4603.2012.00367.x/pdf (accessed 16 August 2012).

97. Hasan F, Shah AA, Hameed A. Industrial applications of microbial lipases. Enzyme and Microbial Technology 2006;39(2) 235–251.

98. Galliard T, Gallagher DM. The effects of wheat bran particle size and storage period on bran flavour and baking quality of bran/flour blends. Journal of Cereal Science 1988;8(2) 147-154.

99. Qi Si J. Synergistic effect of enzymes for breadmaking. Cereal Foods World 1997;42(10) 802-807.

100. Olesen T, Qi Si J, Donelyan V. Use of lipase in baking. US Patent Application; 2000. US 6110508.

101. Primo-Martin C, Hamer RJ, de Jongh, HHJ. Surface layer properties of dough liquor components: Are they key parameters in gas retention in bread dough? Food Biophysics 2006;1(2) 83–93.

102. Stojceska V, Ainsworth P. The effect of different enzymes on the quality of high-fibre enriched brewer's spent grain breads. Food Chemistry 2008;110(4) 865–872.

103. Johnson RH, Welch EA. Baked goods dough and method. US Patent Application; 1968. US 3,368,903.

104. Siswoyo TA, Tanaka N, Morita N. Effects of lipase combined with α-amylase on retrogradation of bread. Food Science and Technology Research 1999;5(4) 356-361.

105. Gandra KM, Del Bianchi M, Godoy VP, Queiroz FPC, Steel CJ. Application of lipase and monoglyceride in fiber enriched pan bread. Ciência e Tecnologia de Alimentos 2008;28(1): 182-192.

106. Colakoglu AS, Ozkaya H. Potential use of exogenous lipases for DATEM replacement to modify the rheological and thermal properties of wheat flour dough. Journal of Cereal Science 2012;55(3) 397-404.

107. Baysal Taner, Demirdoven A. Lipoxygenase in fruits and vegetables: A review. Enzyme and Microbial Technology 2007;40(4) 491–496.

108. De Simone V, Menzo V, De Leonardis AM, Ficco DBM, Trono D, Cattivelli L, De Vita P. Different mechanisms control lipoxygenase activity in durum wheat kernels. Journal of Cereal Science 2010;52(2) 121-128.

109. Casey R, Westy SI, Hardy D, Robinson DS, Wu Z, Hughes RK. New frontiers in food enzymology: recombinant lipoxygenases. Trends in Food Science and Technology 1999;10(9) 297-302.

110. Levavasseur L, Rakotozafy L, Manceau E, Louarme L, Robert H, Baret J-L, Potus J, Nicolas J. Discrimination of wheat varieties by simultaneous measurements of oxygen consumption and consistency of flour dough during mixing. Journal of the Science of Food and Agriculture 2006;86(11) 1688–1698.

111. Feussner I, Kuhn H, Wasternack C. Lipoxygenase-dependent degradation of storage lipids. Trends in Plant Science 2001;6(6) 268-273.

112. Boussard A, Cordella CBY, Rakotozafy L, Moulin G, Buche F, Potus J, Nicolas J. Use of chemometric tools to estimate the effects of the addition of yeast, glucose-oxidase, soybean or horse bean flours to wheat flour on biochemical bread dough characteristics. Chemometrics and Intelligent Laboratory Systems 2012;113 68–77.

113. McDonald CE. Lipoxygenase and lutein bleaching activity of durum wheat semolina. Cereal Chemistry 1979;56(2) 84–89.

114. Stauffer, CE. Enzymes. In: Stauffer CE. (ed.) Functional Additives for Bakery Foods. New York: Van Nostrand Reinhold; 1990. p. 148-152.

115. Cumbee B, Hildebrand DF, Addo K. Soybean flour lipoxygenase isozymes effects on wheat flour dough rheological and breadmaking properties. Journal of Food Science 1997;62(2) 281-283.

116. Tilley KA, Benjamin RE, Bagorogoza KE, Okot-Kotber BM, Prakash O, Kwen H. Tyrosine Cross-Links: Molecular Basis of Gluten Structure and Function. Journal of Agricultural and Food Chemistry 2001;49(5) 2627-2632.

117. Garssen GJ, Vliegenthart JFG, Boldingh J. An anaerobic reaction between lipoxygenase, linoleic acid and its hydroperoxides. Biochemical Journal 1971;122(3) 327-332.

118. Bright HJ, Appleby M. The pH Dependence of the Individual Steps in the Glucose Oxidase Reaction. The Journal of Biological Chemistry 1969; 244(18)3625-3634.

119. Weibel MK, Bright HJ. The glucose oxidase mechanism. The Journal of Biological Chemistry 1971; 246(9) 2734-2744.

120. Kalisz HM, Hecht HJ, Schomburg D, Schmid RD. Effects of carbohydrate depletion on the structure, stability and activity of glucose oxidase from Aspergillus niger. Biochimica et Biophysica Acta 1991;1080(2) 138–142.

121. Hatzinikolaou DG, Hansen OC, Macris BJ, Tingey A, Kekos D, Goodenough P, Stougaard P. A new glucose oxidase from Aspergillus niger characterization and regulation studies of enzyme and gene. Applied Microbiology and Biotechnology 1996; 46(4)371–381.

122. Eryomin AN, Droshdenyuk AP, Zhavnerko GK, Semashko TV, Mikhailova RV. Quartz sand as an adsorbent for purification of extracellular glucose oxidase from Penicillium funiculosum 46.1. Applied Microbiology and Biotechnology 2004;40(2): 178–185.

123. Sukhacheva MV, Davydova ME, Netrusov AI. Production of Penicillium funiculosum 433 glucose oxidase and its properties. Applied Microbiology and Biotechnology 2004;40(1) 25-29.

124. Sisak C, Csanádi Z, Rónay E, Szajáni B. Elimination of glucose in egg white using immobilized glucose oxidase. Enzyme and Microbial Technology 2006;39(5) 1002–1007.

125. Isaksen A, Adler-Nissen J. Antioxidative effect of glucose oxidase and catalase in mayonnaises of different oxidative susceptibility. I. Product trials. LWT - Food Science and Technology 1997;30(8)841–846.

126. Kirk O, Borchert TV, Fuglsang CC. Industrial enzyme applications. Current Opinion in Biotechnology 2002;13(4) 345–351.

127. Pickering GJ, Heatherbell DA, Barnes MF. Optimizing glucose conversion in the production of reduced alcohol wine using glucose oxidase. Food Research International 1998;31(10) 685–692.

128. International Agency for Research on Cancer (IARC) Some naturally occurring and synthetic food components. In: Furocoumarines and Ultraviolet Radiation. IARC Monographs on Evaluation of the Carcinogenic Risk of Chemicals to Humans, vol. 40. IARC Scientific Publication, Lyon, France (1986) p. 207–220. http://monographs.iarc.fr/ENG/Monographs/vol40/volume40.pdf (accessed 23 August 2012).

129. Moore MM, Chen T. Mutagenicity of bromate: Implications for cancer risk assessment. Toxicology 2006;221(2-3) 190–196.

130. Bonet A, Rosell CM, Caballero PA, Gomez M, Perez-Munuera I, Lluch MA. Glucose oxidase effect on dough rheology and bread quality: a study from macroscopic to molecular level. Food Chemistry 2006;99(2) 408–415.

131. Rasiah IA, Sutton KH, Low FL, Lin HM, Gerrard JA. Crosslinking of wheat dough proteins by glucose oxidase and the resulting effects on bread and croissants. Food Chemistry 2005;89(3) 325–332.

132. Lagrain B, Thewissen BG, Brijs K, Delcour JA. Mechanism of gliadin–glutenin cross- linking during hydrothermal treatment. Food Chemistry 2008;107(2) 753–760.

133. Angioloni A, Rosa MD. Effects of cysteine and mixing conditions on white/whole dough rheological properties. Journal of Food Engineering 2007;80(1) 18–23.

134. Bekes F, Gras PW, Gupta RB. Mixing properties as a measure of reversible reduction and oxidation of doughs. Cereal Chemistry 1994;71(1) 44–50.

135. Bonet A, Rosell CM, Perez-Munuera I, Hernando I. Rebuilding gluten network of damaged wheat by means of glucose oxidase treatment. Journal of the Science of Food and Agriculture 2007;87(7) 1301–1307.

136. Anese M, Quarta B, Frias J. Modelling the effect of asparaginase in reducing acrylamide formation in biscuits. Food Chemistry 2011;126(2) 435–440.

137. Capuano E, Ferrigno A, Acampa I, Serpen A, Açar ÖÇ, Gökmen V, Fogliano V. Effect of flour type on Maillard reaction and acrylamide formation during toasting of bread crisp model systems and mitigation strategies. Food Research International 2009;42(9) 1295–1302.

138. Bråthen E, Knutsen SH. Effect of temperature and time on the formation of acrylamide in starch-based and cereal model systems, flat breads and bread. Food Chemistry 2005; 92(4) 693–700.

139. Wriston Jr. JC, Yellin TO. L-asparaginase: a review. Advances in Enzymology and Related Areas of Molecular Biology 1973;39 185–248.

140. International Agency for Research on Cancer IARC. IARC monographs on the evaluation of carcinogenic risks to humans, Vol. 60. International Agency for Research on Cancer, Lyon, France (1994) http://monographs.iarc.fr/ENG/Monographs/vol60/ mono60.pdf (accessed 23 August 2012).

141. Zyzak DV, Sanders RA, Stojanovic M, Tallmadge DH, Eberhart BL, Ewald DK, Gruber DC, Morsch TR, Strothers MA, Rizzi GP, Villagran MD. Acrylamide formation mechanism in heated foods. Journal of Agricultural and Food Chemistry 2003;51(16) 4782-4787.

142. Borek D, Jaskolski M. Sequence analysis of enzymes with asparaginase activity. Acta Biochimica Polonica 2001;48(4)893–902.

143. DSM. DSM publishes DNA sequence of Aspergillus niger genome in "Nature Biotechnology". In: DSM Press Release (2007). http://www.dsm.com/en_US/cworld/public/media/downloads/01e_07_Aspergilles_Niger.pdf?fileaction=openFile Accessed 25 August 2012).

144. Food and Agriculture Organization of the United Nations. FAO. Compendium of Food Additive Specifications. No. 4 FAO JECFA Monographs, Joint FAO/WHO Expert Committee on Food Additives (2007). ftp://ftp.fao.org/docrep/fao/010/a1447e/ a1447e.pdf (Accessed 25 August 2012).

145. Renzetti S, Dal Bello F, Arendt EK. Microstructure, fundamental rheology and baking characteristics of batters and breads from different gluten-free flours treated with a microbial transglutaminase. Journal of Cereal Science 2008;48(1) 33-45.

146. Steffolani ME, Ribotta PD, Perez GT, Leon AE. Effect of glucose oxidase, transglutaminase, and pentosanase on wheat proteins: Relationship with dough properties and bread-making quality. Journal of Cereal Science 2010;51(3) 366-373.

147. Wang F, Huang W, Kim Y, Liu R, Tilley M. Effects of transglutaminase on the rheological and noodle-making characteristics of oat dough containing vital wheat gluten or egg albumin. Journal of Cereal Science 2011;54(1) 53-59.

148. Renzetti S, Behr J, Vogel RF, Barbiroli A, Iametti S, Bonomi F, Arendt EK. Transglutaminase treatment of brown rice flour: A chromatographic, electrophoretic and spectroscopic study of protein modifications. Food Chemistry 2012;131(4) 1076–1085.

149. Caballero PA, Gomez M, Rosell CM. Improvement of dough rheology, bread quality and bread shelf-life by enzymes combination. Journal of Food Engineering 2007;81(1) 42–53.

150. Collar C, Bollaín C, Angioloni A. Significance of microbial transglutaminase on the sensory, mechanical and crumb grain pattern of enzyme supplemented fresh pan breads. Journal of Food Engineering 2005; 70(4) 479–488.

151. Gerrard JA, Fayle SE, Wilson AJ, Newberry MP, Ross M, Kavale S. Dough properties and crumb strength of white pan bread as affected by microbial transglutaminase. Journal of Food Science 1998;63(3) 472–475.

152. Autio K, Kruus K, Knaapila A, Gerber N, Flander L, Buchert J. Kinetics of transglutaminase-induced cross-linking of wheat proteins in dough. Journal of Agricultural and Food Chemistry 2005;53(5) 1039–1045.

153. Bauer N, Koehler P, Wieser H, Schieberle P. Studies on effects of microbial transglutaminase on gluten proteins of wheat. I. Biochemical analysis. Cereal Chemistry 2003;80(6) 781–786.

154. Bourbonnais R, Paice MG, Reid ID, Lanthier P, Yaguchi M. Lignin oxidation by laccase isozymes from Trametes versicolor and role of the mediator 2',2'-azinobis (3- ethylbenzthiazoline-6-sulphonate) in kraft pulp depolymerisation, Applied and Environmental Microbiology 1995;61(5), 1876-1880.

155. Selinheimo E, Kruus K, Buchert J, Hopia A, Autio K. Effects of laccase, xylanase and their combination on the rheological properties of wheat doughs. Journal of Cereal Science 2006;43(2) 152–159.

156. Qi Si J Use of laccase in baking. PCT Int. Appl.WO 9428728 A1 (1994). http://www.freepatentsonline.com/6296883.pdf (Accessed 23 August 2012).

157. Caballero PA, Gomez M, Rosell CM. Bread quality and dough rheology of enzyme- supplemented wheat flour. European Food Research and Technology 2007;224(5) 525–534.

158. Almeida EL, Chang YK. Effect of the addition of enzymes on the quality of frozen pre-baked French bread substituted with whole wheat flour. LWT - Food Science and Technology 2012;49(1) 64-72.

159. Katina K, Salmenkallio-Marttila M, Partanen R, Forssell P, Autio K. Effects of sourdough and enzymes on staling of high-fibre wheat bread. LWT Food Science and Technology 2006;39(5) 479–491.

160. Saxena RK, Gupta R, Saxena S, Gulati R. Role of fungal lipases in food processing. Applied Mycology and Biotechnology 2001;1(9) 353-386.

161. Leon AE, Duran E, Barber CB. Utilization of enzyme mixtures to retard bread crumb firming. Journal of Agricultural and Food Chemistry 2002;50(6) 1416-1419.

162. Faccio G, Flander L, Buchert J, Saloheimo M, Nordlund E. Sulfhydryl oxidase enhances the effects of ascorbic acid in wheat dough. Journal of Cereal Science 2012;55(1) 37-43.

163. Minussi RC, Pastore GM, Duran N. Potential applications of laccase in the food industry. Trends in Food Science and Technology 2002;13(6-7) 205–216.

164. Renzetti S, Courtin CM, Delcour JA, Arendt EK. Oxidative and proteolytic enzyme preparations as promising improvers for oat bread formulations: Rheological, biochemical and microstructural background. Food Chemistry 2010;119(4) 1465–1473.

165. Damen B, Pollet A, Dornez E, Broekaert WF, Van Haesendonck I, Trogh I, Arnaut F, Delcour JA, Courtin CM. Xylanase-mediated in situ production of arabinoxylan oligosaccharides with prebiotic potential in whole meal breads and breads enriched with arabinoxylan rich materials. Food Chemistry 2012;131(1) 111–118.

166. Waters DM, Ryan LAM, Murray PG, Arendt EK, Tuohya MG. Characterisation of a Talaromyces emersonii thermostable enzyme cocktail with applications in wheat dough rheology. Enzyme and Microbial Technology 2011;49(2) 229–236.

167. Zhang C, Tao T, Ying Q, Zhang D, Lu F, Bie X, Lu Z. Extracellular production of lipoxygenase from Anabaena sp. PCC 7120 in Bacillus subtilis and its effect on wheat protein. Applied Microbiology and Biotechnology 2012;94() 949–958.

168. Gerday C, Aittaleb M, Bentahir M, Chessa J.-P, Claverie P, Collins T, D'Amico S, Dumont J, Garsoux G, Georlette D, Hoyoux A, Lonhienne T, Meuwis M-A, Feller G.

169. Cold-adapted enzymes: from fundamentals to biotechnology. Trends in Biotechnology 2000;18(3) 103-107.

170. Dornez E, Verjans P, Arnaut F, Delcour JA, Courtin CM. Use of psychrophilic xylanases provides insight into the xylanase functionality in bread making. Journal of Agricultural and Food Chemistry 2011;59(17) 9553-9562.

171. Zheng H, Guo B, Chen X-L, Fan S-J, Zhang Y-Z. Improvement of the quality of wheat bread by addition of glycoside hydrolase family 10 xylanases. Applied Microbiology and Biotechnology 2011; 90(2) 509–515.

172. Wang M, Si T, Zhao H. Biocatalyst development by directed evolution. Bioresource Technology 2012;115 117–125.

173. Bottcher D, Bornscheuer UT. Protein engineering of microbial enzymes. Current Opinion in Microbiology 2010;13(3) 274–282.

174. Jones A, Lamsa M, Frandsen TP, Spendler T, Harris P, Sloma A, Xu F, Nielsen JB, Cherry JR. Directed evolution of a maltogenic α-amylase from Bacillus sp. TS-25. Journal of Biotechnology 2008;134(3-4) 325–333.

175. Kim JH, Maeda T, Morita N. Effect of fungal α-amylase on the dough properties and bread quality of wheat flour substituted with polished flours. Food Research International 2006; 39(1) 117–126.

176. Derde LJ, Gomand SV, Courtin CM, Delcour JA. Characterisation of three starch degrading enzymes: Thermostable β-amylase, maltotetraogenic and maltogenic α-amylases. Food Chemistry 2012; 135(2) 713–721.

177. Trogh I, Courtin CM, Andersson AAM, Aman P, Sørensen JF, Delcour JA. The combined use of hull-less barley flour and xylanase as a strategy for wheat/hull-less barley flour breads with increased arabinoxylan and (1→3, 1→4)-β-D-glucan levels. Journal of Cereal Science 2004;40(3) 257–267.

178. Lebesi DM, Tzia C. Use of endoxylanase treated cereal brans for development of dietary fiber enriched cakes. Innovative Food Science and Emerging Technologies 2012;13 207–214.

179. Rani KU, Prasada Rao, UJS, Leelavathi K, Haridas Rao P. Distribution of Enzymes in Wheat Flour Mill Streams. Journal of Cereal Science 2001;34(3) 233–242.

180. Moroni AV, Bello FD, Arendt EK. Sourdough in gluten-free bread-making: An ancient technology to solve a novel issue? Food Microbiology 2009;26(7) 676–684.

181. Huttner EK, Arendt EK. Recent advances in gluten-free baking and the current status of oats. Trends in Food Science and Technology 2010;21(6) 303-312.

182. Bankar SB, Bule MV, Singhal RS, Ananthanarayan L. Glucose oxidase - An overview. Biotechnology Advances 2009;27(4) 489–501.

183. Borrelli GM, Ficco DBM, Di Fonzo N, Fares C. Effects of lipoxygenase and of chemical oxidising agent potassium iodate on rheological properties of durum dough. International Journal of Food Science and Technology 2006;41(6) 639–645.

15 Chapter

GRINDING PROPERTIES OF WHEAT IN INDUSTRIAL MILLS

1. INTRODUCTION: AN OVERVIEW

The grinding of cereal seeds results from the mechanical action of several forces: compression, shearing, crushing, cutting, friction, and impact, which the seeds endure based on the design of the grinding mill employed (roller mill, hammer mill, stone mill, or ball mill). Upon the application of these forces, when the mechanical resistance of the particles is surpassed, they undergo division into numerous smaller particles of varying sizes, geometric forms, masses, and volumes.

An industrial wheat mill comprises multiple technological stages, commencing with the coarse grinding of seeds and culminating in the fine grinding of the resultant milling products, following their classification into various size fractions. The initial technological step of the grinding process in wheat mills is the gristing or coarse grinding phase, which comprises multiple technological passages.

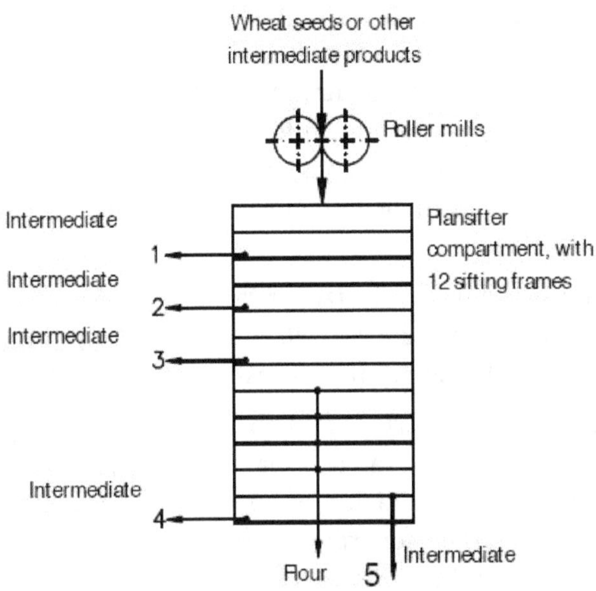

Figure 1. Schematic diagram of a grinding passage

A technical tunnel comprises a grinding machine (roller mill), a machine for sifting and sorting the resultant milling fractions (plansifter compartment), and, optionally, an equipment for conditioning the semi-final product (semolina machine or bran finisher). Intermediate fractions are produced during a technical process, which, with subsequent grinding, yield high-quality flour in milling stages (fine grinding).Wheat processing requires a long and gradual transformation into flour. This process takes place after a gradual crushing schedule, from fine to finer, from machine to machine, of wheat seed, respectively of the crushed particles resulting from it. Each grinding operation is immediately followed by a sorting operation by sifting (fig.1) because during grinding, a wide variety of grinded seed particles is obtained.

Before the grinding process is started, grains must undergo the cleansing process. This is followed by a conditioning process that ensures a uniform moisture content for the entire lot of grains, helping endosperm softening and cover harshening, which improves the separation process.

One of the fractions resulting from a plansifter compartment is composed of flour particles (with sizes under 160 μm), in a higher or lower percentage of the total flour that can be withdrawn in the industrial mill. To extract the full amount of flour from the wheat berries, multiple passes (passages) are required. Some passages are part of coarse grinding phase (gristing), where the milling rollers have fluted surface, while other passages are part of milling phase (fine grinding), where the milling rollers have smooth surface.

Intermediate milling products are, mainly, grists (seed particles with various sizes), semolina (large, average and small) and dunsts (harsh and smooth). They all return in the grinding process for flour extraction, but the grists are grinded by mills with fluted rollers (gristing passages), while semolina and dunsts are grinded by mills with smooth rollers (milling passages). Semolina and dunsts, as intermediate milling products, are particles of clean endosperm or with a small percentage of cohesive coat.

Particles obtained by grinding have sizes in a fairly wide range (1200-160 μm, within the mentioned fractions), average size of the particles of resulted fraction being determined by granulometric analysis using sieve classifier.

In roller mills, wheat seeds are grinded in the gristing phase by pairs of fluted rollers, thus being obtained a wide range of particles with sizes from < 200 μm to > 2000 μm, [1], consisting in coat particles (of larger sizes) and endosperm particles (of smaller sizes), to be further separated with plansifters. The milling process aims to grind the endosperm into finer particles of flour and semolina, while the coating and the seed particles must remain in large sizes to be separated by sifting, [2]. In gristing passages, milling rollers with fluted surface are used, and in milling passages, rollers with smooth surface are used. The quality of wheat milling process is influenced by the physical and mechanical properties of seeds and of the intermediate products (size distribution, seeds hardness, moisture content) and by the design and functional parameters of the roller mill (mutual arrangement of the rollers, differential speed, distance between the rollers, flutes profile, mutual position of the flutes), [3,4]. Effects of these factors are manifested in the size distribution of material particles, compositional distribution of the material, wear degree of the rollers, energy consumed for grinding, [4].

Fang, Campbell et al. (2002) showed that if the distance between rollers increases from 0.3 mm to 0.7 mm, wheat seeds breakage in the gristing phase has a lower intensity, resulting in more particles of large sizes and less particles of smaller sizes. Distance between rollers indirectly influences the specific surface and energy consumption per mass unit and directly influences the specific energy, [5]. Different flutes arrangements on the rollers lead to the obtaining of different size distributions. If the roller flutes are arranged in blade/blade position results in a relatively uniform size distribution, and back to back arrangement lead to a deep parabolic distribution, [1].

Differential speed of milling rollers has a significant effect on the grinding of semolina, flour and wheat bran. With the increase of differential speed of rollers, it also increases the amount of semolina and decreases the amount of flour and wheat bran, [6]. This is due to the difference between shearing and compression forces which are applied on the particles.

It is very important to know the size distribution of the material subjected to grinding, as well of the grist, so that appropriate adjustments can be made to roller mills, and also to choose the fabrics for the sieving frames of plansift compartments. Particles size distribution of the granular material can be determined using superposed sieve classifiers (sieve shakers), with different sizes of sieve holes. This can be assessed by various mathematical functions, from which, most used is the Rosin-Rammler function.

Experiments were performed on the material subjected to grinding (before and after grinding) and cumulative distribution curves were drawn for the sieved material, by computer aided regression analysis of the experimental data with Rosin-Rammler function. Based on the data obtained from particle size distribution were also determined other physical characteristics of the analyzed material: average particle size, grist modulus, specific surface of the granular material, surface increasement resulted from grinding within a passage (break), bulk density and specific mass.

Within this chapter are presented the flow diagrams for two wheat mills of different capacities, one of 100 tons / 24 hours and one of 220 tons / 24 hours, from which it can be estimated the movement of products within the mill.

The experimental results regarding the particle size distribution of the material subjected to grinding and the resultant grist are presented for both technological phases and the two mills, along with the particle size distribution of the material for various grinding machines of the analyzed mills.

Understanding the mechanical properties of wheat seeds and grist particles, along with their dimensional parameters, volume, and mass, is essential for determining the energy needed for crushing.

This study presents the results of experimental research on the behavior of wheat seeds during uniaxial compression experiments conducted between parallel plates. The charts illustrating the fluctuation of crushing force and energy absorbed till the seeds reach the crushing point are also given.

The provided conclusions and acquired data are of significant relevance to roller mill designers, manufacturers, and users of these equipment.

2. TECHNOLOGICAL DIAGRAMS FOR WHEAT GRINDING

The technological passage consists of one or two pairs of milling rollers, both processing the same product, combined with one or more plansifter compartments for sieving.

Gristing is the technological phase aiming to fragment the wheat seed in particles of different sizes and to remove the endosperm from the coating. Particles resulted from first, second and third grinding phase vary in size, from breakages like half seeds to flour particles with very fine granulometry. As gristing is repeated, particles will get increasingly finer, the amount of white flour decreases, and seeds coating reaches the penultimate and last phase as fine dust, [7]. Thus, grist is the intermediate product obtained in the milling industry, by grinding grains by mean of roller mills with fluted surface.

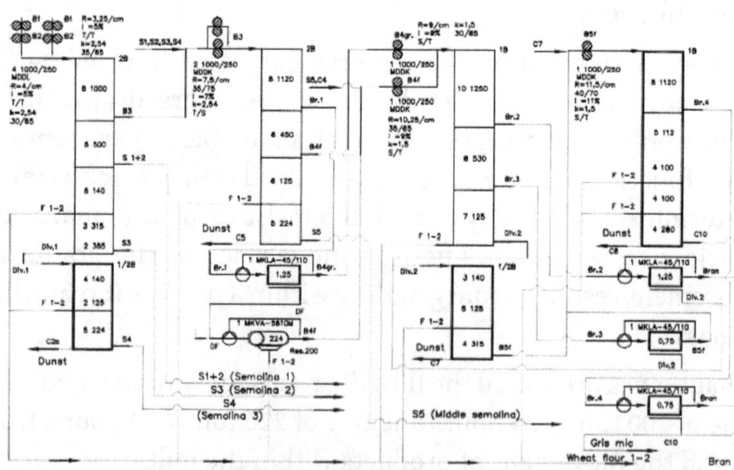

Figure 2. Technological diagram of gristing phase for a Bühler mill with capacity of 220 t / 24 h, [8]

Fig. 2 presents the technological diagram of gristing phase of the wheat in an industrial mill with the capacity of 220 t/24 h.

Milling unit consists of 9 double roller mills, of which the first processes, in both sections, the same material (whole seeds), two plansifters, together amounting 14 compartments, three double semolina machines and five brushes and bran finishers. The three phases of the process (gristing, milling, sorting) can be observed in fig.3 – fig.5.

Gristing phase consists of six simple mills with fluted rollers, four full and two half's of plansifter compartments and four bran finishers which process the coatings resulted from multiple grinding operations. The seeds are processed in a mill with double rollers placed in horizontal plane, noted by B1–B2.

The first grist is processed in passage B3, and the fractions obtained here will follow different routes, to the milling passages, or to the semolina machines or bran finishers, passages B4gr and B5f being responsible for the processing of material particles with high coating content, and passage B4f processes the second refuse from gristing passage B3, with fractions having the same characteristics processed in plansifter compartments. The development of gristing phase in directly connected to the type of meal and the degree of flour extraction. Products resulted from gristing are named intermediate products and they consist of: big grist, fine grist, big semolina, middle semolina, fine semolina, big dunst, soft dunst, flour and bran, [7].

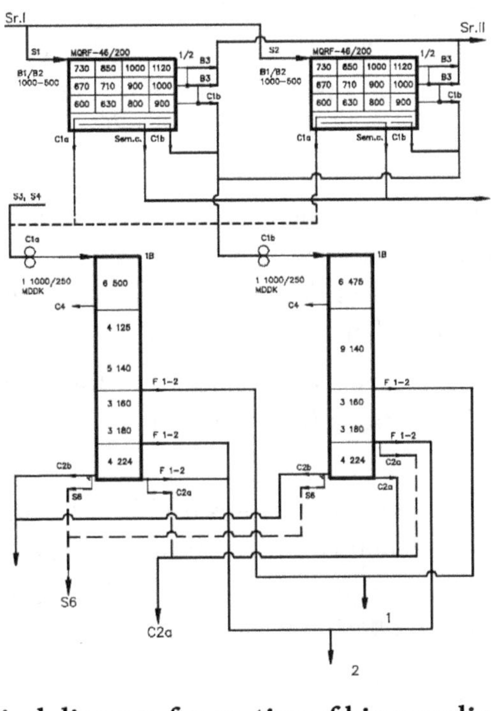

Figure 3. Technological diagram for sorting of big semolina in Bühler mill, [7]

Particles size of these components, resulted from sieving process, is determined by the size of the sieve holes used in sieving compartments. Depending on the particles size, semolina and dunsts can be classified as: big semolina with average size of 1200-630 μm; middle semolina 630–400 μm; fine semolina 400–310 μm; big dunsts 310–245 μm; soft dunsts 245–160 μm. Semolina sorting is dome in sorting phase (fig.2.) A clear delineation between soft dunsts and flour can not be practically achieved, and therefore, are cases when soft dunsts (dm = 220 μm) are considered to be flour (flours granulosity is given by the sieves, with mean equivalent size of the particles below 160 μm).

Particles of intermediate products can be highlited not only by their size, but also by shape, volume, specific mass, aerodynamic properties. Particles with rich coating have irregular shape in the form of foils with rolled or folded edges. Particles of clean endosperm have polyhedral shape with sharp edges and convex lateral surfaces.

Semolina is an intermediate product obtained in percentage of 25...30% in industrial wheat milling, is found as small granules and after cleaning is further milled to obtain flour or a food product known as "kitchen semolina". This is obtained in percentage of 2...3 % at wheat milling and it is cleaned in special semolina machines in order to remove coating particles by the combined action of sieving and airflows. Dunst is a fine semolina obtained as intermediate product from the grinding of wheat or semolina.

After gristing phase it is important to sort the milling products using a wide range of sizes for sieve holes (1000...224 μm), followed by the cleaning of semolina and dunsts, the phase of semolina opening being no longer necessary, since most coating was already removed in the gristing phase (fig.3).

The unit is fully automated, all mill equipments starting and stopping from the computer, starting with the equipments from the final technological phases (bagging, flour homogenization, sieving with plansifters, semolina cleaning, bran finishers, etc.) from the circuit of flour or intermediate products, while stopping begins with the first pair of rollers, i.e. reverse of start up.

In fig.4 and fig.5 is presented the technological flow for a wheat mill with capacity of 100 t/24 h, in grinding phases (fig.4) and in the milling (breakage) phase of semolina (fig.5), [10].

The technological flow of wheat mill is ensured by 12 processing passages, with 12 pairs of milling rollers (6 double rollers of Buhler type) from which 5 gristing passages and 7 milling passages. In addition, the technological flow is fitted with a sorting passage (separate compartment of plansifter).

Apart from the 12 technological passages, each consisting in a pair of roller mills and one plansifter compartment, the mill also has a double machine for semolina, three bran finishers and other auxiliary equipments (detachers, wheat brushes, filters and cleaning cyclones, etc.), as well as the proper elements for the pneumatic transport system from one equipment to another, according to the technological flow.

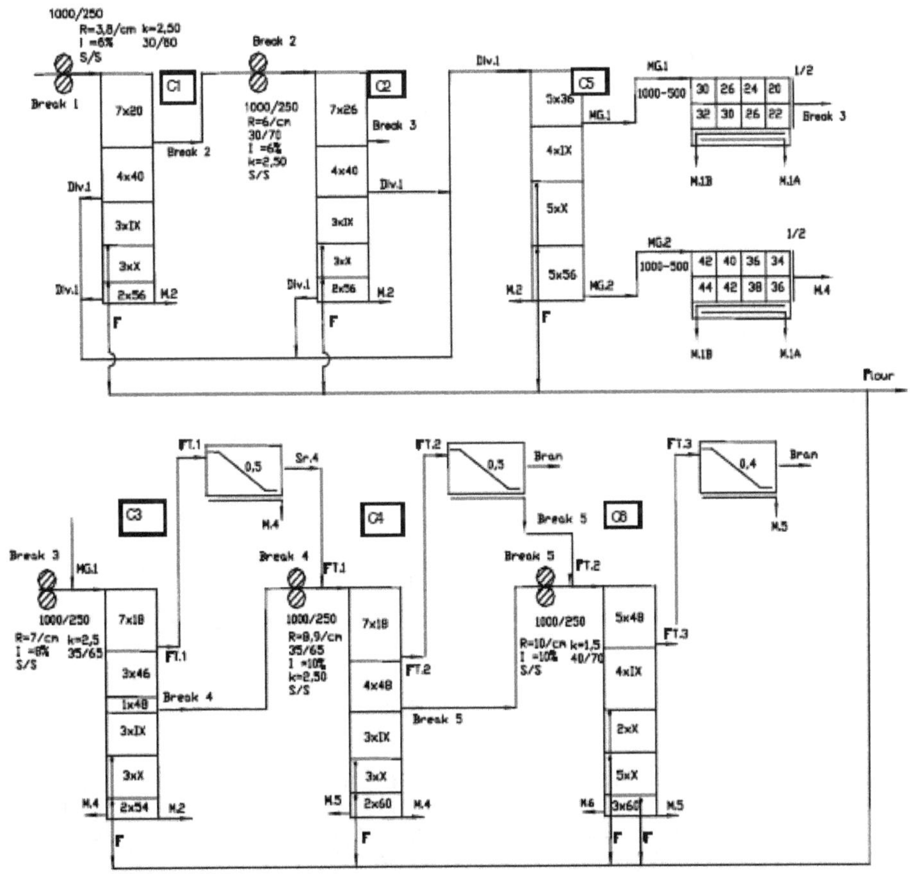

Figure 4. Technological flow of semolina grinding phase in wheat mill, capacity of 100 t/24 h, [11]

In breakage phase the technological diagram of mill contains five pairs of rolls, filled with one compartment of plane sieve, two semolina machines and three wheat bran finishers. The technological breakage phase is completed with one compartment of plane sieve without grinding machine, in which the material is sorted by fractions of different sizes as well as the other compartments of plane sieve.

The first grist, obtained from seeds processing with the pair of fluted rollers Sr.1, is processed in passage Sr.2, and from here the fractions follow various routes, to grinding passages, to semolina machines to wheat bran finishers. The sifting material from the second and the last set of gristing passage Sr.1, is send to a plansifter compartment for division in fractions (Div.1), which next reach the MG1 and MG2 semolina machines. The refuse from the last set of frames in the first passage is then sent to the M2 grinding passage.

The circulation of grist intermediate products in the technological diagram is shown in fig.4 and fig.5.

In the grinding phase (fig.5), the technological diagram of milling unit consists of seven simple roller mills, each fitted with one plansifter compartment for sorting in fractions of the grinded products and the extraction of flour from these products.

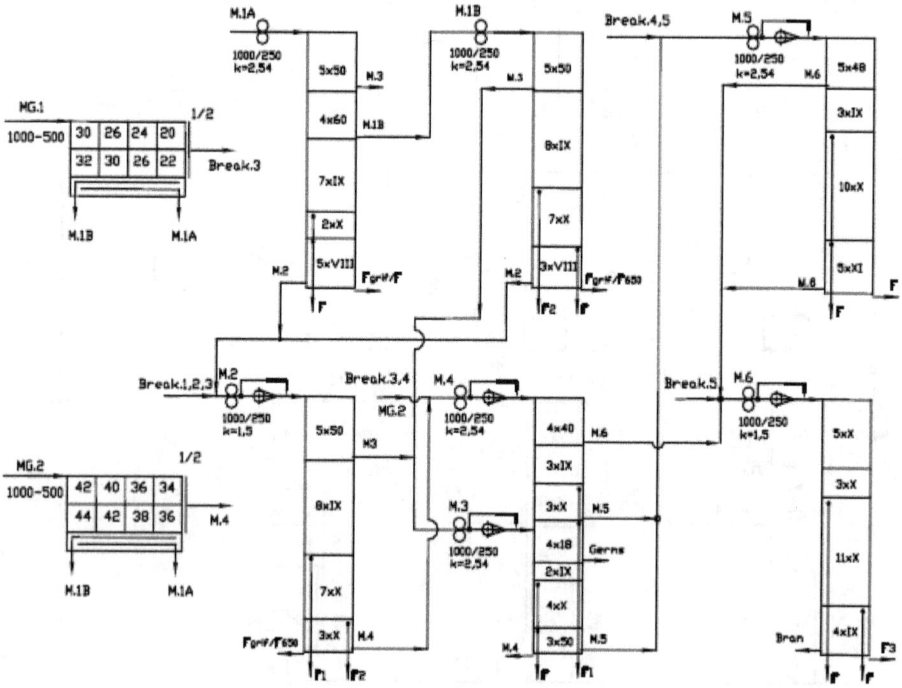

Figure 5. Technological flow of the semolina grinding phase in a wheat mill with 100 t/24 h, [9]

All roller mills of both technological phases have the length of 1000 mm and diameter of 250 mm, with fluted surface, in the gristing phase, respectively smooth surface without flutes in the grinding phase. In the gristing phase, the ratio of the tangential speeds of fluted rollers is k=2.54, and in the grinding phase, for five pairs of rollers, k=2.54, and for two pairs of rollers k=1.5.

As shown in fig.5, the products to be grinded into the grinding phase are products arriving from gristing phase (or breakage phase), inclusive from grists (Sr.1-6) or from semolina machines and bran finishers. The siftings from MG1 and MG2 semolina machines, which are semolinas with sizes below 0.8-1.0 mm, are grinded in the first technological passages M1A and M1B, while the siftings from FT1 and FT3 bran finishers go to the last two grinders M4 and M5, which processed products with higher content of bran. In diagram, the first refusal from M1A and M1B grinders is led to M3 grinder, working with half compartment of plane sieve. It is noted that to grinders which grinded smaller particles of endosperm (about 0.40 mm), after the mill rollers in technological flow are placed detached of material, due to agglomerations arising from the compression of smaller particles of endosperm in the action zone of grinding rolls.

In fig.6 is shown the arrangement of rollers to a mill with 100 t /24 h capacity, where the samples for our determinations were collected.

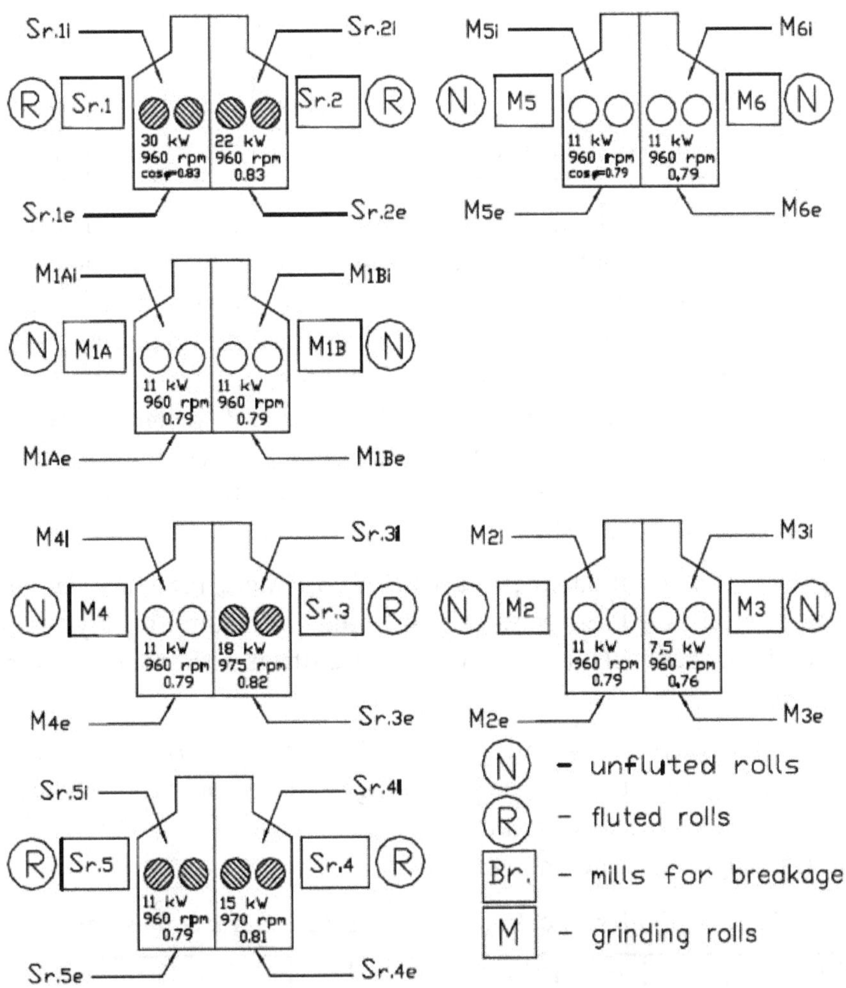

Figure 6. Arrangement of roller mills for the mill with capacity of 100 t /24 h

Plansifters are driven by electric motors of 4 kW, cos φ = 0,81 and speed of 960 rot/min. Double machine for semolina is driven by two moto-vibrators of 400 W and speed of 960 rpm. Characteristics of driving motors for mill rollers are given in table 1.

Table 1. Characteristics of electric motors for the drive of mill rollers, for wheat mill with capacity of 100 t/24 h [9]

Passage	I (A)		P, kW	n, rpm	cos φ	Passage	I (A)		P, kW	n, rpm	cos φ
	No-load	Load					No-load	Load			
Sr. 1	19	45	30	960	0.83	M1 B	10	15	11	960	0.79
Sr. 2	21	37	22	960	0.83	M2	11,9	16	11	960	0.79
Sr. 3	23	32	18	975	0.82	M3	10,9	18	7,5	960	0.76
Sr. 4	13	30	15	970	0.81	M4	10,9	15	11	960	0.79
Sr. 5	13	17	11	960	0.79	M5	12	16	11	960	0.79
M1 A	10	15	11	960	0.79	M6	12	17	11	960	0.79

According to relevant regulations, on the technological diagram (fig.4 or fig.5) should be written the characteristics of grinding rollers: length, diameter (ex.1000x250, in mm), number of flutes and their inclination (ex.7/cm, I=8%), flute angles (ex.35/65), mutual arrangement of the flutes (ex.S/S), speed ratio (ex.k=2.5), and the characteristics of fabrics used in plansifter frames (ex.3x46 – 3 frames with 46 wires per inch or 3xX for flour frames), at semolina machines (ex.42, which represents the number of wires per inch or 1000-500, which is frame size) or at bran finishers (ex.0.5 – size of fabric hole).

Sieve frames from top of compartments are fitted with metal mesh as they separate seed brokens of relatively large sizes (which would wear quite quickly the textile fabrics), while flour frames from the lower set are fitted with frames with plastic or textile fabrics.

Lately, textile fabrics have been replaced with sieve frames with meshes of plastic fabric. According to literature, fot the technological diagram of the analyzed mill, the equivalence between the sieve number and the size of its holes, as they are specified in the diagram, is shown in table 2.

Table 2. Equivalence between sieve number and hole sizes

Sieve no.	18	20	26	36	40	46	48	50	54	56	60	VIII	IX	X	XI
Hole size (μm)	1170	1050	780	520	470	390	370	350	320	310	280	180	170	150	130

3. PHYSICAL AND GRANULOMETRIC CHARACTERISTICS OF SEEDS AND GRINDING PRODUCTS

In the grinding process is necessary to know the physico-mechanical characteristics of the material at the entry and exit from a processing machine, in this case, roller mills.

Main factors influencing the process of grain grinding are the physico-mechanical properties of seeds and of the grinding products, the constructive and functional characteristics of the grinding machines as well as the technological regime, most of those factors having a random character.

As a result of grinding it is obtained a mass of particles with various smaller sizes and different geometrical shapes (grist).

Granulometric distribution of the grinded material and of the material leaving the grinding process can be assessed by the cumulative weight (%) of material passing through the sieve holes of classifier T(x) or which are refused by its sieves R(x), calculated on base of mass weight (%) of the fractions from the sieve. (R(x)+T(x)=100). The mathematical expression of granulometric distribution in case of grinded biological materials, is based on laws of mathematical statistical method of small particles, [11-14].

There will be defined three usual types of laws of cumulative granulometric distribution.

- The Rosin-Rammler distribution, for material particles with larger sizes than sieve holes, is expressed by the relation:

$$R(x) = 100 \cdot e^{-bx^n} \quad (1)$$

where: R(x) is the mass percentage weight of fraction with larger particles than x (which remained on the sieve with meshes with size x); x – is the sieves meshes size by which the particles rest; b and n are the own coefficients of grinding material.

- The Schuhman distribution is defined by the relation:

$$R(x) = 100 \cdot \left\{1 - (x/k)^a\right\} \quad (2)$$

where: R(x) and x have the significance from to relationship (1), k - the module product particles size (the size of sieve mesh through which, theoretical, pass all the sample particles (100%)), a - the distribution module.

- The logistics type distribution with two parameters is defined by the relation:

$$R(x) = 100 \cdot \frac{e^{\alpha+\beta x}}{1+e^{\alpha+\beta x}} \quad (3)$$

where: R(x) and x have the significance from relationships (1-3) α and β are logistical con stants.

Of these characteristics are important: the bulk density, ϱv (kg/m3), of the material to be processed, the density of the material, ϱ (kg/m3); the equivalent sizes of material particle at entry and exit of the grinding machine, dm (mm); angle of internal friction of particles appreciated by natural slope angle, ψ (o); angle of material friction with the surfaces working components, φ (o); material porosity, ε (%) and others.

Of particular importance is the equivalent size of seeds subjected to grinding in the first technological passage.

The density is the ratio between the sample mass and the volume of the particle in it. To determine the densities of wheat seeds, respectively the grinding products, the pycnometrical method was used (xylene 0.8254 kg/cm3).

The porosity is the property of granular materials, respectively of the grains, to not occupy the entire volume of storage, with an intergranular space. Knowing the values of bulk density and material density, the porosity was evaluated using the following relation, [15]:

$$\varepsilon(\%) = (1 - \rho_v / \rho) \cdot 100\ (\%) \tag{4}$$

The static friction coefficient. The most common method for determining the coefficient of static friction is inclined plane method which was used in this paper. It was used a device with adjustable incline plane, [15]. Two sets of determinations were realized on three types of surfaces: glossy fiberglass, steel sheet and cotton canvas.

Assessing parameters of the grinding process are: grinding degree, grinding finesse and specific energy consumption at grinding.

Grinding degree and grinding finesse are determined by granulometric analysis, using a sieve overlay classifier with oscillatory movement.

Grinding degree is defined by the λ index and represents the ratio between equivalent sizes of particles before and after grinding, De, respectively dm, or the ratio between the outer surface of the particles resulted in the grinding process and the initial surface of the particle subjected to grinding, Sf, respectively Si:

$$\lambda = D_e / d_m = S_f / S_i \tag{5}$$

Absolute value of the increase of particles outer surface in the grinding process ΔS, is given by:

$$\Delta S = S_f - S_i = S_i(\lambda - 1) \tag{6}$$

The grinding finesse has been appreciated by the geometric mean diameter dm of the grinding particles which was determined by the size distribution analysis, using the relation of weighted average:

$$d_m = (1/100) \cdot \sum_{i=0}^{n} p_i d_i, \tag{7}$$

where: pi is mass weight of fraction remaining on the sieve i of the classifier, di, is diameter (average value) of fractions particle on the sieve i, considered the arithmetic average of the sieves holes size that contain fraction i.

The surface areaand the surface increase. Knowing the mean diameter of particles of a granular mixture, their specific surface Se.m is determined with the relation, [10,15]:

$$S_{e.m.} = 6 / \rho \cdot d_m \ (m^2 / kg), \tag{8}$$

where: ϱ is the density of the particles.

There are presented the results of some experimental research on the physical characteristics of grinding products on the technological flow of gristing phase of wheat from a mill with capacity of 100 t / 24 h (SC Spicul Rosiori de Vede, Teleorman, Romania).

The material tested in the experiments was taken from the entry, respectively from the exit of each pair of milling rollers (from the five pairs of the phase).

The experimental data characterizing the physical properties of the grist obtained are shown in table 3. Also, in table 4 and table 5 are presented the results of size distribution analysis on mixtures of material entering and leaving the rolls placed in the technological grinding phase.

Table 3. The values of static friction coefficient and natural slope angles [15]

Break		Static friction coefficient μ			
		Cotton canvas	Glossy fiberglass	Steel sheet	Natural slope angle ψ
M1A	E	1.74 - >1.76	68,7	0.61 – 0.85	39,43
M1B	I	1.24 – 1.82	56,4	0.45 – 0.67	30,19
M2	I	>1.76	61	0.6 – 0.86	43,97
M3	E	>1.76	72,4	0.58 – 0.82	38,35
M4	E	>1.76	67,5	0.58 – 0.73	41,96
M5	I	>1.76	64,3	0.60 – 0.88	46,62
M6	E	>1.76	69,5	0.54 – 0.71	40,17

From table 3 it is noted that the static coefficient values, on the glossy fiberglass and metal are within the limits set in various specialized papers, while the values obtained in the experiments on cotton canvas fall in broad limits, probably due to material fractions moisture, but also because of its granularity, this phenomenon is observed, especially, to flours and relatively small particle fractions of endosperm.

Table 4. The ponder values (pi) of the fractions from the sieving machine classifier sieves and of the cumulative weights Ti(%) for the collected gritting, at entrance "I" and exit "E" from the mentioned rolls (only M1A, and M1B), [9]

li (mm)	M1A -		M1A – E		li (mm)	M1B – I		M1B - E	
	pi (%)	Ti (%)	pi (%)	Ti (%)		pi (%)	Ti (%)	pi (%)	Ti (%)
0,00	0,90	0,00	34,40	0,00	0,00	0,40	0,00	7,70	0,00
0,18	1,00	0,90	16,30	34,40	0,13	0,60	0,40	8,00	7,70
0,25	3,70	1,90	10,30	50,70	0,18	3,90	1,00	23,80	15,70
0,32	33,30	5,60	15,80	61,00	0,25	22,10	4,90	35,60	39,50
0,50	49,10	38,90	12,50	76,80	0,32	48,90	27,00	21,90	75,10
0,71	12,00	88,00	10,70	89,30	0,40	24,10	75,90	3,00	97,00
dM1AI = 0,55		dM1AE = 0,33			dM1BI = 0,36		dM1BE = 0,26		

Table 5. The values of grinding degree, specific surface, surface increase and porosity

Break	Equivalent size(I/E)	Grinding degree	Bulk density	True density	Specific surface	Surface Increase	Porosity
	mm	λ	g/dm3	g/dm3	x103 m2/ kg	x103 m2/ kg	%
M1A	0,55-0,33	1,68	560,0-389,5	1344,9-1247,1	8,13-14,72	6,595	58,3-68,7
M1B	0,36-0,26	1,385	583,5-499,0	1338,7-1372,0	12,34-16,72	4,38	56,4-63,6
M2	0,19-0,17	1,113	480,5-437,5	1233,3-1313,4	26,04-27,55	1,513	61-66,7
M3	0,35-0,45	0,788	363,5-308,5	1252,9-1119,8	13,55-11,95	-1,605*	71-72,4
M4	0,22-0,24	0,940	452,5-419,5	1290,6-1290,6	21,02-19,76	-1,252*	64,9-67,5
M5	0,22-0,24	0,924	430,5-419,5	1205,4-1210,2	22,74-20,89	-1,854*	64,3-65,3
M6	0,24-0,27	0,903	416,0-373,0	1274,4-1224,5	19,40-18,23	-1,164*	67,4-69,5
The sign * in table 3, for negative values of specific surface increases, means that at the passage through milling rollers with smooth surface, agglomeration of gritting particles occurs.							

Based on the data obtained from the experiments and presented in table 6, were mapped graphics, using MS Excel version 7.0 program (fig.6), the variations of mean diameter and bulk density to technological breakage passage of milling unit.

Table 6. Physico-mechanical characteristics of grinding products at gristing passages of wheat, from the mill with capacity of 100 t / 24 h, [10]

Physical characteristic	Sr.1-I	Sr.1-E	Sr.2-I	Sr.2-E	Sr.3-I	Sr.3-E	Sr.4-I	Sr.4-E	Sr.5-I	Sr.5-E
Bulk density, ϱv (kg/m3)	713.0	381.5	482.0	346.5	267.8	292.0	255.0	257.0	269.0	266.0
Density, ϱ (kg/m3)	1239	1250	1219	1200	1100	1063	1016	1130	1100	1191
Equivalent size, (mm)	3.76	2.13	2.23	1.22	1.51	0.90	1.06	0.84	0.65	0.63
Grinding degree, λ	1.76		1.83		1.67		1.26		1.03	
Specific surface, (m2/kg)	1.29	2.25	2.21	4.10	3.61	6.27	5.57	6.32	8.39	8.00*
Surface increase, ΔS (m2/kg)	0.96		1.89		2.66		0.75		-0.39*	
Natural slope angle, ψ (gr.)	21.8	37.8	37.1	37.5	44.6	39.0	41.1	39.2	42.6	44.4
Porosity, ε (%)	42.5	69.5	60.5	71.1	75.7	72.5	74.9	77.3	75.5	77.7

Correlation between individual volume of the seeds, calculated with the relation: $V = (1/6)\pi l w t$ (where: l, w, t represent the measured length, width, and thickness of each seed, and the seeds are assimilated with ellipsoid geometrical bodies) and their weight is presented in fig.7.

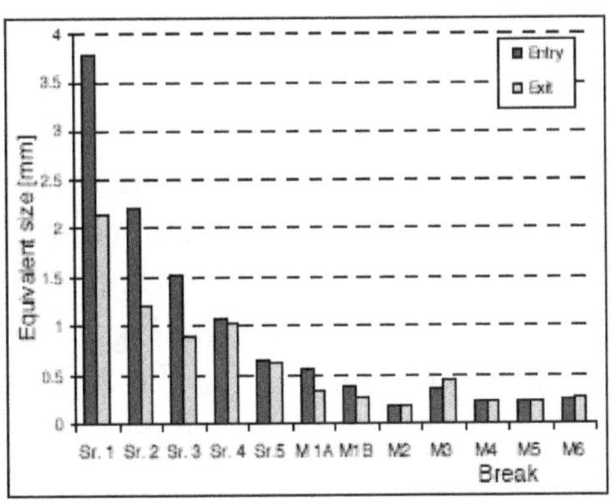

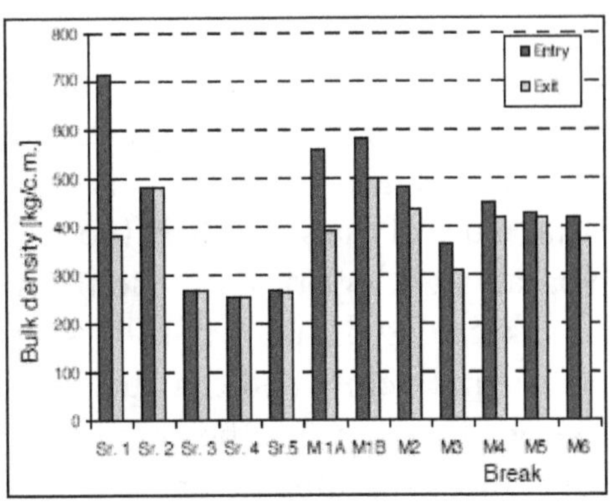

Figure 7. Variation of mean diameter and bulk density of grinding intermediate products on the grinding technological flow with grinding rollers [15]

Sieves used in granulometric analysis with sieve classifier and the results obtained by analysis are given in table 7, for each of the five technological passages, at the entry end exit from the respective mill rollers.

Table 7. Values of weights (%) pi for the fractions on the shaker sieves of the sifter machine and of the cumulative percentages Ri (%) for the collected grinded products, at entry "I" and exit "E" from pairs of mentioned rollers (Sr.1... Sr.5), [10]

li (mm)	Sr.1 - E pi (%)	Ri (%)	li (mm)	Sr.2 - I pi (%)	Ri (%)	Sr.2 - E pi (%)	Ri (%)	li (mm)	Sr.3 - I pi (%)	Ri (%)	Sr.3 - E pi (%)	Ri (%)
0.00	24.20	0.00	0.00	2.00	0.00	34.70	0.00	0.00	13.40	0.00	43.10	0.00
1.00	8.40	24.20	0.71	6.00	2.00	11.50	34.70	0.71	22.50	13.40	20.60	43.10
1.40	15.10	32.60	1.00	19.20	8.00	22.20	46.20	1.00	22.80	35.90	23.00	63.70
2.00	20.10	47.70	1.40	13.90	27.20	11.50	68.40	1.40	12.00	58.70	5.10	86.70
2.80	27.00	67.80	2.00	29.50	41.10	14.90	79.90	2.00	20.10	70.70	7.20	91.80
4.00	5.20	94.80	2.80	29.40	70.60	5.20	94.80	2.80	9.20	90.80	1.00	99.00
d1E = 2.13 mm			d2I = 2.23 mm			d2E = 1.22 mm		d3I = 1.51 mm			d2E = 0.90 mm	

li (mm)	Sr.4 - I pi (%)	Ri (%)	Sr.4 - E pi (%)	Ri (%)	li (mm)	Sr.5 - I pi (%)	Ri (%)	Sr.5 - E pi (%)	Ri (%)
0.00	26.00	0.00	42.60	0.00	0.00	5.20	0.00	5.80	0.00
0.71	25.20	26.00	27.50	42.60	0.25	3.90	5.20	3.90	5.80
1.00	28.30	51.20	20.70	70.10	0.32	21.10	9.10	22.60	9.70
1.40	11.90	79.50	3.60	90.80	0.50	28.20	30.20	29.70	32.30
2.00	7.80	91.40	5.00	94.40	0.71	35.30	58.40	32.70	62.00
2.80	0.80	99.20	0.60	99.40	1.00	6.30	93.70	5.30	94.70
d4I = 1.06 mm		d4E = 0.84 mm			d5I = 0.65 mm		d5E = 0.63 mm		

Based on the results obtained by granulometric analysis with the sieve classifier were tested by nonlinear regression analysis, the three laws of cumulative distribution for the refuse of the sieves R(x) (Rosin-Rammler function, Schuhman function and two parameters logistical function), for products entering the process, and for the products leaving the pairs of rollers, in the gristing phase of the grinding process. Experimental points and the curves of cumulative distribution for the refuse of the sieves (R(x)), using the three functions (eq.1, eq.2, eq.3), for some grinding products are presented in fig.8.

The coefficient values k, a, b, n, α and β, from the cumulative distribution relations Rosin– Rammler, Schuhman and the two parameters logistical function, as well as the R2 correlation coefficient values (which verifies the distribution adequacy degree expressed through the (1), (2), (3) relations), correspondent for the nine analyzed probes (from the five roll pairs) are presented in table 8.

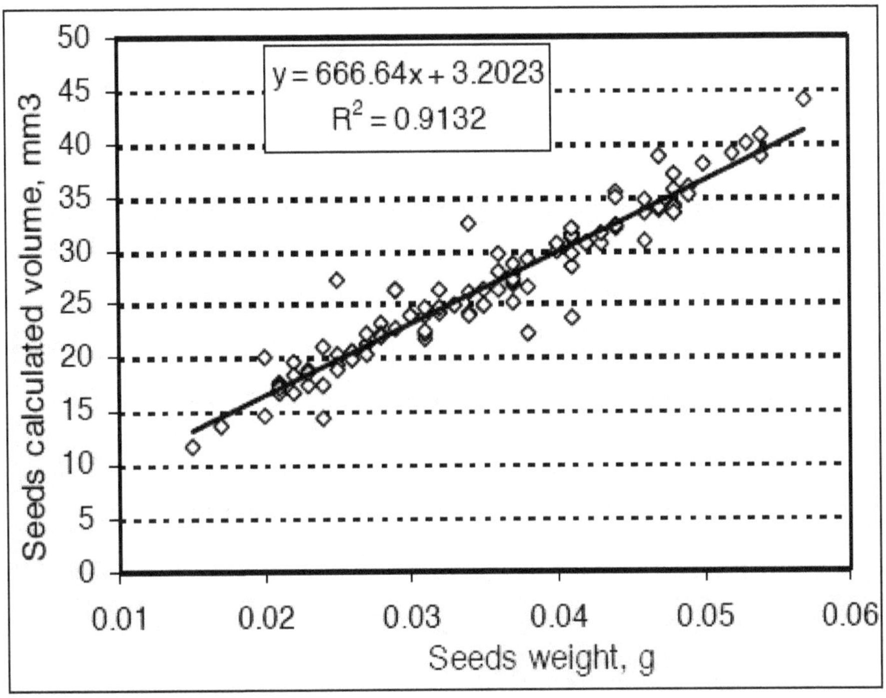

Figure 8. Correlation between volume and the mass of wheat seeds in an technological mixture (before grinding) [16]

From the analysis and interpretation of the obtained data for the 9 probes, which come from the mill rolls with rifles (for the coarse gritting in the breaking passages) (fig.9), following conclusions were found:

- For the vast analyzed material probes, from the mills flux, the best law of cumulative distribution is the Rosin-Rammler (1) with a correlation coefficient R2≥0.982, time in which the Schuhman type distribution law with a correlation coefficient R2≥0.933 (usually R2≥0.956) can be used with satisfactory results, in these cases;
- For the two parameter distribution law, the R2 correlation coefficient presents close values from the ones obtained through the Rosin-Rammler function, R2≥0,963, at half the probes being very close;
- The total grinding degree of the wheat breakage phase at the analyzed mill is approximately $\lambda = 7$, correspondent to a coarse gritting (crushing);
- It is appreciated that, in all cases, at seeds wheat grinding in the complex roller mills, we can consider that the best law of distribution is the Rosin-Rammler (1), (R2≥0,982), but the other methods, Schuhman and two parameter logistic, also can be used with satisfactory results.

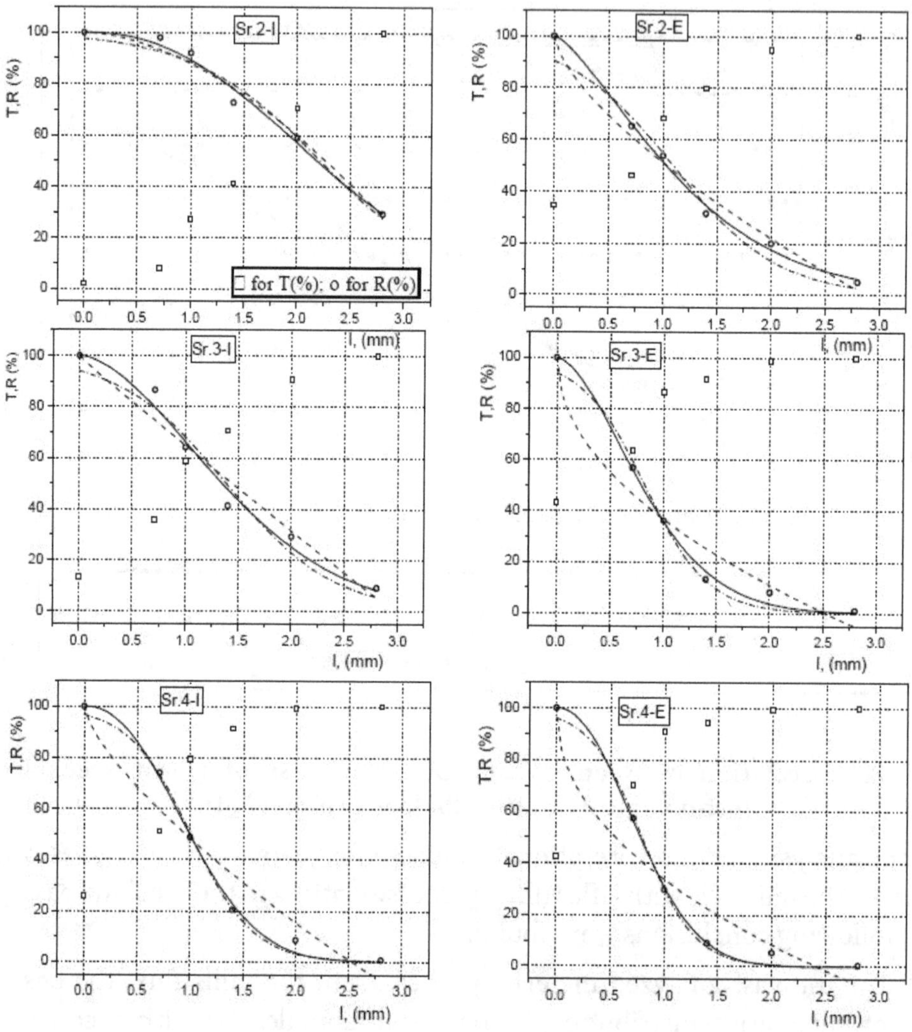

Figure 9. The curves described by the cumulative distribution laws (1), (2), (3) towards the experimental points R(%) for the gritting product from the five roll pairs (Sr.2...Sr.5) [10]; (I-entrance; E exit); Rosin-Rammler; Schuman; —• — • logistical function)

Table 8. The coefficient values a, k, b, n, α and β and of the R2 correlation coefficients, for the three size distribution laws tested, for the gritted products from the „I" entry to the „E" exit between the mentioned roll pairs (Sr.1...Sr.5), [10]

Law type	Coeff.	Sr.1-E	Sr.2-I	Sr.2-E	Sr.3-I	Sr.3-E	Sr.4-I	Sr.4-E	Sr.5-I	Sr.5-E
Rosin-Rammler (eq.1)	b	0.224	0.114	0.665	0.411	1.025	0.701	1.169	2.472	2.652
	n	1.659	2.302	1.382	1.747	1.682	2.220	2.093	2.852	2.817
	R2	0,988	0,987	0,996	0,982	0,996	0,996	0,996	0,998	0,999

Schuhman (eq.2)	k	4.201	3.398	2.893	2.966	2.531	2.532	2.431	1.025	1.016
	a	0.996	1723	0.674	0.960	0.495	0.711	0.464	1.710	1.639
	R2	0,999	0,981	0,985	0,956	0,958	0,933	0.940	0,991	0.987
Logistic with two parameters (eq.3)	α	2.573	3.701	2.243	2.744	2.760	3.397	3.216	4.347	4.303
	β	-1.245	-1.666	-2.053	-1.981	-3.345	-3.380	-4.056	-6.739	-6.878
	R2	0.984	0,974	0.972	0.963	0.988	0.995	0.994	0.997	0997

In plansifter compartments, material fractions are separated and sorted, as any granular material is made of particles with sizes between a minimum and a maximum value, in the interior of the mixture the size distribution being characterised by various distribution laws.

It must be mentioned that material particles, being extracted from various areas of the seed (from exterior to interior) have different mechanical characteristics and composition. This, and the different sizes of particles gives a different behaviour of the particles during grinding.

Thus is important to study and to know the size distribution of the particles of each fraction obtained in each frame set of the six plansifter compartments.

Size of sieve holes used for the experiments and the amount of material fractions on each sieve (individual and cumulative) for the separated material are presented in table 9.

In every fraction there is a percentage of material with sizes smaller than the size of the sieve hole, which means that sieving is incomplete, even if the number of frames is quite high. However, the average particle size of fraction C1–Break 2 is 2.27 mm, much larger than the opening of sieve holes of the package (1.05 mm). This shows that here are obtained the parts of seed with quite large sizes, which must be reintroduced in the grinding process at the passage Break 2.

At the second set of sieving frames of plansifter compartment C1, the opening of fabric holes is 470 μm (no. 40), but mean size of particles of fraction C1-DIV1' is 0.58 mm, slightly larger than the opening of the holes. It is noticed (Table 9) that there are particles with sizes smaller than the size of holes which remain unseparated (at least 8.4%). This phenomenon is valid for all sets of sieves in the plansifter with six compartments, as can be seen from the analysis of the results presented in table 9.

Composition of fraction C1-DIV1" of plansifter compartment C1 consists of the refuse of sieve frames no. 56 (with holes opening 0.31 mm), after the sieved of the second set, consisting of particles that passes through sieve no. 40 (with holes opening 0.47 mm) was extracted flour F (mean size of particles 0.08 mm). This fraction with fraction C1-DIV1' and with the two fractions C2-DIV1 of the second plansifter compartment are directed to the sorting-dividing compartment DIV1 (compartment C5). Mean particle sizes of fraction DIV1", from compartment C1, are 0.31 mm (equal to the opening of sieve holes which refused them, proving that here also the sieving is incomplete).

Table 9. Values of weights pi(%) of sieved fractions and of the cumulative weights Ti (%) for products collected at the entrance, respectively exit of plansifter compartments, C1 and C2

l_i (mm)	C1 Entrance		l_i (mm)	C1 Break 2		l_i (mm)	C1 DIV1'		l_i (mm)	C1 F		l_i (mm)	C1 DIV1"		l_i (mm)	C1 M2	
	p_i (%)	T_i (%)		p_i (%)	Ti (%)		p_i (%)	Ti (%)		p_i (%)	Ti (%)		p_i (%)	Ti (%)		p_i (%)	Ti (%)
0.000	24.20	0.00	0.000	10.20	0.00	0.000	1.10	0.00	0.000	4.20	0.00	0.000	6.00	0.00	0.000	0.60	0.00
1.000	8.20	24.20	1.000	21.30	10.20	0.180	2.30	1.10	0.045	45.10	4.20	0.125	8.00	6.00	0.090	1.90	0.60
1.400	15.10	32.40	1.400	14.60	31.50	0.250	5.00	3.40	0.063	24.30	49.30	0.180	12.80	14.00	0.125	41.50	2.50
2.000	20.20	47.50	2.000	21.60	46.10	0.400	51.70	8.40	0.090	18.80	73.60	0.250	24.50	26.80	0.180	15.00	44.00
2.800	27.10	67.70	2.500	20.70	67.70	0.630	28.60	60.10	0.125	6.30	92.40	0.315	32.40	51.30	0.200	30.10	59.00
4.000	5.20	94.80	4.000	11.60	88.40	0.710	11.30	88.70	0.160	1.30	98.70	0.400	16.30	83.70	0.250	10.90	89.10
d_{1E} = 2.13 mm			$d_{1Break2}$ = 2.27 mm			$d_{1DIV1'}$ = 0.58 mm			d_{1F} = 0.08 mm			$d_{1DIV1"}$ = 0.31 mm			d_{1M2} = 0.19 mm		

l_i (mm)	C2 Entrance		l_i (mm)	C2 Break 3		l_i (mm)	C2 DIV1'		l_i (mm)	C2 F		l_i (mm)	C2 M2		l_i (mm)	C2 DIV1"	
	p_i (%)	Ti (%)		p_i (%)	Ti (%)		p_i (%)	Ti (%)		p_i (%)	Ti (%)		p_i (%)	Ti (%)		p_i (%)	Ti (%)
0.000	34.40	0.00	0.000	23.50	0.00	0.000	5.60	0.00	0.000	19.10	0.00	0,000	17.70	0.00	0.000	0.30	0.00
0.710	11.50	34.40	1.000	30.10	23.50	0.180	4.00	5.60	0.045	34.90	19.10	0,090	9.40	17.70	0.125	1.60	0.30
1.000	22.20	45.90	1.400	13.50	53.60	0.250	7.70	9.60	0.063	21.60	54.00	0,125	22.30	27.10	0.180	5.60	1.90
1.400	11.60	68.10	2.000	17.60	67.10	0.400	26.10	17.30	0.090	16.10	75.60	0,180	9.60	49.40	0.250	19.20	7.50
2.000	15.00	79.70	2.500	12.40	84.70	0.500	49.20	43.70	0.125	6.70	91.70	0,200	25.10	59.00	0.315	44.50	26.70
2.800	5.30	94.70	3.150	2.90	97.10	0.710	7.40	92.60	0.160	1.60	98.40	0,250	15.90	84.10	0.400	28.80	71.20
d_{2E} = 1.22 mm			$d_{2Break3}$ = 1.56 mm			$d_{2DIV1'}$ = 0.52 mm			d_{2F} = 0.07 mm			d_{2M2} = 0.17 mm			$d_{2DIV1"}$ = 0.37 mm		

The last components of plansifter compartments in gristing passage shave higher content of coating particles which are found in the upper layers of material on the frames, thus being recommended that they do not separate through the holes, even if their sizes are about the size of endosperm particles, to be further removed in semolina machines (sieving motion leads to the layering of mixture components by density). Flour particles have mean sizes under 0.18 mm in all plansifter compartments, while particles of last refuse from the five passages fitted with pairs of rollers have mean sizes over 0.37 mm (see Table 9). Values of coefficients b and n in the equation of relationship Rosin–Rammler cumulative distribution law (eq.1), for the material which passed through sieve holes in granulometric analysis, and the correlation coefficients R2 and χ2 have high values which show the adequacy degree of the given function with the experimental data. In all cases, for all fractions obtained during gristing phase of wheat in the studied mill, the correlation is very good, appreciated by values of coefficient R2 ≥ 0.926.

As it can be noticed from fig.10, there are fractions having most particles of sizes close to the minim value of sieve classifier holes, but there are also components with particles with sizes from the mean size to the maximum size of the sieve holes used for granulometric analysis.

However, most components show mean profile (with central inflection point) of the separation curves which demonstrates the correct choosing of sieve classifier sizes (made from a set of 30 sieves by trying to take into consideration the arrangement in geometric distribution with holes ratio of $\sqrt{2}$).

From the analysis of coefficients b and n from Rosin–Rammler law (eq.1) it is noticed that values of coefficient b are 0.2–1.5 103 for most analyzed fractions, generally with high values, for the small size components of the particles (flour or dunsts), 1 106 – 5 107, giving the size characteristics of such particles (Table 10).

Table 10. Values of coefficients b and n and correlation coefficient R2 for Rosin – Rammler granulometric distribution, for the granulometric distribution law for fractions of the two plansifter compartments

Plansifter Compartment		b	n	R^2	χ^2	Plansifter Compartment		b	n	R^2	χ^2
C1	C1 Entrance	0.222	1.663	0.988	17.039	C4	C4 Entrance	0.621	1.958	0.988	21.914
	C1 Break 2	0.169	1.964	0.987	18.890		C4 F	2.06 103	5.665	0.997	1.144
	C1 DIV1"	37.812	3.412	0.993	9.147		C4 M4	1.15.103	3.967	0.976	36.414
	C1 DIV1'	15.782	6.033	0.997	6.245		C4 FT2	0.027	4.557	0.996	9.071
	C1 F	2.2 103	3.051	0.938	142.827		C4 Break 5	2.590	2.983	0.999	1.091
	C1 M2	2.86 103	5.028	0.988	20.440		C4 M5	1.6 103	6.054	0.999	3.370

Values of exponent n indicate the uniformity or the irregularity degree of particles from the analyzed fractions.

The analysis of this exponent values for the fractions of each plansifter compartment (Table 9) shows that they have a wide range of values, even for the same type of grinding product (for example flour – F), which shows the irregularity of particles, both for a given fraction and between fractions.

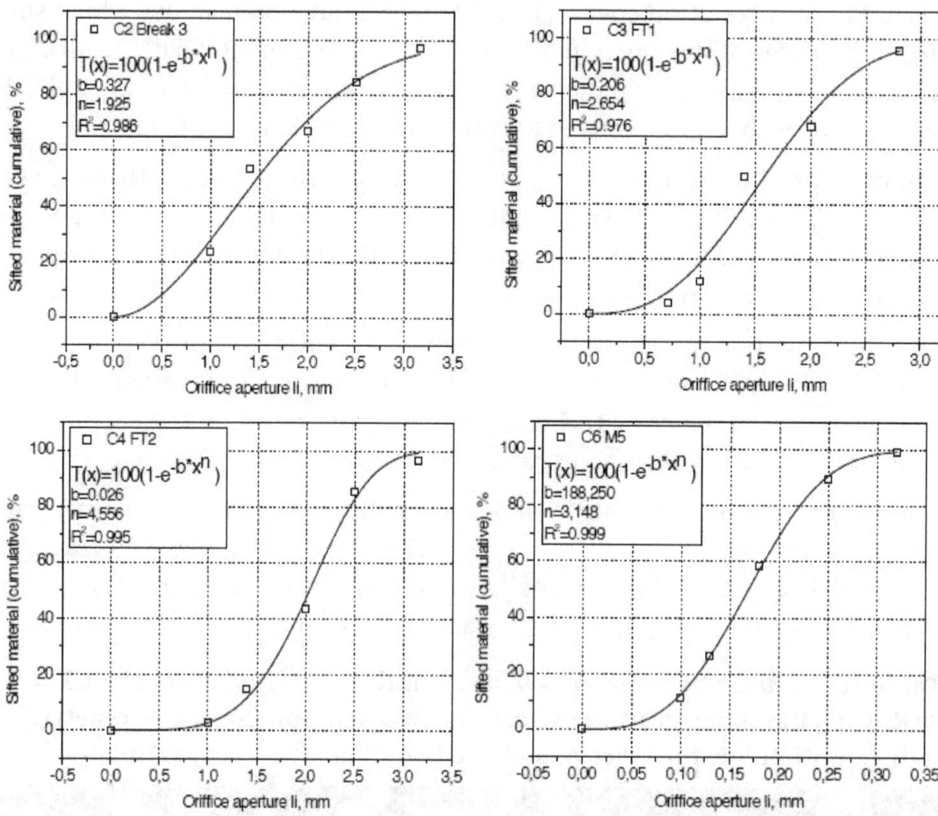

Figure 10. Curves of granulometric distribution given by eq. (1) in correlation with experimental data for grinding fractions in plansifter compartments during gristing phase of wheat in a mill with capacity of 100 t /24 h

4. SOME MECHANICAL CHARACTERISTICS OF WHEAT SEEDS IN UNIAXIAL COMPRESSION TESTS

Main stress to which seeds are subjected, while passing through mill rollers, is given by the type of rollers surface, namely smooth or fluted. Regardless the surface type, one of the main stress during grinding is compression (or crushing), especially if the mill rollers have smooth surface. To estimate the behaviour of seeds while passing through the rollers, experimental research is required on the compression stress of seeds from various wheat varieties, knowing that not all varieties have similar mechanical characteristics. Even seeds from the same variety have different behaviour, due to the irregular development stage in the ear, and also from one ear to another.

The compression of wheat seeds is performed in three different stages: the first stage is elastically deformation, characterized by the proportionality between the compression force and the deformation; the second stage is plastic deformation, characterized by large increases of seed deformation at small increases of compression force; the last stage consists in cracking or rupture, being characterized by seed crushing when reaching a certain value of compression force, [17-20].

Compression test is an objective method for determining the mechanical properties of cereal seeds and also one of the best techniques for determining the modulus of elasticity by the study of their behaviour at compression stress, using force-deformation curve, [21,22].

By performing uniaxial compression tests on wheat seeds, force-deformation curve is obtained, giving the possibility to determine hardness, apparent modulus of elasticity, crushing resistance, force and deformation and energy consumption in various specific points of the curve (i.e. rupture point) and maximum stress in the material, [21,23].

Cereal seeds have a different behaviour under the action of compression forces, depending on their moisture content, [17,20], variety, development stage, geometric sizes, individual mass, glassiness, (soft cereals and hard cereals) etc.

In fig.1 is presented a typical force-deformation curve for compressed Flamura wheat seed.

The bioyield point is the point on the force – deformation curve at which the force decreases or remains constant with increasing deformation. Force in the rupture point (rupture force) is the minimum required force for the wheat seed to break (rupture). Deformation at bioyield and rupture points is the deformation at loading direction, [24,25]. Values of force and deformation to bioyield and rupture points are directly read from force-deformation curve and recorded by machine used for compression test, [21].

Energy absorbed in bioyield and rupture points could be determined from the area under the force-deformation curve between the initial point and the bioyield and rupture point, respectively, using equation [24,25]:

$$W = \frac{FD}{2} (\mathrm{mJ}) \tag{9}$$

where: W is energy absorbed (mJ), F is force in bioyield or rupture point (N), D is deformation in bioyield or rupture point (mm), (see fig.11).

Based on a standard method (ASAE 2008, [21]), for a seed placed between two parallel plates, the modulus of elasticity could be calculated with following equation, [20,21,21]:

$$E = \frac{0.338 k_u^{3/2} F\left(1-\mu^2\right)}{D^{3/2}} \left[\frac{1}{R'} + \frac{1}{R_1{}'}\right]^{1/2} \tag{10}$$

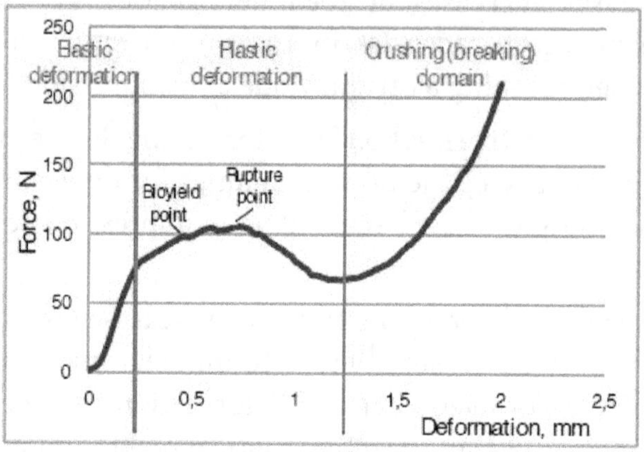

Figure 11. A typical force –deformation curve of wheat grain (type Flamura), [20]

where: E – modulus of elasticity for cereal seeds, (MPa); ku – coefficient which depends on the geometrical properties of wheat seeds (ku = 1,303 - adapted from calculus tables of Kozma and Cunningham, 1962); F – compression force, (N); D – seed deformation (m); μ - Poisson ratio, (μ = 0,3 for wheat seeds); R' and R'1 – small and large radius of the curvature of convex surface seed in contact with the flat surface, (m), (see fig.12, left).

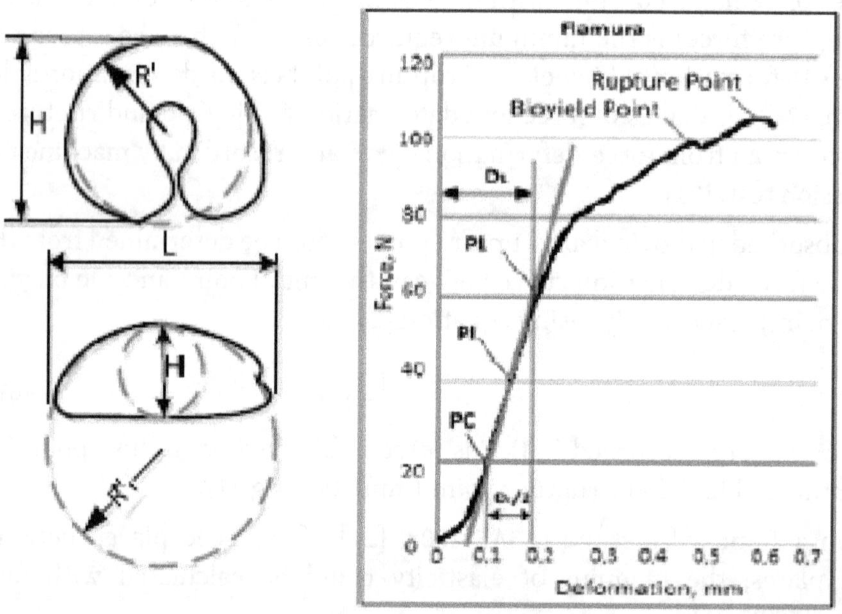

Figure 12. Estimation of curvature radius and force-deformation curve of wheat seed, (adapted from [25,26]) PL – proportional limit; PI – point of inflection; Pc – point of calculation

According to the standard method (ASAE 2008, [21]), also presented by Mohsenin in [25,26], curvature radius of convex surface, R' and R1' (fig.12) can be calculated using relations (11) and (12):

$$R' = \frac{H}{2} \tag{11}$$

$$R_1' = \frac{H + L^2/4}{2H} \tag{12}$$

where: H is seed thickness, (m), and L is seed length, (m), in undistorted state.

This method was used by many researchers to determine the modulus of elasticity for different agricultural products, [27-30].

According to the standard method (ASAE 2008), values of force F and deformation D, from equation (2) are calculated for the proportionality area of force-deformation curve in the point of calculation Pc (fig.12). The position of this point is estimated visually, as the point is located halfway between curve origin and proportionality limit PL (fig.12, right). It was found that the point of calculation Pc is located lower than the point of inflection, also established visually, [21].

To determine the variation of mechanical resistance characteristics of wheat seeds from the same variety, compression tests were performed for sets of 100 seeds of three varieties of Romanian wheat (Flamura, Glosa and Trivale – soft wheat), using Hounsfield mechanical testing machine, at a constant speed of the crushing device of 5 mm min-1, using a force cell of 1000 N. Were graphically plotted the force-deformation curves for each seed, and from each diagram were collected data about: force, deformation and energy absorbed in the bioyield point (F1, ε1, W1), respectively in the final point (rupture), (F2, ε2, W2).

The analysis of measured data showed that the seeds of Flamura variety were larger than Trivale variety, for all three main sizes, and for their volume. The same goes for seeds mass. Flamura variety seeds were more uniform as size and mass. Regarding the mechanical characteristics of wheat seeds, it was found that compression forces, for bioyield point and for final seeds crushing, were smaller for Trivale variety than Flamura. The same goes for energy absorbed to the bioyield point, respectively to crushing. Since the sizes of Trivale seeds were smaller than Flamura seeds, the deformations carried to the bioyield point, respectively to crushing, were smaller for Trivale than Flamura, but the standard deviation of the values was smaller for Flamura for deformations, showing that Flamura seeds were more regular in terms of deformations (until crushing).

In fig.13 are presented two examples of force-deformation curves for two varieties of wheat, and in fig.14 are presented the histograms of bioyield force and energy absorbed for seed crushing.

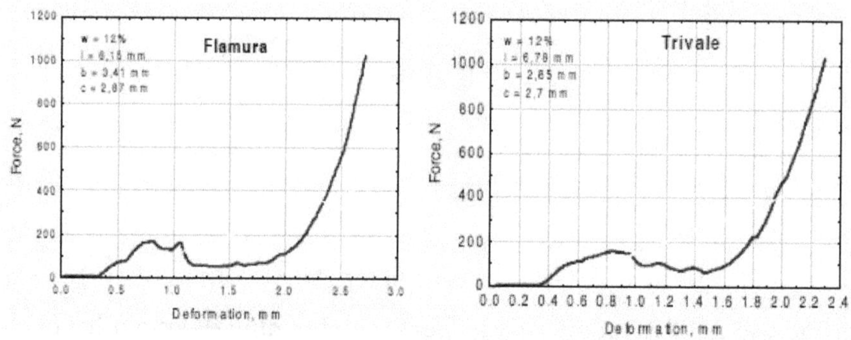

Figure 13. Examples of force-deformation curves for the two wheat varieties, [23]

On the histograms were traced the variation curves for the analyzed parameters by regression analysis of the values given by the histogram, using the normal function presented in equation (13), [23]:

$$p_x = 100\,a \frac{1}{\sqrt{2\pi}\,b} e^{-\frac{1}{2}\left(\frac{x-c}{b}\right)^2} \tag{13}$$

where: px(%) is the percentage weight of each class interval (number of seeds with values in the considered class interval); a – class interval for each analyzed parameter; b and c are regression coefficients of the analyzed function (b is the standard deviation, c is the values mean).

Values of coefficients for the regression function (eq.13) used in statistical analysis and values of correlation coefficient R2 for data given by histograms are presented in Table 11.

Table 11. Values of coefficients for regression equation (eq.13) and its correlation with experimental data [20]

Measured parameters of wheat	Flamura wheat variety				Trivale wheat variety			
seeds	a	b	c	R2	a	b	c	R2
Length l, (mm)	0.20	0.404	6.443	0.989	0.20	0.501	6.186	0.971
Width w, (mm)	0.10	0.202	3.429	0.974	0.20	0.284	2.994	0.981
Thickness t, (mm)	0.20	0.248	3.058	0.975	0.20	0.315	2.664	0.983
Mass m, (g)	0.01	0.008	0.051	0.988	0.01	0.009	0.037	0.981
Volume V, (mm3)	5.00	5.870	35.57	0.968	5.00	6.207	26.23	0.985
Bioyield force F1, (N)	20.0	41.36	122.64	0.921	20.0	39.48	104.70	0.888
Bioyield energy W1, (J)	0.01	0.024	0.036	0.923	0.01	0.017	0.026	0.884

Analysis of histograms and variation curves, as well as of data in table 11, shows that all analyzed parameters have almost normal distribution, assessed by values of correlation coefficient R2.

Using standard method (ASAE 2008, [21]) and equations (10), (11) and (12) were determined the values of modululs of elasticity for wheat seeds of Flamura, Trivale and Glosa varieties, in this paper being presented their mean values, (table 12).

Fig.15 shows the machine used for uniaxial compression tests between parallel plates of weat seeds and their position.

From the sample of 100 determinations for each variety of wheat, were selected the 50 most representative determinations, being kept the values found for force and absolute deformation of the seed.

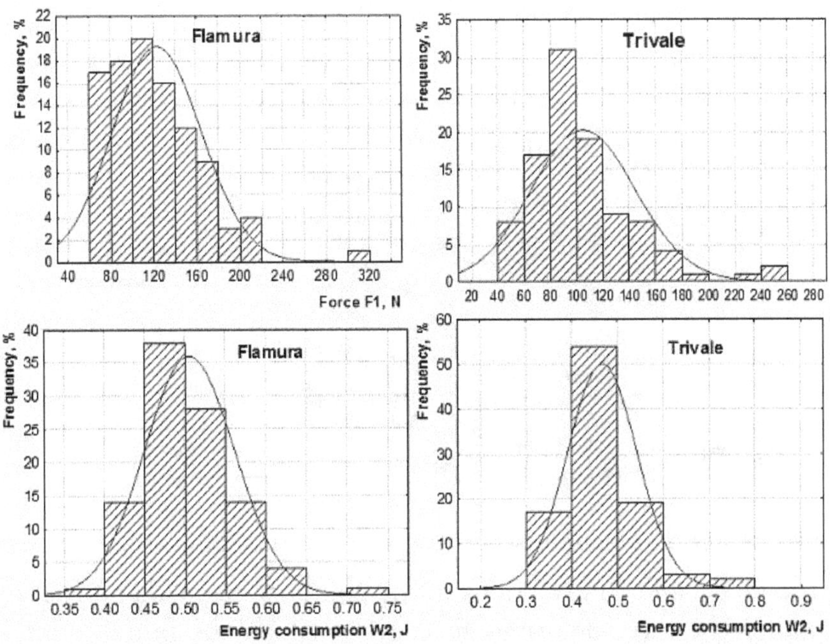

Figure 14. Histograms and variation curves for the force in bioyeld point and the energy consumption in rupture point for wheat seeds [23]

Force-deformation curves, for each of the 50 determinations (of a variety) were processed so that each has the same origin (same starting point), and the intervals of reading (recorded) to be the same. Values for the parameter on the ordinate (forces in the mentioned points) were averaged (arithmetic average for the 50 determinations was calculated) for the same value of deformation (parameter on the abscissa), and these values were used to retrace the force-deformation curve, which respresents the curve of mean values of compression force (fig.16). Using the approximately normal distribution, were statistically estimated the limits within which the mean force-deformation curve is found, for a confidence interval of 95%. For normal distribution, the confidence interval corresponding to 95% confidence level ranges between +/- 1,96, considered standard deviations. Thus, the confidence interval of mean curve was calculated using the following equation:

$$\mu = m \pm 1.96\frac{\sigma}{\sqrt{n}} \tag{14}$$

where: μ is the confidence interval, and m is the mean value of the analyzed parameter (in this case, the compression force) and $\sigma / \sqrt{n}$ = Sm is the standard error of the mean, σ - standard deviation, and n –number of seeds from each variety of wheat (in this paper, n = 50).

On the curve of mean values (fig.16), were determined the values of mechanical characteristics mentioned before (forces and deformations in the characteristic points) and it was calculated the value of modulus of elasticity using the standard method (ASAE 2008, [21]), for mean curve (for the three varieties of wheat).

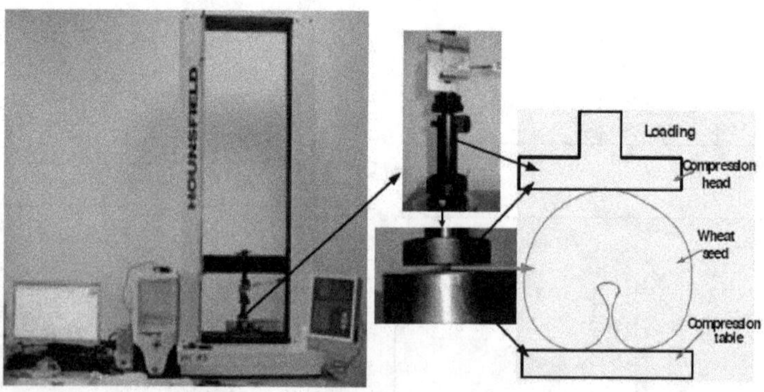

Figure 15. Hounsfield - Mechanical testing machine used in compression test [20]

Knowing the forces and deformations in the points of bioyield and rupture, from the area under the force-deformation curve between the initial point and the bioyield and rupture point, respectively, using equation (1), energy absorbed in bioyield and rupture point was determined.

Table 12. Values of measured and determined parameters in uniaxial compression test [20]

Measured parameters of wheat seeds		Mean of parameters values			Values of parameters read from the mean curve		
		Flamura	Trivale	Glosa	Flamura	Trivale	Glosa
	Bioyield force Fb, (N)	93.2	83.1	98.0	98.4	81.1	94.0
	Bioyield energy Wb, (J)	-	-	-	0.028	0.018	0.016

	Rupture force Fr, (N)	107.8	90.5	103.6	104.2	83.2	94.7
	Rupture energy Wr, (J)	-	-	-	0.038	0.018	0.016
	Relative deformation, δb	0.138	0.092	0.077	-	-	-
deformation	Absolute deformation, Db (mm)	0.304	0.267	0.260	0.464	0.348	0.292
	Relative deformation, δr	0.099	0.109	0.086	-	-	-
deformation	Absolute deformation, Dr (mm)	0.419	0.320	0.290	0.576	0.400	0.360
Modulus of elasticity, (MPa)		313	364	486	298	369	468

Analysis of data presented in table 12 showed that the values of bioyield force, respectively values of the force in the point of rupture of wheat seeds, determined from the mean curve are very close to the values of these forces obtained from the force-deformation curves for each particular seed.

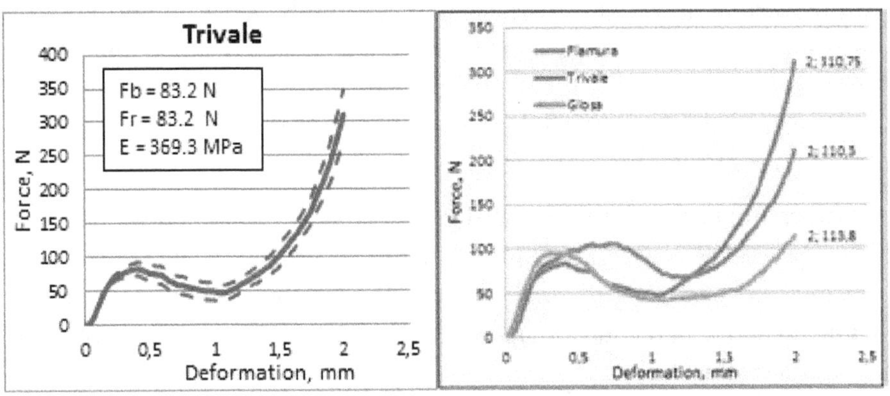

Figure 16. Mean curves force-deformation for three wheat varieties and 95 % confidence interval, [20]

Analysis of curves presented in figure 15 shows that they have similar shapes for the three varieties of wheat, and also within each of them and the force-deformation curves for each individual seed analyzed from each variety of wheat.

As absolute values of the force in the bioyield point, respectively in the rupture point, they are found in between 83.1 N for Trivale variety and 98.0 N for Glosa variety regarding the bioyield force, respectively 90.5 N for Trivale and 107.8 N for Flamura (values calculated with arithmetic average of the 50 determinations). These values are very close to the values presented in literature [31], where is stated that crushing force (rupture) of wheat seeds is of approximately 100 N.

On the relative deformation of seeds, during the compression tests, for the force in the bioyield point (bioyield force), respectively rupture, data in table 12 also show relatively close values for the wheat seeds of the three varieties.

5. CONCISE SUMMARY

The advancement of the technical gristing process for wheat in a mill is crucial for the overall technological workflow, significantly impacting the flour extraction rate while minimizing the excessive grinding of seed coats.

Laboratory analysis of material samples collected from the entrance and exit of each pair of milling rollers can ascertain the equivalent average particle sizes, grinding degree during passage, and specific surface area of the material particles.

Granulometric analysis of the material to be grinded or of the grinded material at mill rollers, and of the sorted fractions in plansifter compartments show a distribution after multiple known laws, from which most used is Rosin-Rammler distribution function, with high correlation coefficient R2.

However, it is shown that there can also be used with good results the Schuhman and logistical two parameters distribution laws, the finding suggest that the type of granulometric distribution law which best describes the size of grinded biological materials depends on material nature and the place and role of roller mill used for grinding in the general technological flow. Knowledge of adequate mathematical models describing the size distribution of grinded materials is useful in all engineering activities related to the processes on the flow of complex roller mills of last generation.

Values of mechanical characteristics of wheat seeds (regardless the variety) are It is essential to assess the energy expended during the grinding process in grain mills. The grinding energy is significantly influenced by the crushing force and their relative and absolute deformation, as established through experimental studies on uniaxial compression.

Certain wheat types discussed in this chapter exhibit a rupture point compression force, derived from force-deformation curves, ranging from 100 to 110 N, with a seed moisture content of approximately 12%.

The crushing energy ranges from 0.02 to 0.04 J per wheat seed, influenced by seed wetness and arrangement during compression, either width-wise or thickness-wise.

The modulus of elasticity ranges from 313 to 487 MPa, with higher values corresponding to decreased moisture content. It was determined that reduced moisture content led to increased modulus of elasticity and decreased rupture energy, confirming that wetter seeds exhibit greater flexibility than drier seeds, resulting in higher energy consumption.

REFERENCES

1. Fang Ch., Campbell G.M. Stress-Strain Analysis and Visual Observation of Wheat Kernel Breakage During Roller Milling Using Fluted Rolls. Cereal Chemistry 2002; 79(4) 511–517.
2. Mazlina S., Kamal M. Evaluation of the potential role recycle within the flour milling break system. Universiti Putra Malaysia, TLN 2006; 42.
3. Fang Q., Hanna M.A., Haque E., Spillman C.K. Neural Network Modelling of Energy Requirements for Size Reduction of Wheat, Transaction of the ASAE 2000; 43(4) 947-952.
4. Campbell G.M. Roller Milling of Wheat, Handbook of Powder Technology 2007, 12, 383-421.
5. Pasikatan M.C., Milliken G.A., Steele J.L., Spillman C.K., Haque E. Modelling the Size Properties of First-Break Ground Wheat. Transaction of the ASAE 2001; 44(6): 1727-1735.
6. Hareland G.A., Effects of Break-Roll Speed Differential on Product Yield and Semolina Granulation in a Durum Pilot Mill System. Cereal Chemistry, 1998; 75(6) 836-840.
7. Căsăndroiu T., Voicu G., Tutuianu G.D. Cu privire la legile distribuției granulometrice al produselor mărunțite într-o moară cu valțuri, Sesiunea jubiliară de comunicări ştiinţifice ISBTeh-2002, Politenica University of Bucharest 2002; 194-199.
8. Voicu G., David M., Tudosie E.M., Constantin G. Researches on the grindings charactersitics in technological breakage phase at a fluring mill with capacity of 220 t / 24 h. Romanian Technical Sciences Academy, Alma Mater, Modelling and optimization in the machines building field MOCM 2009; 15(3) 85-90.
9. Voicu G., Tudosie E.M., Tarcolea C., Voicu P. Particle Size Distribution and Sieving Characteristics of Semolina in the Grinding Passages of a Wheat Mill, 11th International Congress on Mechanization and Energy in Agriculture TRAKAGENG 2011, Istanbul, Turkey, 457-462;
10. Voicu G., Căsăndroiu T., Târcolea C., Tudosie E.M., Voicu P. Experimental research on the physical characteristics of products obtained from wheat milling, Actual Tasks on agricultural engineering 2010; Opatija, Croatian, 399-410.
11. Headley V., Pfost H. A comminution equation relating energy to surface area by log probability method, Transactions of the ASAE 1968; 11(3) 331-334 and 338.

12. Henderson S., Hansen R. Farm grain comminution: Hammer mill and burr mill performance analyzed, Transactions of the ASAE 1968, 11(3) 339-402.
13. Voicu G., Căsăndroiu T. Utilizarea funcției logistice pentru descrierea proceselor de separare pe sitele sistemelor de curățire de la combine, Lucrări ştiinţifice "Concepte, tehnologii şi echipamente tehnice moderne pentru agricultură şi industria alimentară" INMATEH- III 2004, Bucharest, 107-112.
14. Voicu G., Căsăndroiu T., Toma L.M. A multiple logistic regression statistical model to estimate grain losses on sieve cleaning system from combine, Actual tasks on agricultural engineering 2006; Croatian, 481-491.
15. Voicu G., Tudosie E.M., Casandroiu T., Biris S.S. Some aspects of physical properties of seeds and wheat products obtained from a roller mill. Buletinul Institutului Politehnic din Iasi, Secția Constructii de masini, Editura Politehnium, LVI (LX), 4B, 2010; 477–488.
16. Voicu G., Tudosie E.M., Paraschiv G., Voicu P., Ipate G. Testing certain distribution laws regarding some physical characteristics of grinded wheat seed mixture inside milling units and the connection between them, Actual tasks on agricultural engineering 2011, Opatija, Croatian, 427-436.
17. Clarke B., Rottger A. Small mills in Africa. Selection, installation and operation of equipment. Food and Agriculture Organization of the United Nations, Roma 2006.
18. Nedeff V., Panainte M., Mosnegutu E., Savin C. Echipamente de proces în industria alimentară. Măruntirea produselor agroalimentare, Editura Meronia, 2005.
19. Romański L., Stopa R., Niemiec A. Distribution of surface pressure on grains crushed between flat plates. Inzynieria Rolnicza 2005; 11(71) 413-421.
20. Voicu G., Stefan E.M., Ungureanu N. Constantin G. Some mechanical characteristics of wheat seeds obtained by uniaxial compression tests, Scientific Bulletin of University Politehnica of Bucharest, 2012 (sent for review).
21. ***ASAE Standards 2000 (R2008), Compression Test of Food Materials of Convex Shape. American Society of Agricultural and Biological Engineers, S368.4.
22. Khodabakhshian R., Emadi B. Determination of the Modulus of Elasticity in Agricultural Seeds on the Basis of Elasticity Theory. Middle-East Journal of Scientific Research 2011; 7(3) 367-373.
23. Tudosie E.M., Voicu G., Căsăndroiu T., Târcolea C. Variation of crushing characteristics by compression of Romanian wheat seeds. Research People and Actual Tasks on Multidisciplinary Sciences 2011; vol.2, 95-99.

24. Karaj S., Muller J. Determination of physical, mechanical and chemical properties of seeds and kernels of Jatropha curcas L. Industrial Crops and Products 2010; http:// www.sciencedirect.com (accessed May 2011).
25. Mohsenin N.N. Physical properties of plant and animal materials, vol I (Structure, physical characteristics and mechanical properties). Gordon and Breach Science publishers, N.Y., 1970.
26. Shelef L. Mohsenin N.N. Evaluation of the Modulus of Elasticity of Wheat Grain. Journal Series of the Pennsylvania Agricultural Experiment Station, no. 3174, 1967; http://www.aaccnet.org/publications/cc/backissues/1967/ Documents/ Chem44_392.pdf (accessed 20 January 2011).
27. Afkari Sayyah A.H., S.Minaei S. Behavior of Wheat Kernels under Quasi-Static Loading and its Relation to Grain Hardness, Journal Agricultural Science and Technology 2004; 6 11-19.
28. Baslar M., Kalkan F., Mazhar Kara M., Ertugay M.F., Correlation between the protein content and mechanical properties of wheat, Turkish Journal of Agriculture and Forestry 2012; 36, http://mistug.tubitak.gov.tr (accessed May 2012).
29. Moya M., Ayuga F., Guaita M., Aguado P.J. Mechanical properties of granular agricultural materials considered in silos designe. 15th ASCE Engineering Mechanics Conference, Columbia University, New York, (2002), http://www. civil.columbia.edu/ em2002/ proceedings/papers/271.pdf (accessed January 2010).
30. Abbaspour-Fard M.H., Khodabakhshian R., Emadi B., Sadrnia H. Evaluation the Effects of Some Relevant Parameters Elastic Modulus of Pumpkin Seed and its Kernel. International Journal of Biomaterials 2012; ID 271650, http:// www.sciencedirect.com (accessed May 2012).
31. Emadi B., Khodabakhshian R., Abbaspour Fard M.H., Sadrnia H. Experimental comparison of applying different theories in elasticity for determination of the elasticity modulus of agricultural produce. Pumpkin seed as a case study. Journal of Agricultural Technology 2011; 7(6) 1495-1508.

[illegible]. Karal S, Müller [illegible] determination of physical, mechanical and chemical properties of seeds and kernels of [illegible] [illegible] to Industrial [illegible] [illegible] [illegible] [illegible] [illegible] accessed [illegible]

[illegible] [illegible] properties of plant and animal materials [illegible] [illegible] [illegible] [illegible] [illegible]

[illegible]

27. [illegible] [illegible] [illegible] [illegible] [illegible]

[illegible] [illegible] [illegible] properties of [illegible] [illegible] Accessed May [illegible]

[illegible] [illegible] M. Arnaud F. [illegible] properties [illegible] [illegible] conference [illegible] design [illegible] [illegible] Mechanics Conference, [illegible] University, New York (2002) http://www.[illegible] 2002/proceedings/papers/224.pdf (accessed January [illegible]).

[illegible] Abbaspour-Fard M.H., Khodabakhshian R., Emadi B., Sadrnia H. Evaluation the effects of some physical parameters Elastic Modulus of pumpkin seed and its kernel. International Journal of [illegible] [illegible] [illegible] http://www.sciencedi[illegible] (accessed May 2012).

[illegible] [illegible], Khodabakhshian R., Abbaspour-Fard M.H., Sadrnia H. [illegible] comparison of applying different theories in elasticity for determination of the elasticity modulus of agricultural produce, Pumpkin seed as a case study. Journal of Agricultural Technology 2011; 7(6) 1495-1508.

16 Chapter
TECHNOLOGICAL PACKAGING SOLUTIONS FOR FOOD QUALITY CONTROL

1. INTRODUCTION: AN OVERVIEW

The shelf life of perishable foods such as meat, poultry, fish, fruits, vegetables, and fresh cereal-based goods is constrained by several variables that typically result in alterations in odor, flavor, color, and texture until they become entirely unacceptable. Packaging serves as the primary mechanism to avert product degradation and extend its shelf life. The package safeguards the food from physical, chemical, and biological harm. It serves as a physical barrier against oxygen, moisture, volatile chemical compounds, and microbes that are harmful to food. The packaging must be regarded as a fundamental component of the preservation system, as it offers a barrier between the food and the external environment. It is typically a composite entity addressing multiple distinct requirements [1]. The preservation function is an essential aspect of food packaging, as it is directly linked to consumer safety. The performance of packaging is influenced by various factors, including the quality of the starting food, processing methods, dimensions, design, distribution techniques, and disposal practices. Generally speaking, the characteristics that define their suitability for meeting performance criteria can be classified into the following categories: mechanical, thermal, optical, and mass transport properties. Mass transport phenomena are crucial in food packing using plastics, as a polymeric matrix allows permeability to moisture, oxygen, carbon dioxide, nitrogen, and other low molecular weight molecules. Glass and metal packaging materials exhibit impermeability to low molecular weight compounds, while paper-based materials demonstrate excessive permeability. Consequently, these recent materials do not allow the designer to enhance the barrier qualities for diverse applications.The polymers can provide a wide range (by three or four orders of magnitude) of permeability for different applications, thus justifying studies aimed to ensure adequate barrier protection. Therefore, in situations where food deteriorations are driven by either gas or moisture permeation to the ambient environment, an accurate choice of packaging mass transport properties may bring about an increase to product shelf life. Each category food has its specificity in quality attributes, storage conditions, expected shelf life and packaging tools applied. Together with transport properties of film, another valid packaging option to maintain product quality is represented by the proper selection of headspace conditions. Vacuum packaging and modified atmosphere packaging (MAP) are two widely used strategies for food preservation [2]. The first strategy means a complete lack of gas in the package whereas, under MAP, headspace environment may change during storage but there is no additional manipulation of the internal environment. Packaging under these conditions can protect products against deteriorative effects, which may include discoloration, off-flavor and off-odor development, nutrient loss, texture changes, pathogenicity, and other measurable factors. With the increasing demand for fresh and natural products without addition of dangerous chemicals, MAP or vacuum seem to be ideal methods of preservation for many foods, being simple and cheap to be applied. The few disadvantages are related to need of equipments and proper packaging materials and, in the specific case of MAP, to the limitation on retail for the increased pack volume of bags.

Today the efforts to improve the performances of packaging with clear effects on food quality can be directed towards two many working areas, green polymers and active packaging. The performance expected from bio-plastic materials used in food packaging application is containing the food and protecting it from the environment while maintaining food quality. It is obvious that to perform these functions is important to control and modify their mechanical and barrier properties, that consequently depend on the structure of the polymeric packaging material. In addition, it is important to study the change that can occur on the characteristics of the bioplastics during the time of interaction with the food. Studies of the literature show up that only a limited amount of biopolymers are used for food packaging application [3, 4]. Unlike the usual wrap, films, labels and laminates came from fossil fuel resources, the use of biodegradable polymers represents a real step in the right direction to preserve us from environmental pollution. This kind of packaging materials needs more research, more added value like the introduction of smart and intelligent molecules able to give information about the properties of the food inside the package and nutritional values. It is necessary to make researches on this kind of material to enhance barrier properties, to ensure food properties integrity, to incorporate intelligent labeling, to give to the consumer the possibility to have more detailed product information than the current system [5, 6].

Active packaging is the most relevant innovative idea applied for consumer satisfaction. It has been defined as a system in which the product, the package and the environment interact in a positive way to extend shelf life of product or to achieve some characteristics that cannot be obtained otherwise [7]. In many present-day active packaging technologies the active agent is placed in the package with the food, in a small sachet, pad or device manufactured from a permeable material which allows the active compound to achieve its purpose but prevents direct contact with the food product, protecting the food from contamination or degradation. Active packaging developments are now focusing on incorporating the agents into the polymeric matrices which constitute the package walls; the resulting materials act by releasing substances which have a positive effect on the food or by retaining undesired substances from the food or the internal atmosphere of the package. The migration of a substance may be achieved by direct contact between food and packaging material or through gas phase diffusion from packaging layer to food surface. Although the former is the packaging situation usually meets, the latter solution has exerted interesting effects due to simple and wide applications. Among the migratory agent categories, a further division would be made between controlled and uncontrolled release systems. Even though uncontrolled delivery packages intended for food applications are more abundant, controlled release systems are of industrial relevance due to their aptitude to prevent sensorial or toxicological problems or inefficiency of the system, caused by a too high or a too low concentration of delivered substance [8]. The active packaging technology provides several advantages compared to direct addition of active compounds, such as lower amounts of active substances required, localisation of the activity to the surface, migration from film to food matrix and elimination of additional steps within a standard process intended to introduce the active compounds at the industrial processing level such as mixing, immersion, or spraying. New regulations, the Commission Regulation (EC) No 450/2009 (EC, May 2009), and the Question Number EFSA-Q-2005-041 (EFSA, July 2009), together with Regulation 1935/2004 (EC, October 2004) make active packaging possible within the European Union [9].

This article seeks to provide an overview of the primary technology alternatives in packaging for the regulation of food quality. The emphasis will be on the utilization of appropriate headspace conditions and the implementation of active packaging technologies. Case studies are provided for primary food categories, including dairy products, meat, fish, fruits, and vegetables.

2. HEADSPACE CONDITIONS

Vacuum, gas flushing or controlled permeability of the pack are valid techniques to control biochemical, enzymatic and microbial degradations so as to avoid or decrease the main degradations that might occur in food. This allows the preservation of fresh state of the food product without temperature or chemical treatments used by competitive preservation techniques, such as canning, freezing, dehydration and other processes. MAP is the replacement of air in a pack with a single gas or mixtures of gases; the proportion of each component is fixed when the mixture is introduced. No control is exerted over the initial composition, and the gas composition is likely to change with time owing to the diffusion of gases into and out of the product, the permeation of gases into and out of the pack, and the effects of the product and microbial metabolism [10]. MAP was first recorded in 1927 as an extension of shelf life of apples by storing them in atmosphere with reduced oxygen and increased carbon dioxide concentrations. In 1930s it was used to transport fruit in the holds of ships. Increasing the carbon dioxide concentration surrounding beef carcasses transported long distances an increase in shelf life by up to 100% was shown [11]. Marks and Spenser introduced MAP for meat in 1979; the success of this product led, two years later, to the introduction of MAP for bacon, fish, sliced cooked meats and cooked shellfish. MAP techniques are now used on a wide range of fresh or chilled foods, including raw and cooked meats and poultry, fish, fresh pasta, fruit and vegetables and more recently coffee, tea and bakery products. The advantage of MAP for the consumer are:

- increased shelf life allowing less frequent loading of retail display shelves;
- reduction in retail waste;
- improved presentation-clear view of product and all round visibility;
- hygienic stackable pack, sealed and free from product drip and odor;
- easy separation of sliced products;
- little or no need of chemical preservatives;
- increased distribution area and reduced transport costs due to less frequent deliveries;
- centralized packaging and portion control;
- reduction in production and storage costs due to better utilization of labor, space and equipment.
- The disadvantages of MAP are:
- capital cost of gas packaging machinery;
- cost of gases and packaging materials;
- cost of analytical equipment to ensure that correct gas mixtures are used;
- cost of quality assurance systems to prevent distribution of leakers;
- increase of pack volume which will adversely affect transport costs and retail display space;
- benefits of MAP are lost once the pack is opened or leaks.

MAP does not increase significantly the shelf life of every type of food since some products that undergo processes such as smoking, curing, etc., already have extended shelf lives because of pre-packaging treatments. In these cases MAP may improve other quality aspects such as color stability or slice separation.

The safety and the stability of foods depend on microorganisms initially present, being unable to overcome various adverse factors, both extrinsic and intrinsic to the food. Modification of the atmosphere surrounding the food may provide one condition to inhibit microbial growth. In table 1 there is a description of the principal degradations that take part in a common food product and the role of MAP in contrasting these factors. The combination of chilled temperatures and MAP generally results in a more effective and safer storage regime and longer shelf life [12]. Atmospheres within the product are influenced by the type of material used in the package and the initial gas mixture used. Some materials allow diffusion of gases in and/or out of the package during storage while if the film is fully permeable, the atmosphere inside the pack becomes the same as the air outside. In semi-permeable films, the atmosphere in the pack rise to a gas equilibrium. Packs for MAP are made from one or more polymers: polyvinylchloride (PVC), polyethylene terephthalate (PET), polyethylene (PE), and polypropylene (PP), depending on the characteristics desired for the final use. There are many factors which must be taken into account:

- **barrier properties:** permeability to various gases and water vapor transmission rate;
- **machine capability:** capacity of trouble-free operation (resistance to tearing, possibility to be heat-formed)
- **sealing reliability:** ability to seal to itself and to container;
- **anti-fog properties:** good product visibility;
- **special characteristics:** possibility of heating without removing product from the packaging and easy-peel seals for convenient opening.

Table 1. Main degradation factors and roles of MAP

	Spoilage browning	Rancidity	Enzymatic
Effects on products	change in smell, taste, texture, appearance, toxicity	less nutritional value, rancid taste	browning of vegetables
Type of products	vegetables, bakery cooked products	products containing vitamins and fats	vegetables
Role of MAP	gas mixture mainly constituted by CO2	N2 or other neutral gas replacement of air	N2 or other neutral gas replacement of air

The main gases used in MAP are: oxygen, nitrogen and carbon dioxide. These three gases are used in different combination according to the product and the needs of manufacturer and consumer. The choice for a particular combination is influenced by the microbiological flora and the sensitivity of the product to gases and color stability requirements. The basic concept of MAP of fresh foods is the replacement of the air surrounding the food in the package with a mixture of atmospheric gases different in proportion from that of air. Oxygen is the most important gas being used by both aerobic spoilage microorganisms and plant tissues and taking part in some enzymatic reactions responsible for food deterioration. For these reasons, under MAP, oxygen is either excluded or set as low as possible. This gas is generally set at low levels to reduce oxidative deterioration of foods, particularly in high fat product. Oxygen generally stimulates growth of aerobic bacteria, inhibiting growth of anaerobic bacteria, although there is a wide variation in the sensitivity of anaerobes to this gas. The exceptions occur when oxygen is needed for fruit and vegetable respiration, color retention as in the case of red meat or to avoid anaerobic conditions in white fish [13]. One of the main function of oxygen is the maintenance of myoglobin in its oxygenated form, oxymyoglobin because this is the form responsible for the bright red color, which most consumers associate to fresh meat. Carbon dioxide is both water and lipid soluble and although is not bactericide or fungicide, it has a bacteriostatic and fungistatic properties. The effect on microorganisms consists in the extension of the lag phase and a decrease of growth rate. The effectiveness of this gas is influenced by its original and final concentrations, the storage temperature, the partial pressure of carbon dioxide, the initial bacterial population, the microbial growth phase, the growth medium used, the acidity, the water activity and the type of product being packaged [10, 14-16]. Yeasts which produce carbon dioxide during growth are stimulated by high levels of carbon dioxide and thus for some products where they are potentially a major cause of spoilage, MAP may not be an advisable option. Also Clostridium perfrigens and botulinum are not affected by the presence of carbon dioxide and their growth is encouraged by anaerobic conditions. In general, carbon dioxide is most effective in foods where the normal spoilage microorganisms consist of aerobic and gram negative psychrotropic bacteria [17]. For maximum antimicrobial effect, the storage temperature of the product should be kept as low as possible, because the solubility of carbon dioxide decreases dramatically with increasing temperature, thus improper temperature could eliminate the beneficial effects of carbon dioxide. The absorption of carbon dioxide is dependent on the moisture and fat content of the product. If product absorbs excess carbon dioxide the total volume inside the package will be reduced, giving a vacuum package look known as pack collapse. Excess carbon dioxide absorption in combination with package collapse can also reduce water holding capacity of meats, resulting in unsightly drip. Genigeorgis [18] suggested that the antimicrobial activity of carbon dioxide was a result of the gas being absorbed onto the surface of the product forming carbonic acid, subsequent ionization of the carbonic acid and a reduction in pH. Other theories have been summarized:

- alteration of cell membrane function including effects on nutrient uptake and absorption;
- direct inhibition of enzyme systems or decreases in rate of enzyme reactions;
- penetration of membranes resulting in changes of intracellular pH;
- direct changes to physic-chemical properties of proteins.

Nitrogen is an inert gas which has been used as a packaging filler for many years to prevent pack collapse because of its low solubility in water and lipid. In MAP products, especially fresh meat packed in high concentrations of carbon dioxide, pack collapse occurs because of the solubility of carbon dioxide in meat tissue. Nitrogen is also used to replace oxygen in MAP products, to prevent rancidity and inhibit growth of aerobic organisms. The gas combination depends on product characteristics. Table 2 reports a comparison between storage in MAP and in air for different products. Some relevant case- studies that highlight the benefits and limits of MAP for main food categories are reported hereinafter

Table 2. Examples of food shelf life under MAP and air

Product	Temperature	Shelf life under MAP	Shelf life in AIR
Toast bread	Room	2-3 months	10 days
Cake	Room	40-60 days	-a
Croissant, milk bread	Room	6 weeks	Several days
Pizza	4-5°C	30 days	Several days
Hamburger, hot dog rolls	4-5°C	30 days	1 week
Cakes with cream	Room	25-30 days	-
Emmenthal	2-4°C	4-5 weeks	A few days
Bovine mozzarella	2-4°C	6-8 days	3 days
Robiola, Crescenza	2-4°C	3-4 weeks	1 week
Cheese slices	2-4°C	2-3 months	2-3 months
Gorgonzola	2-4°C	30 days	10 days
Parmesan in pieces	2-4°C	40-60 days	-
a - means that air was never used to store product.			

The properties of meat that are important in determining shelf life include water binding capacity, color, microbial quality, lipid stability and palatability. The variables that influence the shelf life of packaged fresh meat are: product, package and headspace, packaging equipment, storage temperature, and additives. Plastic film properties, shrinkage, strength, oxygen transmission, moisture transmission, and anti-fog agents are important for meat package materials [19]. Fresh meat packaging is only minimally permeable to moisture to prevent desiccation, while gas permeability varies with the applications. MAP (commonly 70-80% O_2 and 20-30% CO_2) and vacuum packaging are widely used methods for packaging meat. Packaging under high oxygen concentration, however, may cause an increase in the lipid and protein oxidation. These reactions affect the functional, sensory and nutritional quality of meat products. Lipid oxidation leads to discoloration, increase drip-loss, off-odors and production of toxic compounds. In addition, these modifications can negatively affect the sensory quality of meat products in terms of texture, tenderness and color [20]. Protein oxidation can also result in the loss of enzyme activity and protein solubility, as well as in the formation of protein complexes and non enzymatic browning products. In a recent study [21], the effects of MAP (70% O_2 and 30% CO_2) and vacuum skin packaging on protein oxidation and texture of pork were investigated. Packaging under MAP containing high level of O_2 resulted in protein cross-linking, which reduced tenderness and juiciness of pork. Rowe et al. [20] proposed that the oxidation of muscle proteins may have a negative effect on beef tenderness that was attributed to an inactivation of μ-calpain with a subsequent decrease in proteolysis. Zakrys et al. [22]compared the effects of high levels of oxygen (80% O_2 20% CO_2) with vacuum packaging and showed that high O_2 levels lead to high myosin inter molecular cross-links, low free thiol groups and high carbonyl content, demonstrating that a significant level of protein oxidation occurred. This protein oxidation was found to have a negative effect on meat tenderness. Results from this study suggested that high oxygen induced changes in myosin and intermolecular cross-linking, increased disulphide bond formation, protein oxidation and drip loss compared to vacuum packaged. Color of meat is a very important quality attribute that influences consumer acceptance of meat. The surface color of meat depends on the quantity of myoglobin present, on its chemical state and also on the chemical and physical conditions of other components. Meat showing a bright red color is assumed to be fresh, while oxidation of heme iron to form methamyoglobin produces the brown color which consumers find undesirable. An interesting study conducted by Mastromatteo et al. [23] evaluated the combination of different MAPs (from 20% to 40% of CO_2 ; from 5% to 20% of O_2 and from 75% to 40% of N_2) with natural essential oils on shelf life of reduced pork back fat content sausages. They found that lemon and thymol recorded the highest sensory score while all the investigated MAPs showed an antimicrobial effect; moreover, low carbon dioxide concentrations caused low color variations during storage. The combination of MAP and thymol was able to further improve the shelf life of meat, in fact the microbial threshold was never reached. A shelf life of more than 5 days for thymol- MAP samples was obtained, respect to the other investigated samples (2 days). To sum up, integration of meat characteristics with available packaging materials, equipment into current cold chain logistical and information systems have resulted in a sufficiently high state of complexity that has caused uncertainty and confusion among industry, regulatory agency, and consumer segments [13]. Meat and packaging industry must continue to work on systems that will ensure safe and palatable products. The review of Belcher [24] well summarized packaging developments that are resulting from numerous trends taking place in the meat industry and in the retail sector. Moreover, alternative non-thermal preservation technologies such as high hydrostatic pressure, super chilling, natural biopreservatives and active packaging have been proposed because they are also effective against spores. To increase their efficacy, a combination of several preservation technologies under the so-called hurdle concept has to be investigated [25].

As regard fresh-cut fruits and vegetables, process increases respiration rate and causes major tissue disruption as enzymes and substrates normally sequestered within the vacuole. Processing also increases wound-induced ethylene, water activity and surface area per unit volume, which, may accelerate water loss and enhance microbial growth since sugars also become readily available. These physiological changes may be accompanied by flavor loss, cut surface discoloration, color loss, increased rate of vitamin loss, shrinkage and shorter shelf life. MAP is largely used for minimally processed fruit and vegetables. It relies on the modification of the atmosphere inside the package in order to extend the food shelf life by reducing the respiration of the product and consequently its degradation rate. The effect of MAP on quality of many fresh-cut products has been studied and successful applications include mushroom [26], apples [27], tomato [28], butterhead lettuce [29], potato [30], or kiwifruit [31]. These products are metabolically active for long periods after harvesting due to both endogenous activity, such as respiration, and external factors such as physical injury, microbial flora, water loss and storage temperature. Respiration may result in anaerobiosis, being quickly established if the produce is sealed in an impermeable film with low initial O2 concentration. Subsequently anaerobic respiration of the produce will be initiated at very low O2 concentrations, resulting in the accumulation of ethanol, acetaldehyde and organic acids and deterioration of organoleptic properties. Rate of respiration is influenced by the initial gas concentration so that, for example, reducing the oxygen to 2% and increasing the carbon dioxide concentration to 5%, results in more than 10-fold reduction of respiration rate in vegetables such as broccoli [32]. The maintenance of color is important and in red peppers, MAP has been shown to increase carotenoid retention and reduce browning [33]. The combination of storage time and temperature has been shown to be important in extending the shelf life of fruit in terms of texture, weight loss, pH and other nutritional changes. Temperature is also a factor in determining respiration rates of fruits. In freshly harvested beansprouts for example, which not only have a high respiration rate but also are characterized by high initial microbial populations, Varoquaux et al. [34], observed a 10-fold increase in respiration rate at 16.5 °C. In the case of fruit and vegetables with high respiration rates, there is an optimum initial atmosphere concentration to ensure minimal growth of aerobic spoilage bacteria together with an optimal film permeability to delay the development of anaerobic respiration and necrosis of vegetable. One of the obvious ways in which produce may be assessed for freshness is in terms of wilting and shriveling which are due to loss of moisture. Fruits and vegetables lose moisture when the relative humidity in the packaging is less than 80-95% of saturation and reduction in quality occurs if 3-6% of the produce moisture is lost [35]. Most films used for MAP of fruit and vegetables are relatively good water vapor barriers and are able to maintain a high relative humidity inside the pack. The relative humidity within a pack is influenced by the rate at which the product loses water vapor and by vapor transmission rate of the packaging film. Successful applications include broccoli florets, cauliflower florets, carrots, peeled garlic [36]. LDPE was found a good alternative to PVC for wrapping these vegetables. Comparative evaluation of the effect of storage temperature fluctuation on MAP of selected fruit and vegetables like mushrooms and mature green tomatoes was also studied by Tano et al. [37]. The quality of the products stored under temperature fluctuating regime was severely affected as indicated by extensive browning, loss of firmness, weight loss increase, ethanol level in plant tissue and infection, due to physiological damage and excessive condensation, compared to products stored at constant temperature. It was clear that temperature fluctuation can seriously compromise the benefits of MAP and safety of the product. Temperature is the most effective environmental factor in the prevention of fruit ripening. Both ripening and ethylene production rates increase with increase in temperature. To delay fruit ripening, temperature should be held as close to 0 °C as possible. The use of MAP as a supplement to proper temperature maintenance in the effort to delay ripening is effective for all fruits. Reducing oxygen concentration below 8% and/or elevating Carbon dioxide concentration above 1% retards fruit ripening. Successful applications include broccoli slaw, coleslaw, dry slaw, casserole mix, and mixed salads. Degradation of cut vegetables in terms of appearance was delayed by N2 gas packaging and vegetables remained acceptable at temperature below 5 °C after 5 days. MAP may have the effect of increasing shelf life of some vegetables in terms of sensory properties but does not reduce growth of some microorganisms such as L. monocytogenes and Salmonella Enterica. Therefore, the use of appropriate pre-harvest and postharvest sanitation practices to prevent contamination remains the most important measure for ensuring the microbiological safety of ready-to-eat fresh-cut products.

Shelf life of milk and milk-based products is limited because of their high water content and favorable pH for microbial growth [38-40]. The rapid spoilage adversely affects flavor and texture along with visual color changes of refrigerated raw and pasteurized milk, cottage cheese and other similar products. The responsible microorganisms include psychrotropic Gram negative bacteria, yeasts and moulds. These organisms produce extracellular protease and lipase activity, which reduce the functionality of milk proteins and fat and often produce undesirable aromas. Gram positive bacteria particularly those producing lactic and acetic acids, can spoil dairy foods, but the number of organisms required are generally higher than for Gram-negative bacteria, and the changes can be less noticeable. It has been reported that the product shelf life increases by low oxygen atmospheres because of the reduction in aerobic microorganisms. The antimicrobial effect of Carbon dioxide occurs near 10% level, and further increase in carbon dioxide affects growth of Pseudomonas and Moraxella. The largest inhibition by carbon dioxide occurs with Gram negative psychrotrophs bacteria [41]. The protective role of carbon dioxide is also important for mould proliferation; its function in creating an anaerobic environment with the displacement of existing molecular oxygen, its extra and intracellular pH decreasing effect and its destroying effect on the cell membrane make carbon dioxide an inhibitory substance towards microorganisms. The antimicrobial effect of carbon dioxide is dependent on many factors, including the partial pressure, application time, concentration of gas, temperature of the medium [42], volume of headspace, acidity, water activity of the medium and type of organism present [43]. The applied composition for packaging of dairy products can vary from 10% to 100% carbon dioxide, balanced with N2 as inert gas filler, to prevent package collapse as a result of carbon dioxide solubilization in the cheese. MAP has been applied to the packaging of cheese. The packaging of each type of cheese needs to be consider separately. Another fact to be considered is that some cheeses are carbon dioxide producers, while other not. It is important that the levels of carbon dioxide are controlled because for certain cheese high levels of carbon dioxide have been found to impart off-flavor [44-46]. Cheese stored under carbon dioxide contained high concentrations of aldehydes and fatty acids and lower concentrations of alcohols and esters than cheeses stored under nitrogen. Hard and semi-soft cheeses, such as cheddar, are commonly packed in 100% carbon dioxide or mixtures of carbon dioxide and nitrogen. Soft cheese have also a limited shelf life. An alternative to conventional packaging is to use MAP. Carbon dioxide acts both directly on moulds and indirectly by displacing oxygen. Vacuum packaging does not remove all of the oxygen and thus, moulds and yeasts can still occur [47]. The gas mixture typically used is 70% N2 and 30% CO2 to inhibit mould growth, to keep the package from collapsing and to prevent shred matting. Alves et al. [48] reported that atmospheres ≥ 50% carbon dioxide were more effective than air or 100% nitrogen in improving shelf life of sliced mozzarella cheese. High carbon dioxide atmospheres have been shown to inhibit growth of lactic and mesophilic bacteria [49]. Piergiovanni et al. [50] compared Taleggio cheese packaged under four modified atmospheres and stored at 6 °C to conventional paper wrapping and found that samples packaged in MAP had satisfactory quality. Gammariello et al. [51] evaluated the shelf life of Stracciatella cheese packaged in four different gas mixtures at 8 °C and showed that MAP 50:50 and 95:5 (O2:CO2) prolonged the sensorial acceptability limit by delayed growth of spoilage bacteria, without affecting the dairy microflora. Del Nobile et al. [52] suggest that MAP of Ricotta with 95% carbon dioxide inhibits microbial growth without effects on lactic acid bacteria, probably due to their facultative anaerobic nature, and also maintains the natural color of Ricotta.

MAP found wide application also for fresh fish. Fish such as herring and haddock benefits from being packaged under MAP since this reduces the production of peroxides which affect fish sensory characteristics and hence shelf-life [53]. However, high levels of carbon dioxide may result in carbon dioxide dissolution into the fish flesh, causing deformation or collapse of the packaging and also affecting the product color. The resulting drop in pH of the tissue may cause a decrease in the flesh's water holding capacity and drip may occur, reducing shelf-life. Fresh hake stored in up to 60% carbon dioxide exerted a shelf life significantly longer than those stored in air. MAP inhibits bacterial growth, reduces the formation of total volatile bases and trimethylamine and delays alterations in protein functionality [54]. In cooked fish such as smoked blue cod and smoked atlantic and silver salmon, a high concentration of carbon dioxide increases shelf life without showing drip or muscle exudate observed in fresh fish. carbon dioxide extents fish shelf life due to the inhibition of Gram negative and lactic acid bacteria. carbon dioxide concentrations in all MAP fishery products should be carefully monitored, especially when stored for long periods of time, because carbon dioxide does not inhibit C. botulinum and the effect of temperature abuse may increase the risk of botulism in those products which contain spores of non-proteolytic C. botulinum [55, 56].

3. ACTIVE PACKAGING

Active packaging has been classified as a subset of smart packaging and referred to as the incorporation of certain additives into packaging film or within packaging containers with the aim of maintaining and extending product shelf life [57-59]. Another definition states that packaging may be termed active when it performs some desired role in food preservation other than providing an inert barrier to external conditions [60]. Hence, active packaging includes components of packaging systems that are capable of scavenging oxygen; absorbing carbon dioxide, moisture, ethylene and/or flavor/odor taints; releasing carbon dioxide, ethanol, antioxidants and/or other preservatives; and/or maintaining temperature control and/or compensating for temperature changes (Table 3). In the food and beverage market, growth of active packaging concepts is being driven by the growing use of packaged food, increasing demand for ready-prepared foods such as microwave meals, and increasing use of smaller package sizes. Although many active packaging technologies are still developmental, there are commercial successes, particularly in oxygen scavengers. Oxygen scavengers are easily oxidizable substances included in the packaging system to remove oxygen by means of a chemical reaction. The substance is usually contained in sachets made of a material highly permeable to air but it can also be included in bottle closures or in the plastic film matrix. Different studies show that the use of scavengers led to faster reduction and to lower levels of residual oxygen, as compared to nitrogen flushing. The most common substances used are iron powder and ascorbic acid. Scavengers also differ in the reaction speed, from immediate action (0.5 to 1 day) to slow action (4 to 6 days), on the application, particularly the moisture content of the food, and on the function, i.e., oxygen scavenging only or dual function, such as absorbing or generating carbon dioxide, besides removing the oxygen. The scientific literature contains a number of references which examine the influence of oxygen scavenger sachets on fresh beef discoloration. Gill and McGinnis [61] performed an oxygen absorption kinetics study with a commercial oxygen scavenger (FreshPaxTM 200R) and reported that discoloration could be prevented in ground beef if large numbers of scavengers were used in each pack to bring residual oxygen to <10 ppm within 2 h at a storage temperature of -1.5 °C. The inclusion of oxygen scavengers (Ageless® SS200) in master packs flushed with 50% carbon dioxide and 50% nitrogen significantly improved color stability of M. longissimus dorsi and M. psoas major, relative to controls [62]. In addition to fresh beef oxygen scavenging technology has also been applied to pork [63] and pork products, where, Martìnez, Djenane, Cilla, Beltràn, and Roncalès [64] reported that fresh pork sausages stored in 20% CO2 and 80% N2 plus an oxygen scavenger (Ageless® FX-40) for up to 20 days at 2 ± 1 °C reduced psychrotrophic aerobe counts and extended shelf life in terms of color and lipid stability. An alternative to sachets involves the incorporation of the oxygen scavenger into the packaging structure itself. This UV light-activated oxygen scavenging film, which structurally is composed of an oxygen scavenger layer extruded into a multilayer film, can reduce headspace oxygen levels from 1% to ppm levels in 4–10 days, compared to oxygen scavenging sachets. The OS2000TM scavenging films found applications in a wide variety of food products including dried or smoked meat products and processed meats [65]. Berenzon and Saguy [66] evaluated the applicability of oxygen absorbers for extending shelf life of military ration crackers packaged in hermetically sealed tin cans and stored at 15, 25 and 35 °C for up to 52 weeks. Sensory evaluations suggested that crackers stored without oxygen absorbers developed oxidative rancid odors after 24 weeks at 25 and 35 °C.

Independently of storage temperatures, no oxidative rancid odors were observed after 44 weeks with oxygen absorbers. Opposed to the currently available chemical oxygen scavengers, systems based upon natural and biological components could have advantages towards consumer perception and sustainability [67]. A model system for a new oxygen scavenging poly(ethylene terephthalate) (PET) bottle is proposed using an endospore-forming bacteria genus [68]. Incorporated spores could actively consume oxygen for minimum 15 days, after an activation period of 1–2 days at 30 °C under high humidity conditions. Although the system shows some clear opportunities, such as being a biological based system and its capability of solving polymer compatibility and recyclability issues, towards the current chemical systems, further investigations are necessary to determine a possible interaction between spores and food product.

Table 3. Examples of active packaging systems

Active packaging	Action mechanisms	Food applications
Oxygen scavengers	Slowed food metabolism, reduced oxidative rancidity, inhibited undesirable oxidation of labile pigments and vitamins, controlled enzymatic discoloration, inhibited growth of aerobic microorganisms.	Bread, cakes, cooked rice, biscuits, pizza, pasta, cheese, cured meats and fish, coffee, snack foods, dried foods and beverages.
Carbon dioxide scavengers/ emitters	Slowed respiration rate, inhibited microbial growth.	Coffee, fresh meats and fish, nuts and other snack food products and sponge cakes.
Ethylene scavengers	Slowed respiration rate, thus slowed softening and ripening	Fruit, vegetables and other horticultural products.
Preservative releasers	Antimicrobial and antioxidant effect.	Cereals, meats, fish, bread, cheese, snack foods, fruit and vegetables.
Ethanol emitters	Effective against mould, can also inhibit the growth of yeasts and bacteria.	Pizza crusts, cakes, bread, biscuits, fish and bakery products.
Moisture absorbers	Inhibited microbial growth and moisture related degradation of texture and flavor	Fish, meats, snack, cereals, dried foods, sandwiches, fruit and vegetables.
Flavor/odor absorbers	Malodorous constituents causing off-flavors are absorbed.	Fruit juices, fried snack foods, fish, cereals, poultry, dairy products and fruit.
Temperature control	Able to maintain chilled temperature.	Ready meals, meats, fish, poultry and beverages.
Temperature compensating	Gas permeability responding to temperature changes to avoid anoxic conditions.	Fruit, vegetables and other horticultural products.

Another important example of scavenger is the packaging with ethylene scavenger property. Ethylene has long been recognized as a problem in post-harvest handling of horticultural products because it is responsible for a wide variety of undesirable effects: it accelerates the respiration of fruits and vegetables, as well as softening and ripening, and it is responsible for a number of specific post-harvest disorders. The removal of this gas from storage chambers and packages of fruits and vegetables is, therefore, of the utmost importance, and it is done as a regular practice in the case of chambers, although is only more recently done in the case of removal from a single package. Ethylene is a very reactive compound that can be altered in many ways, such as chemical cleavage and modification, absorption, adsorption, etc. This creates a diversity of opportunities for commercial applications for the removal of ethylene [69]. Most substances designed to remove ethylene from package are delivered either as sachets that go inside the package or are integrated into the packaging material, usually a plastic polymer film. The most commonly used are based in potassium permanganate, activated carbon and activated earth. Meyer & Terry [70] studied the effect of 1-methylcyclopropene (1-MCP) and a newly developed palladium (Pd)-promoted ethylene scavenger (e + ®Ethylene Remover) on changes in firmness, color, fatty acids and sugar content of early and late season avocado (Persea americana Mill.), cv. Hass, during storage at 5 °C and subsequent ripening at 20 °C. Results have shown that the ®Ethylene Remover is effective at delaying ripening of avocado at low temperature, similarly to 1-MCP; however, subsequent ripening was not impaired. Similarly, but to a lesser extent and concomitant with trends in firmness retention and color changes, ®Ethylene Remover led to greater maintenance of mannoheptulose and perseitol than that of controls. Initial findings have demonstrated for the first time that the presence of a palladium-based scavenger was effective at removing ethylene to below physiologically active levels for preclimacteric green bananas and green avocado fruits. Reduced carbon dioxide production and control of color change from green to yellow was observed for the preclimacteric bananas. Results suggested that the normal and expected climacteric respiratory rise has been disrupted. Therefore, for the first time an ethylene scavenger has been shown to be capable of extending shelf life even when the climacteric respiratory rise has already been initiated [71].

In order to suppress spoilage and remove offensive odors in fresh products carbon dioxide absorbers are used. On the other hand, carbon dioxide emitters are useful in modified atmosphere packaging, because carbon dioxide suppresses the bacteria that cause spoilage. Both carbon dioxide generators and absorbers are available in sachet format [72].

Ethanol emitters are particularly effective in extending the shelf life of high water activity baked products. The use of ethanol generating sachets or strips avoids the ethanol spraying directly onto the product surface prior to packaging [60].

The ethanol is absorbed or encapsulated in a carrier material enclosed in sachets of selective permeability to ethanol to allow for ethanol accumulation in the headspace. The level of ethanol in the packaging headspace depends obviously on the sachet size and on product water activity.

Fragrances incorporated in packaging also found commercial use in food, personal care, pharmaceutical, and nutraceutical packaging. In food packaging, fragrances are being used as a marketing tool to create consumer awareness and to enhance brand image. Because cyclodextrins are able to form inclusion complexes with various compounds, they present a potential interest as agents to retain or scavenge substances such as odors, bitter compounds, lactose, cholesterol, etc., or to add aromas, colors, or functional ingredients whose release could enhance the quality of the packaged product and extend its shelf life [73-75].

Control of moisture is also important for food preservation. In most cases, the packaging material itself is responsible for the control of moisture transfer between the internal and external environment, providing an adequate barrier. There are situations however, where a greater control is needed to avoid the build-up of liquid water inside the package, therefore requiring liquid water control or humidity buffering as in the case of transpiration of fresh produce, melting of ice in fish transportation, temperature fluctuation in high water activity food packages and drip of tissue fluid from cut meats and produce [60]. To this aim, absorbent pads or sheets, anti-fog additives in the polymer film, humectant between two layers of a highly water vapor permeable film or sachets of inorganic desiccant salts, are generally used to accomplish liquid removal or humidity buffering.

Antioxidant food packaging films were produced by incorporation of ascorbic acid, ferulic acid, quercetin, and green tea extract into an ethylene vinyl alcohol copolymer matrix. The efficiency of the films developed was determined by real packaging applications of brined sardines. The evolution of the peroxide index and the malondialdehyde content showed that, in general, the films improved sardine stability. Films with green tea extract offered the best protection against lipid oxidation [76]. A natural citrus extract was also sprayed onto the surface of polyethylene terephthalate trays to delay lipid oxidation of cooked turkey meat slices, stored at 4 °C over 4 days [9]. The high surface roughness, demonstrated by optical profilometry, and the high level of solubility of the antioxidant in water allowed a good effectiveness of the citrus extract coating. Patties made of minced chicken breast and thigh packed in standard vacuum-packaging or in antioxidant active packaging containing rosemary extract were subjected to high pressure treatment (800 MPa, 10 min, 5 °C) and subsequently stored at 5 °C. The active packaging was able to delay surface lipid oxidation up to 25 days. The migration of α-tocopherol from a multilayer active packaging made up of high density polyethylene, ethylene vinyl alcohol and a layer of low density polyethylene containing the antioxidant, was studied. The antioxidant delivering system delayed the lipid oxidation of whole milk powder and it was more effective at temperatures higher than 20 °C [77].

The most investigated active systems are the packaging with antimicrobial properties, even if there has been little commercial activity in North America or Europe. Japan has historically been a leader in antimicrobial use. To date the literature counts various research works and review papers dealing with advances in antimicrobial packaging. Generally specking most of them are focused on in vitro test and can be subdivided in various categories: (i) there are studies aimed to develop new active systems with recently exposed natural compounds; (ii) studies finalized to underline the release kinetic of the active agents from the matrix to the food, with the intent to realize controlled release systems; (iii) studies aimed to develop bio-based active systems and (iv) works that use the nanotechnology approach. Among the abundant list of research articles and reviews available in the scientific literature on this topic, some recent works have been selected. Kanatt et al. [78] studied active films of chitosan and polyvinyl alcohol containing aqueous mint extract/pomegranate peel extract. Ramos et al. [79] studied antimicrobial active films based on polypropylene (PP) and containing thymol and carvacrol at three different concentrations. Trans-2-hexenal was encapsulated into β-cyclodextrins and incorporated into a poly(L-lactic acid) matrix by extrusion and casting [75]. A newly sinthetized polyester (poly-butylene adipate) containing covalently bound quaternary phosphonium groups was developed by Anthierens et al. [80]. The resulting polyester showed great antimicrobial activity through direct contact without any migration of active groups. Among the bio-based active systems, various bio-active packaging were developed to control insect pest in granary weevils [81, 82]. The efficacy of edible films produced from whey protein isolate and glycerol, including incorporation of lactic acid, propionic acid, chitooligosaccharides and natamycin was assessed by Ramos et al. [83]. Suppakul et al. [84] studied the diffusion of linalool and methylchavicol from thin antimicrobial low-density polyethylene-based films. Cellulose acetate-based mono and multilayer films including potassium sorbate were prepared using dry phase inversion technique [85].

Monolayer films, prepared using powdered cellulose and poly(vinyl) alcohol were coated with cellulose membrane to obtain multilayer films and sorbic acid was incorporated as antimicrobial agent [86]. Cerisuelo et al. [87] studied by a mathematical model the release of carvacrol from an ethylene-vinyl alcohol coating on a polypropylene film while Del Nobile et al. [88] studied the release of thymol from zein-based films. Bierhalz et al. [89] studied release behavior in water and diffusion coefficients of single and composite films based on alginate and pectin containing natamycin. Organic/inorganic compounds, essential oils, bacteria originated antibacterial proteins, enzymes and fruit extracts have shown great potential in inhibiting microbial growth in food stuff. However, the development of new resistant strains of bacteria to current antibiotics has led to the search for new bactericides that can effectively reduce the harmful effects of microorganisms. With the emergence of nanotechnology, the search for effective biocidal agents has focused on the development of nanostructures of coinage metals like silver, copper, zinc and gold [90]. ZnO nanoparticles loaded starch-coated polyethylene film were developed by Tankhiwale and Bajpai [91]. Montmorillonite nanoclay and rosemary essential oil were incorporated into chitosan film to improve its physical and mechanical properties as well as antimicrobial and antioxidant behavior [92]. Silver nanoparticles (AgNPs) have been abundantly exploited for technological applications as bactericidal agents. Recently, AgNPs were incorporated with success into bio- based materials [93] and into a hydroxy-propyl methylcellulose matrix [94]. Although numerous antimicrobial systems continue to be investigated in food simulating models, real applications are limited by technical, aesthetic and regulatory barriers. To this regards, a few recent examples can be cited. Microencapsulated beta-cyclodextrin and trans-cinnamaldehyde complex was incorporated into a multilayered edible coating made of chitosan and pectin to coat fresh-cut papaya that was then packaged in Ziploc trays with Ziploc lids for 15 days. The layer-by-layer assembly with incorporation of microencapsulated antimicrobial was effective in extending shelf life and quality of fruit [95]. The antimicrobial proteins lysozyme and lactoferrin were incorporated into paper containing carboxymethyl cellulose [96]. The antimicrobial activity on common food contaminants was also retained in the released protein, and a synergism between the two proteins was evident in tests carried out with paper containing both proteins. Lysozyme was most effective in preventing microbial growth when the system was applied to thin meat slices laid on paper sheets containing either or both antimicrobial proteins. Cellulose/silver nanocomposites were investigated to decrease the microbial loads in minimally processed foods and meat [97]. The active systems were synthesized by means of reduction by UV/heat of silver nitrate adsorbed on fluff pulp cellulose fibres. Minimally processed fruits and meat products were packaged in trays containing commercial absorbent pads or silver loaded absorbers and in contact with silver loaded absorbers, spoilage counts were significantly reduced. Active packaging based on silver nanoparticles, obtained by allowing silver ions from nitrate solutions to replace the Na+ of natural montmorillonite and then reduced by a thermal treatment, were applied to fruit salad [98] and fresh dairy products [99, 100]. The striking feature of these works is the interesting antimicrobial effects, without compromising sensory properties. The antimicrobial effectiveness is usually complicated by several factors, including temperature, moisture levels, chemistry of the antimicrobial agent and release mechanism. Moreover, it is necessary to consider odor or color change that an antimicrobial could provoke in the packaged product. The cost-benefit ratio of antimicrobials is also a limiting factor for commercial growth and rates of return in the food industry are small. All these considerations explain the limited diffusion of active systems, although several antimicrobials have been successful in the laboratory. Applications with good potential are value added products such as pre-sliced and prepared foods.

4. CONCISE SUMMARY

Packaging design is an essential component of a product introduction. Given the significance of packaging in influencing product shelf life, it is essential to regard product development and its packaging system with equal consideration. The essential factor for effective packaging is the choice of materials and designs that optimally reconcile the conflicting demands of product attributes, marketing factors such as distribution and consumer requirements, environmental and waste management concerns, and cost. Food packaging technologies necessitate integration with additional processing and preservation methods, including freezing, irradiation, pulsed electric fields, high-pressure processing, and pulsed light. The pressures of globalization, packaging life cycles, and stringent safety regulations are driving the development of innovative packaging methods that facilitate the transportation of food items while ensuring traceability across the food distribution chain. Given the variety of product attributes and fundamental food packaging requirements and applications, any packaging technologies that provide enhanced product delivery and quality control in a cost-effective and versatile manner would be highly regarded. To address future challenges, extensive research persists in evaluating domains such as active packaging, traceability, sustainable resources, and antimicrobial packaging. Progress in these domains will perpetuate a secure and sustainable food supply. The application of modified atmosphere technology can prolong shelf life. Its application does not obviate the necessity for appropriate regulation of storage conditions, particularly temperature, nor for sufficient training of handlers regarding the sensory attributes and shelf life of various food products, thereby preventing the proliferation of dangerous germs. MAP will persist in future applications, likely with many MAP formats utilized globally. Mechanistic, logistical, and perceptual barriers will necessitate effort and imagination to surmount current packaging and system challenges and facilitate the adoption of novel processing and packaging technologies. Furthermore, the integration of antibacterial and antioxidant chemicals within packaging to mitigate spoilage and microbial proliferation in food will significantly enhance shelf life and food safety. While the evidence indicates that active packaging is a promising technology, its potential cannot be completely actualized unless significant technological challenges are addressed. Further study on the regulation of active agent migration at rates appropriate for various real food systems is necessary. The acknowledgment of the advantages of active packaging technologies by the food sector, the creation of economically feasible packaging solutions, and heightened consumer acceptance present new opportunities for active packaging technology.

REFERENCES

1. da Cruz AG, Faria JAF, Van Dender AGF. Packaging systems and probiotic dairy foods. Food Research International 2007; 40 951–956.
2. Kotsianis IS, Giannou V, Tzia C. Production and packaging of bakery products using MAP technology. Trends in Food Science & Technology 2002; 13 319–324.
3. Sanchez-Garcia MD, Lopez-Rubio A, Lagaron MJ. Natural micro and nanobiocomposites with enhanced barrier properties and novel functionalities for food biopackaging applications. Trends in Food Science & Technology 2010; 21 528-536.
4. Zi-Xuan L, Zhong-Su M, Jing W, Huan L. Preparation and characterization of immobilized lysozyme and evaluation of its application in edible coatings. Process Biochemistry 2012; 47 201–208.
5. Aider M. 2010 Chitosan application for active bio-based films production and potential in the food industry: Review. LWT - Food Science and Technology 2010; 43 837– 842
6. Sorrentino A, Gorrasi G, Vittoria V. Potential perspectives of bio-nanocomposites for food packaging applications. Trends in Food Science & Technology 2007; 18 84-95.
7. Cutter CN. Microbial control by packaging: a review. Critical Reviews in Food Science and Nutrition 2002; 42 151–161.
8. Mastromatteo M, Mastromatteo M, Conte A, Del Nobile MA. Advances in controlled release devices for food packaging applications. Trends in Food Science & Technology 2010; 21 591-598.
9. Bolumar T, Andersen ML, Orlien V. Antioxidant active packaging for chicken meat processed by high pressure treatment. Food Chemistry 2011; 129 1406–1412
10. Churc N. Developments in modified atmosphere packaging and related technologies. Trends in Food Science and Technology 1994; 5 345- 352.
11. Davies AR. Advances in Modified Atmosphere Packaging, New Methods of Food Preservation, ed. by G.W. Gould; 1995 p304-320, Glasgow, UK, Blackie.

12. Leistner L. Principles and applications of hurdle technology. In: New Methods of food Preservation (ed.) G.W. Gould. 1995. p1-21. Glasgow, UK, Blackie.

13. McMillin KW. Where is MAP Going? A review and future potential of modified atmosphere packaging for meat. Meat Science 2008; 80 43–65.

14. Farber JM. Microbiological aspects of modified-atmosphere packaging technology-A Review. Journal of Food Protection 1993; 54(1) 58-70.

15. Phillips CA. Review: Modified atmosphere packaging and its effects on the microbiological quality and safety of produce, International Journal of Food Science and Technology 1996; 31 463-479.

16. Church IJ, Parsons AL. Modified atmosphere packaging technology: A Review. Journal of the Science of Food and Agriculture 1995; 67 143-152.

17. Hotchkiss JH. Microbiological Hazards of Controlled/Modified Atmosphere Food Packaging. Journal of the Science of Food and Agriculture 1989; 53(3) 41-49.

18. Genigeorgis C. Microbial and safety implications of the use of modified atmospheres to extend the storage life of fresh meat and fish. International Journal of Food Microbiology 1985; 1 237-251.

19. Smith BS. The maturation of case-ready technology operational challenges. In Meat industry research conference pag117-118, 15-17 October 2001, Chicago, Illinois, USA.

20. Rowe LJ, Maddock KR, Lonergan SM, Huff-Lonergan E. Influence of early post-mortem protein oxidation on beef quality. Journal of Animal Science 2004; 82 785-793.

21. Lund MN, Lametsch R, Hviid MS, Jensen ON, Skibsted LH. High oxygen packaging atmosphere influences protein oxidation and tenderness of porcine longissimus dorsi during chill storage. Meat Science 2007; 77 295-303.

22. Zakrys-Waliwander PI, O'Sullivan MG, O'Neill EE, Kerry JP. The effects of high oxygen modified atmosphere packaging on protein oxidation of bovine M. longissimus dorsi muscle during chilled storage. Food Chemistry 2012; 131 527-532.

23. Mastromatteo M, Incoronato AL, Conte A, Del Nobile MA. Shelf life of reduced pork back fat content sausages as affected by antimicrobial compounds and modified atmosphere packaging. International Journal of Food Microbiology 2011; 150 1-7.

24. Belcher J.N. Industrial packaging developments for the global meat market. Meat Science 2006; 74 143–148.

25. Zhou GH, Xu XL, Liu Y. Preservation technologies for fresh meat – A review. Meat Science 2010; 86 119–128.

26. Simon A, Gonzales-Fandos E, Tobar V. The sensory and microbiological quality of fresh sliced mushroom (Agaricus bisporus L.) packaged in modified atmospheres. International Journal of Food Science and Technology 2005; 40 943.

27. Soliva-Fortuny RC, Ricart-Coll M, Martin-Belloso O. Sensory quality and internal atmosphere of fresh cut Golden Delicius apples. International Journal of Food Science and Technology 2005; 40 369.

28. Aguayo E, Escalona V, Artes F. Quality of fresh cut tomato as affected by type of cut, packaging, temperature and storage time. European Food Research and Technology 2004; 219 492-499.

29. Escalona VH, Verlinden BE, Geysen S, Nicolai BM. Changes in respiration of fresh cut butter head lettuce under controller atmosphere using low and super atmospheric oxygen conditions with different carbon dioxide levels. Postharvest Biology and Technology 2006; 39 48-55.

30. Beltran D, Selma MV, Tudela IA, Gil MI. Effect of different sanitizers on microbial and sensory quality of fresh cut potato strips stored under modified atmosphere or vacuum packaging. Postharvest Biology and Technology 2005; 37 37-46.

31. Rocculi P, Romani S, Rosa MD. Effect of MAP with argon and nitrous oxide on quality maintenance of minimally processed kiwifruit. Postharvest Biology and Technology 2005; 35 319-328.

32. Zagory D, Kader AA. Modified atmosphere packaging of fresh produce. Food Technology 1988; 42 7-77.

33. Lee DS, Chung SK, Yam KL. Cerotenoid loss in dried red pepper products. International Journal of Food Science and Technology 1992; 27 179-185.

34. Varoquaux P, Albagnac G, The CN, Varoquaux F. Modified atmosphere packaging of fresh beansprouts. Journal of the Science of Food and Agriculture 1996; 70 224-230.

35. Day BPF. Fruit and vegetables. In: Principles and Applications of Modified Atmosphere Packaging (ed.) Parry RT. 1993: 114-133. Glasgow, UK: Blackie.

36. Lee DS, Kang JS, Renault P. Dynamics of internal atmospheres and humidity in perforated packages of peeled garlic cloves. International Journal of Food Science and Technology 2000; 37 255.

37. Tano K., Oulé M. K., Doyon G., Lencki R. W., Arul J. Comparative evaluation of the effect of storage temperature fluctuation on modified atmosphere

packages of selected fruit and vegetables. Postharvest Biology and Technology 2007; 46 212–221.

38. Muir DD. The shelf life of dairy products. 1. Factors influencing raw milk and fresh products. Journal of Society of Dairy Technology 49: 1996a; 24-32.
39. Muir DD. The shelf life of dairy products. 2. Raw milk and fresh products. Journal of Society of Dairy Technology 1996b; 49 44-48.
40. Muir DD. The shelf life of dairy products. 3. Factor influencing intermediate and long life dairy products. Journal of Society of Dairy Technology 1996c; 49 67-72.
41. Hendricks A, Hotchkiss JH. Effect of CO2 on Pseudomonas fluorescens and Listeria monocytogenes growth in aerobic atmospheres. Journal of Food Protection 1997; 60 1548-1552.
42. Blickstad E, Enfors SO, Molin G. Effect of hyperbaric CO2 pressure on the microbial flora of pork stored at 4°C-14°C. Journal of Applied Bacteriology 1981; 50 493-504.
43. Davidson PM, Juneja VK. Antimicrobial agents. In Food Additives 8ed.) Branene, A.L., Davidson, P.M., Salminem, S. New York, NY, USA: Marcel Dekker. 1990: 83-87
44. Mannheim CH, Soffer T. Shelf life extension of cottage cheese by modified atmosphere packaging. LWT-Food Science and Technology 1996; 29 767-771.
45. Gonzalas-Fandos E, Sanz S, Olarte C. Microbiological, physicochemical and sensory characteristics of Cameros cheese packaged under modified atmospheres. Food Microbiology 2000; 17 407-414.
46. Trobetas A, Badeka A, Kontominas MG. Light induced changes in grated Graviera cheese packaged under modified atmospheres. International Dairy Journal 2008; 18 1133-1139.
47. Hocking AD, Faedo M. Fungi causing thread mold spoilage of vacuum packaged cheddar cheese during maturation. International Journal of Food Microbiology 1992; 16 123-130.
48. Alves RMV, Sarantopoulos CIGDL, Dender AGFV, Faria JDAF. Stability of sliced mozzarella cheese in modified atmosphere packaging. Journal of Food Protection 1996; 59 838-844.
49. Eliot SC, Vuillemard JC, Emond JP. Stability of shredded Mozzarella cheese under modified atmosphere. Journal of Food Science 1998; 63 1075-1080.
50. Piergiovanni L, Fava P, Moro M. Shelf life extension of Taleggio cheese by modified atmosphere packaging. Italian Journal of Food Science 1993; 2 115-127.

51. Gammariello D, Conte A, Di Giulio S, Attanasio M, Del Nobile MA. Shelf life of stracciatella cheese under modified atmosphere packaging. Journal of Dairy Science 2008; 92 483-490.

52. Del Nobile MA, Conte A, Incoronato AL, Panza O. Modified atmosphere packaging to improve microbial stability of Ricotta. African Journal of Microbiology Research 2009; 3 137-142.

53. Dhananjaya S, Stroud GD. Chemical and sensory changes in haddock and herring stored under modified atmosphere. International Journal of Food Science and Technology 1994; 29 575-583.

54. Pastoriza L, Sampero G, Herrera JJ, Cabo ML. Effect of modified atmosphere packaging on shelf-life of iced fresh hake slices. Journal of Food and Agriculture 1996; 71 541-547.

55. Ashie INA, Smith JP, Simpson BK. Spoilage and shelf life extension of fresh fish and shell-fish. Critical Reviews in Food Science and Nutrition 1996; 36 21-111.

56. Reddy NR, Armstrong DG, Rhodehamel EJ, Kautter DA. Shelf life extension and safety concerns about fresh fishery products packaged under modified atmospheres: a review. Journal of Food Safety 1992; 12 87-118.

57. Conte A., Buonocore G.G., Bevilacqua A., Sinigaglia M., Del Nobile M.A. Immobilization of lysozyme on polyvinylalcohol films for active packaging applications. Journal of Food Protection 2006; 4(69) 866-870.

58. Day BPF. Active packaging – a fresh approach. Journal of Brand Technology 2001; 1 (1) 32–41.

59. Day B.P.F. Active packaging. In Coles R., McDowell D., Kirwan M. (eds) Food Packaging Technologies Boca Raton, FL, USA, pp. 282–302. (2003).

60. Rooney, M.L. (ed.) (1995) Active Food Packaging. Chapman & Hall, London, UK.

61. Gill CO, McGinnis JC. The use of oxygen scavengers to prevent the transient discoloration of ground beef packaged under controlled, oxygen-depleted atmospheres. Meat Science 1995; 41 19–27.

62. Allen P, Doherty AM, Buckley DJ, Kerry J, O'Grady MN, Monahan FJ. Effect of oxygen scavengers and vitamin E supplementation on color stability of MAP beef. In: Proceedings 42nd international congress of meat science and technology 1996. pp. 88– 89 1–6 September 1996, Lillehammer, Norway.

63. Doherty AM, Allen P. The effect of oxygen scavengers on the colour stability and shelf life of CO2 packaged pork. Journal of Muscle Foods 1998; 9 351–363.

64. Martìnez L, Djenane D, Cilla I, Beltràn JA, Roncalès P. Effect of varying oxygen concentrations on the shelf-life of fresh pork sausages packaged in modified atmosphere. Food Chemistry 2006; 94 219–225.

65. Butler B. L. Polymeric oxygen scavenging systems. 2002 http://www.sealedair.com/ library/articles/article-os2000.html.

66. Berenzon S., Saguy I. S. Oxygen Absorbers for Extension of Crackers Shelf-life. Lebensmittel Wissenschaft und Technologie 1998; 31 1–5.

67. Altieri C., Sinigaglia M., Corbo M. R., Buonocore G. G., Falcone P., Del Nobile M. A. Use of entrapped microorganisms as biological oxygen scavengers in food packaging applications. Lebensmittel Wissenschaft und Technologie 2004; 37(1) 9–15.

68. Anthierens T., Ragaert P., Verbrugghe S., Ouchchen A., G. De Geest B., Noseda B., Mertens J., Beladjal L., De Cuyper D., Dierickx W., Du Prez F., Devlieghere F. Use of endospore-forming bacteria as an active oxygen scavenger in plastic packaging materials. Innovative Food Science and Emerging Technologies 2011; 12 594–599.

69. Zagory D. Ethylene-removing packaging. In Rooney M.L. (ed.) Active Food Packaging. Chapman & Hall, 1995. pp. 38 - 51.

70. Meyer MD., Terry LA. Fatty acid and sugar composition of avocado, cv. Hass, in response to treatment with an ethylene scavenger or 1-methylcyclopropene to extend storage life. Food Chemistry 2010; 121 1203–1210.

71. Smith AWJ., Poulston S, Rowsell L, Terry LA, Anderson JA. A new Palladium-based ethylene scavenger to control ethylene-induced ripening of climacteric fruit. Platinum Metals Review 2009; 53(3) 112–122.

72. Smith JP, Hoshino J, Abe Y. Interactive packaging involving sachet technology. In: Rooney M.L. (ed.) Active Food Packaging. Chapman & Hall, 1995. pp. 143-172.

73. López-de-Dicastillo C, Gallur M, Catalá R, Gavara R, Hernandez-Munoz P. Immobilization of β-cyclodextrin in ethylene-vinyl alcohol copolymer for active food packaging applications. Journal of Membrane Science 2010; 353 184–191.

74. López-de-Dicastillo C, Catalá R, Gavara R, Hernández-Muñoz P. Food applications of active packaging EVOH films containing cyclodextrins for the preferential scavenging of undesirable compounds. Journal of Food Engineering 2011; 104 380–386.

75. Joo MJ, Merkel C, Auras R, Almenar E. Development and characterization of antimicrobial poly(l-lactic acid) containing trans-2-hexenal trapped in cyclodextrins. International Journal of Food Microbiology 2012; 153 297–305.

76. López-de-Dicastillo C, Gómez-Estaca J, Catalá R, Gavara R, Hernández-Muñoz P. Active antioxidant packaging films: Development and effect on lipid stability of brined sardines. Food Chemistry 2012; 131 1376–1384.

77. Granda-Restrepo DM, Soto-Valdez H, Peralta E., Troncoso-Rojas R, Vallejo-Córdoba B, Gámez-Meza N, Graciano-Verdugo AZ. Migration of a-tocopherol from an active multilayer film into whole milk powder. Food Research International 2009; 42 1396– 1402.

78. Kanatt SR, Rao MS, Chawla SP, Sharma A. Active chitosanepolyvinyl alcohol films with natural extracts. Food Hydrocolloids 2012; 29 290- 297.

79. Ramos M, Jiménez A, Peltzer M, Garrigós MC. Characterization and antimicrobial activity studies of polypropylene films with carvacrol and thymol for active packaging Journal of Food Engineering 2012; 109 513–519.

80. Anthierens T, Billiet L, Devlieghere F, Du Prez F. Poly(butylene adipate) functionalized with quaternary phosphonium groups as potential antimicrobial packaging material. Innovative Food Science and Emerging Technologies 2012; In press

81. Germinara SG, Conte A, Lecce L, Di Palma A, Del Nobile MA. Propionic acid in bio- based packaging to prevent Sitophilus granarius (L.) (Coleoptera, Dryophthoridae) infestation in cereal products. Innovative Food Science and Emerging Technologies 2010; 11 498–502.

82. Germinara GS, Conte A, De Cristofaro A, Lecce L, Di Palma A, Rotundo G, Del Nobile MA. Electrophysiological and behavioral activity of (E)-2-Hexenal in the granary weevil and its application in food packaging. Journal of Food Protection 2012; 75 366-370.

83. Ramos ÓL, Silva SI, Soares JC, Fernandes JC, Poças MF, Pintado ME, Malcata FX. Features and performance of edible films, obtained from whey protein isolate formulated with antimicrobial compounds. Food Research International 2012; 45 351–361.

84. Suppakul P, Sonneveld K, Bigger SW, Miltz J. Diffusion of linalool and methylchavicol from polyethylene-based antimicrobial packaging films. LWT - Food Science and Technology 2011; 44 1888-1893.

85. Uz M, Alsoy Altınkaya S. Development of mono and multilayer antimicrobial food packaging materials for controlled release of potassium sorbate. LWT - Food Science and Technology 2011; 44 2302-2309.

86. Jipa IM, Stoica-Guzun A, Stroescu M. Controlled release of sorbic acid from bacterial cellulose based mono and multilayer antimicrobial films. LWT - Food Science and Technology 2012; 47 400-406.

87. Del Nobile MA, Conte A, Incoronato AL, Panza O. Antimicrobial efficacy and release kinetics of thymol from zein films. Journal of Food Engineering 2008; 89 57–63

88. Cerisuelo JP, Muriel-Galet V, Bermúdez JM, Aucejo S, Catalá R, Gavara R, Hernández-Muñoz P. Mathematical model to describe the release of an antimicrobial agent from an active package constituted by carvacrol in a hydrophilic EVOH coating on a PP film. Journal of Food Engineering 2012; 110 26–37.

89. Bierhalz ACK, da Silva MA, Kieckbusch TG. Natamycin release from alginate/pectin films for food packaging applications. Journal of Food Engineering 2012; 110 18–25.

90. Sekhon PS. Food nanotechnology – an overview. Nanotechnology, Science and Applications 2010; 3 1–15.

91. Tankhiwale R, Bajpai SK. Preparation, characterization and antibacterial applications of ZnO-nanoparticles coated polyethylene films for food packaging. Colloids and Surfaces B: Biointerfaces 2012; 90 16– 20.

92. Abdollahi M, Rezaei M, Farzi G. A novel active bionanocomposite film incorporating rosemary essential oil and nanoclay into chitosan. Journal of Food Engineering 2012; 111 343–350.

93. Incoronato AL, Buonocore GG, Conte A, Lavorgna M, Del Nobile MA. Active systems based on silver/montmorillonite nanoparticles embedded into bio-based polymer matrices for packaging applications. Journal of Food Protection 2010; 73: 2256-2262.

94. de Moura MR, Mattoso LHC, Zucolotto V. Development of cellulose-based bactericidal nanocomposites containing silver nanoparticles and their use as active food packaging. Journal of Food Engineering 2012; 109 520–524.

95. Brasil IM, Gomes C, Puerta-Gomez A, Castell-Perez ME, Moreira RG. Polysaccharide-based multilayered antimicrobial edible coating enhances quality of fresh-cut papaya. LWT - Food Science and Technology 2012; 47 39-45.

96. Barbiroli A, Bonomi F, Capretti G, Iametti S, Manzoni M, Piergiovanni L, Rollini M. Antimicrobial activity of lysozyme and lactoferrin incorporated in cellulose-based food packaging. Food Control 2012; 26 387-392.

97. Lloret E, Picouet P, Fernández A. Matrix effects on the antimicrobial capacity of silver based nanocomposite absorbing materials. LWT - Food Science and Technology 2012; 1-6.

98. Costa C, Conte A, Buonocore GG., Del Nobile MA. Antimicrobial silver-montmorillonite nanoparticles to prolong the shelf life of fresh fruit salad. International Journal of Food Microbiology 2011; 148 164–167.

99. Incoronato AL, Conte A, Buonocore GG., Del Nobile MA. Agar hydrogel with silver nanoparticles to prolong the shelf life of Fior di Latte cheese. Journal of Dairy Science 2011; 94 1697-1704.

100. Gammariello D, Conte A, Buonocore GG, Del Nobile MA. Bio-based nanocomposite coating to preserve quality of Fior di latte cheese. Journal of Dairy Science 2011; 94 5298-5304.

www.ingramcontent.com/pod-product-compliance
Lightning Source LLC
LaVergne TN
LVHW081941200926
848484LV00009B/2482